# Bioinformatics Biocomputing and Perl

# Bioinformatics Biocomputing and Perl

An Introduction to Bioinformatics Computing Skills and Practice

**Michael Moorhouse**
*Post-Doctoral Worker from Erasmus MC,
The Netherlands*

**Paul Barry**
*Department of Computing and Networking,
Institute of Technology,
Carlow, Ireland*

John Wiley & Sons, Ltd

Copyright 2004   John Wiley & Sons Ltd, The Atrium, Southern Gate, Chichester,
West Sussex PO19 8SQ, England

Telephone (+44) 1243 779777

Email (for orders and customer service enquiries): cs-books@wiley.co.uk
Visit our Home Page on www.wileyeurope.com or www.wiley.com

Reprinted September 2004

All Rights Reserved. No part of this publication may be reproduced, stored in a retrieval system or transmitted in any form or by any means, electronic, mechanical, photocopying, recording, scanning or otherwise, except under the terms of the Copyright, Designs and Patents Act 1988 or under the terms of a licence issued by the Copyright Licensing Agency Ltd, 90 Tottenham Court Road, London W1T 4LP, UK, without the permission in writing of the Publisher. Requests to the Publisher should be addressed to the Permissions Department, John Wiley & Sons Ltd, The Atrium, Southern Gate, Chichester, West Sussex PO19 8SQ, England, or emailed to permreq@wiley.co.uk, or faxed to (+44) 1243 770620.

This publication is designed to provide accurate and authoritative information in regard to the subject matter covered. It is sold on the understanding that the Publisher is not engaged in rendering professional services. If professional advice or other expert assistance is required, the services of a competent professional should be sought.

*Other Wiley Editorial Offices*

John Wiley & Sons Inc., 111 River Street, Hoboken, NJ 07030, USA

Jossey-Bass, 989 Market Street, San Francisco, CA 94103-1741, USA

Wiley-VCH Verlag GmbH, Boschstr. 12, D-69469 Weinheim, Germany

John Wiley & Sons Australia Ltd, 33 Park Road, Milton, Queensland 4064, Australia

John Wiley & Sons (Asia) Pte Ltd, 2 Clementi Loop #02-01, Jin Xing Distripark, Singapore 129809

John Wiley & Sons Canada Ltd, 22 Worcester Road, Etobicoke, Ontario, Canada M9W 1L1

Wiley also publishes its books in a variety of electronic formats. Some content that appears in print may not be available in electronic books.

*British Library Cataloguing in Publication Data*

A catalogue record for this book is available from the British Library

ISBN 0-470-85331-X

Typeset in 9.5/12.5pt Lucida Bright by Laserwords Private Limited, Chennai, India
Printed and bound in Great Britain by Antony Rowe Ltd, Chippenham, Wiltshire
This book is printed on acid-free paper responsibly manufactured from sustainable forestry in which at least two trees are planted for each one used for paper production.

For my parents, who taught me the value of knowledge – MJM

For three great kids: Joseph, Aaron and Aideen – PJB

# Contents

Preface   xv

## 1 Setting the Biological Scene   1

    1.1   Introducing Biological Sequence Analysis   1
    1.2   Protein and Polypeptides   4
    1.3   Generalised Models and their Use   5
    1.4   The Central Dogma of Molecular Biology   6
            1.4.1   Transcription   6
            1.4.2   Translation   7
    1.5   Genome Sequencing   10
            1.5.1   Sequence assembly   11
    1.6   The Example DNA-gene-protein system we will use   12
          Where to from Here   13

## 2 Setting the Technological Scene   15

    2.1   The Layers of Technology   15
            2.1.1   From passive user to active developer   16
    2.2   Finding `perl`   17
            2.2.1   Checking for `perl`   17
          Where to from Here   18

# I Working with Perl   19

## 3 The Basics   21

    3.1   Let's Get Started!   21
            3.1.1   Running Perl programs   22
            3.1.2   Syntax and semantics   23
            3.1.3   Program: run thyself!   25
    3.2   Iteration   26
            3.2.1   Using the Perl `while` construct   26
    3.3   More Iterations   30
            3.3.1   Introducing variable containers   31
            3.3.2   Variable containers and loops   32

viii    *Contents*

|  |  |  |  |
|---|---|---|---|
| | 3.4 | Selection | 34 |
| | | 3.4.1 Using the Perl `if` construct | 35 |
| | 3.5 | There Really is MTOWTDI | 36 |
| | 3.6 | Processing Data Files | 41 |
| | | 3.6.1 Asking `getlines` to do more | 43 |
| | 3.7 | Introducing Patterns | 44 |
| | | Where to from Here | 46 |
| | | The Maxims Repeated | 46 |

## 4 Places to Put Things — 49

|  |  |  |  |
|---|---|---|---|
| | 4.1 | Beyond Scalars | 49 |
| | 4.2 | Arrays: Associating Data with Numbers | 49 |
| | | 4.2.1 Working with array elements | 51 |
| | | 4.2.2 How big is the array? | 51 |
| | | 4.2.3 Adding elements to an array | 52 |
| | | 4.2.4 Removing elements from an array | 54 |
| | | 4.2.5 Slicing arrays | 54 |
| | | 4.2.6 Pushing, popping, shifting and unshifting | 56 |
| | | 4.2.7 Processing every element in an array | 57 |
| | | 4.2.8 Making lists easier to work with | 59 |
| | 4.3 | Hashes: Associating Data with Words | 60 |
| | | 4.3.1 Working with hash entries | 61 |
| | | 4.3.2 How big is the hash? | 61 |
| | | 4.3.3 Adding entries to a hash | 62 |
| | | 4.3.4 Removing entries from a hash | 62 |
| | | 4.3.5 Slicing hashes | 63 |
| | | 4.3.6 Working with hash entries: a complete example | 64 |
| | | 4.3.7 Processing every entry in a hash | 66 |
| | | Where to from Here | 68 |
| | | The Maxims Repeated | 68 |

## 5 Getting Organised — 71

|  |  |  |  |
|---|---|---|---|
| | 5.1 | Named Blocks | 71 |
| | 5.2 | Introducing Subroutines | 73 |
| | | 5.2.1 Calling subroutines | 73 |
| | 5.3 | Creating Subroutines | 74 |
| | | 5.3.1 Processing parameters | 76 |
| | | 5.3.2 Better processing of parameters | 78 |
| | | 5.3.3 Even better processing of parameters | 80 |
| | | 5.3.4 A more flexible `drawline` subroutine | 83 |
| | | 5.3.5 Returning results | 84 |
| | 5.4 | Visibility and Scope | 85 |
| | | 5.4.1 Using private variables | 86 |
| | | 5.4.2 Using global variables properly | 88 |
| | | 5.4.3 The final version of `drawline` | 89 |
| | 5.5 | In-built Subroutines | 90 |
| | 5.6 | Grouping and Reusing Subroutines | 92 |
| | | 5.6.1 Modules | 93 |
| | 5.7 | The Standard Modules | 96 |
| | 5.8 | CPAN: The Module Repository | 96 |
| | | 5.8.1 Searching CPAN | 97 |
| | | 5.8.2 Installing a CPAN module manually | 98 |

|        |       |       | 5.8.3 Installing a CPAN module automatically | 99 |
|---|---|---|---|---|
|  |  |  | 5.8.4 A final word on CPAN modules | 99 |
|  |  |  | Where to from Here | 100 |
|  |  |  | The Maxims Repeated | 100 |

## 6 About Files — 103

- 6.1 I/O: Input and Output — 103
  - 6.1.1 The standard streams: STDIN, STDOUT and STDERR — 103
- 6.2 Reading Files — 105
  - 6.2.1 Determining the disk-file names — 106
  - 6.2.2 Opening the named disk-files — 108
  - 6.2.3 Reading a line from each of the disk-files — 110
  - 6.2.4 Putting it all together — 110
  - 6.2.5 Slurping — 114
- 6.3 Writing Files — 116
  - 6.3.1 Redirecting output — 117
  - 6.3.2 Variable interpolation — 117
- 6.4 Chopping and Chomping — 118
- Where to from Here — 119
- The Maxims Repeated — 119

## 7 Patterns, Patterns and More Patterns — 121

- 7.1 Pattern Basics — 121
  - 7.1.1 What is a regular expression? — 122
  - 7.1.2 What makes regular expressions so special? — 122
- 7.2 Introducing the Pattern Metacharacters — 124
  - 7.2.1 The + repetition metacharacter — 124
  - 7.2.2 The | alternation metacharacter — 126
  - 7.2.3 Metacharacter shorthand and character classes — 127
  - 7.2.4 More metacharacter shorthand — 128
  - 7.2.5 More repetition — 130
  - 7.2.6 The ? and * optional metacharacters — 130
  - 7.2.7 The any character metacharacter — 131
- 7.3 Anchors — 132
  - 7.3.1 The \b word boundary metacharacter — 132
  - 7.3.2 The ^ start-of-line metacharacter — 133
  - 7.3.3 The $ end-of-line metacharacter — 133
- 7.4 The Binding Operators — 134
- 7.5 Remembering What Was Matched — 135
- 7.6 Greedy by Default — 137
- 7.7 Alternative Pattern Delimiters — 138
- 7.8 Another Useful Utility — 139
- 7.9 Substitutions: Search and Replace — 140
  - 7.9.1 Substituting for whitespace — 141
- 7.10 Finding a Sequence — 142
- Where to from Here — 146
- The Maxims Repeated — 146

## 8 Perl Grabbag — 147

- 8.1 Introduction — 147
- 8.2 Strictness — 147

x    Contents

|  |  |  |
|---|---|---|
| 8.3 | Perl One-liners | 149 |
| 8.4 | Running Other Programs from `perl` | 152 |
| 8.5 | Recovering from Errors | 153 |
| 8.6 | Sorting | 155 |
| 8.7 | HERE Documents | 159 |
|  | Where to from Here | 160 |
|  | The Maxims Repeated | 161 |

## II  Working with Data  163

## 9  Downloading Datasets  165

|  |  |  |
|---|---|---|
| 9.1 | Let's Get Data | 165 |
| 9.2 | Downloading from the Web | 165 |
|  | 9.2.1  Using `wget` to download PDB data-files | 167 |
|  | 9.2.2  Mirroring a dataset | 168 |
|  | 9.2.3  Smarter mirroring | 168 |
|  | 9.2.4  Downloading a subset of a dataset | 169 |
|  | Where to from Here | 171 |
|  | The Maxims Repeated | 171 |

## 10  The Protein Databank  173

|  |  |  |
|---|---|---|
| 10.1 | Introduction | 173 |
| 10.2 | Determining Biomolecule Structures | 174 |
|  | 10.2.1  X-Ray Crystallography | 174 |
|  | 10.2.2  Nuclear magnetic resonance | 176 |
|  | 10.2.3  Summary of protein structure methods | 177 |
| 10.3 | The Protein Databank | 177 |
| 10.4 | The PDB Data-file Formats | 179 |
|  | 10.4.1  Example structures | 180 |
|  | 10.4.2  Downloading PDB data-files | 181 |
| 10.5 | Accessing Data in PDB Entries | 182 |
| 10.6 | Accessing PDB Annotation Data | 183 |
|  | 10.6.1  Free R and resolution | 184 |
|  | 10.6.2  Database cross references | 186 |
|  | 10.6.3  Coordinates section | 188 |
|  | 10.6.4  Extracting 3D coordinate data | 191 |
| 10.7 | Contact Maps | 192 |
| 10.8 | STRIDE: Secondary Structure Assignment | 196 |
|  | 10.8.1  Installation of STRIDE | 197 |
| 10.9 | Assigning Secondary Structures | 197 |
|  | 10.9.1  Using STRIDE and parsing the output | 200 |
|  | 10.9.2  Extracting amino acid sequences using STRIDE | 204 |
| 10.10 | Introducing the `mmCIF` Protein Format | 205 |
|  | 10.10.1  Converting `mmCIF` to PDB | 206 |
|  | 10.10.2  Converting `mmCIF`s to PDB with `CIFTr` | 206 |
|  | 10.10.3  Problems with the `CIFTr` conversion | 208 |
|  | 10.10.4  Some advice on using `mmCIF` | 208 |
|  | 10.10.5  Automated conversion of `mmCIF` to PDB | 208 |
|  | Where to from Here | 210 |
|  | The Maxims Repeated | 210 |

## 11  Non-redundant Datasets — 211

| | | |
|---|---|---|
| 11.1 | Introducing Non-redundant Datasets | 211 |
| | 11.1.1  Reasons for redundancy | 211 |
| | 11.1.2  Reduction of redundancy | 212 |
| | 11.1.3  Non-redundancy and non-representative | 212 |
| 11.2 | Non-redundant Protein Structures | 213 |
| | Where to from Here | 217 |
| | The Maxims Repeated | 217 |

## 12  Databases — 219

| | | |
|---|---|---|
| 12.1 | Introducing Databases | 219 |
| | 12.1.1  Relating tables | 220 |
| | 12.1.2  The problem with single-table databases | 222 |
| | 12.1.3  Solving the one-table problem | 222 |
| | 12.1.4  Database system: a definition | 224 |
| 12.2 | Available Database Systems | 224 |
| | 12.2.1  Personal database systems | 225 |
| | 12.2.2  Enterprise database systems | 225 |
| | 12.2.3  Open source database systems | 225 |
| 12.3 | SQL: the Language of Databases | 226 |
| | 12.3.1  Defining data with SQL | 226 |
| | 12.3.2  Manipulating data with SQL | 227 |
| 12.4 | A Database Case Study: MER | 227 |
| | 12.4.1  The requirement for the MER database | 231 |
| | 12.4.2  Installing a database system | 232 |
| | 12.4.3  Creating the MER database | 233 |
| | 12.4.4  Adding tables to the MER database | 235 |
| | 12.4.5  Preparing SWISS-PROT data for importation | 238 |
| | 12.4.6  Importing tab-delimited data into `proteins` | 245 |
| | 12.4.7  Working with the data in `proteins` | 246 |
| | 12.4.8  Adding another table to the MER database | 248 |
| | 12.4.9  Preparing EMBL data for importation | 249 |
| | 12.4.10 Importing tab-delimited data into `dnas` | 253 |
| | 12.4.11 Working with the data in `dnas` | 253 |
| | 12.4.12 Relating data in one table to that in another | 254 |
| | 12.4.13 Adding the `crossrefs` table to the MER database | 255 |
| | 12.4.14 Preparing cross references for importation | 256 |
| | 12.4.15 Importing tab-delimited data into `crossrefs` | 259 |
| | 12.4.16 Working with the data in `crossrefs` | 259 |
| | 12.4.17 Adding the `citations` table to the MER database | 263 |
| | 12.4.18 Preparing citation information for importation | 265 |
| | 12.4.19 Importing tab-delimited data into `citations` | 268 |
| | 12.4.20 Working with the data in `citations` | 268 |
| | Where to from Here | 269 |
| | The Maxims Repeated | 269 |

## 13  Databases and Perl — 273

| | | |
|---|---|---|
| 13.1 | Why Program Databases? | 273 |
| 13.2 | Perl Database Technologies | 274 |
| 13.3 | Preparing Perl | 275 |
| | 13.3.1  Checking the `DBI` installation | 275 |

|       |        |                                                          |     |
|-------|--------|----------------------------------------------------------|-----|
| 13.4  |        | Programming Databases with `DBI`                         | 276 |
|       | 13.4.1 | Developing a database utility module                     | 279 |
|       | 13.4.2 | Improving upon `dump_results`                            | 280 |
| 13.5  |        | Customising Output                                       | 282 |
| 13.6  |        | Customising Input                                        | 285 |
| 13.7  |        | Extending SQL                                            | 289 |
|       |        | Where to from Here                                       | 292 |
|       |        | The Maxims Repeated                                      | 292 |

# III Working with the Web — 295

## 14 The Sequence Retrieval System — 297

| 14.1 | An Example of What's Possible | 297 |
|------|-------------------------------|-----|
| 14.2 | Why SRS?                      | 298 |
| 14.3 | Using SRS                     | 298 |
|      | Where to from Here            | 300 |
|      | The Maxims Repeated           | 300 |

## 15 Web Technologies — 303

|       |        |                                            |     |
|-------|--------|--------------------------------------------|-----|
| 15.1  |        | The Web Development Infrastructure         | 303 |
| 15.2  |        | Creating Content for the WWW               | 305 |
|       | 15.2.1 | The static creation of WWW content         | 308 |
|       | 15.2.2 | The dynamic creation of WWW content        | 308 |
| 15.3  |        | Preparing Apache for Perl                  | 310 |
|       | 15.3.1 | Testing the execution of server-side programs | 312 |
| 15.4  |        | Sending Data to a Web Server               | 315 |
| 15.5  |        | Web Databases                              | 320 |
|       |        | Where to from Here                         | 327 |
|       |        | The Maxims Repeated                        | 327 |

## 16 Web Automation — 329

| 16.1 | Why Automate Surfing?       | 329 |
|------|-----------------------------|-----|
| 16.2 | Automated Surfing with Perl | 330 |
|      | Where to from Here          | 335 |
|      | The Maxims Repeated         | 336 |

# IV Working with Applications — 337

## 17 Tools and Datasets — 339

|       |        |                                        |     |
|-------|--------|----------------------------------------|-----|
| 17.1  |        | Introduction                           | 339 |
| 17.2  |        | Sequence Databases                     | 340 |
|       | 17.2.1 | Understanding EMBL entries             | 343 |
|       | 17.2.2 | Understanding SWISS-PROT entries       | 346 |
|       | 17.2.3 | Summarising sequences databases        | 347 |
| 17.3  |        | General Concepts and Methods           | 347 |
|       | 17.3.1 | Predictions and validation             | 348 |
|       | 17.3.2 | True/False/Negative/Positive           | 348 |

|  |  | 17.3.3 Balancing the errors | 351 |
|---|---|---|---|
|  |  | 17.3.4 Using multiple algorithms to improve performance | 352 |
|  |  | 17.3.5 tRNA-ScanSE, a case study | 353 |
|  | 17.4 | Introducing Bioinformatics Tools | 357 |
|  |  | 17.4.1 ClustalW | 358 |
|  |  | 17.4.2 Algorithms and methods | 359 |
|  |  | 17.4.3 Installation and use | 360 |
|  |  | 17.4.4 Substitution/scoring matrices | 361 |
|  | 17.5 | BLAST | 362 |
|  |  | 17.5.1 Installing NCBI-BLAST | 364 |
|  |  | 17.5.2 Preparation of database files for faster searching | 365 |
|  |  | 17.5.3 The different types of BLAST search | 369 |
|  |  | 17.5.4 Final words on BLAST | 371 |
|  |  | Where to from Here | 371 |
|  |  | The Maxims Repeated | 371 |

## 18  Applications                                                           373

|  | 18.1 | Introduction | 373 |
|---|---|---|---|
|  | 18.2 | Scientific Background to Mer Operon | 374 |
|  |  | 18.2.1 Function | 374 |
|  |  | 18.2.2 Genetic structure and regulation | 374 |
|  |  | 18.2.3 Mobility of the Mer Operon | 375 |
|  | 18.3 | Downloading the Raw DNA Sequence | 377 |
|  | 18.4 | Initial BLAST Sequence Similarity Search | 378 |
|  | 18.5 | GeneMark | 380 |
|  |  | 18.5.1 Using BLAST to identify specific sequences | 382 |
|  |  | 18.5.2 Dealing with false negatives and missing proteins | 386 |
|  |  | 18.5.3 Over-predicted genes and false positives | 387 |
|  |  | 18.5.4 Summary of validation of GeneMark prediction | 388 |
|  | 18.6 | Structural Prediction with SWISS-MODEL | 388 |
|  |  | 18.6.1 Alternatives to homology modelling | 390 |
|  |  | 18.6.2 Modelling with SWISS-MODEL | 390 |
|  | 18.7 | DeepView as a Structural Alignment Tool | 396 |
|  | 18.8 | PROSITE and Sequence Motifs | 401 |
|  |  | 18.8.1 Using PROSITE patterns and matrices | 402 |
|  |  | 18.8.2 Downloading PROSITE and its search tools | 403 |
|  |  | 18.8.3 Final word on PROSITE | 407 |
|  | 18.9 | Phylogenetics | 407 |
|  |  | 18.9.1 A look at the HMA domain of MerA and MerP | 407 |
|  |  | Where to from Here? | 410 |
|  |  | The Maxims Repeated | 411 |

## 19  Data Visualisation                                                     413

|  | 19.1 | Introducing Visualisation | 413 |
|---|---|---|---|
|  | 19.2 | Displaying Tabular Data Using HTML | 415 |
|  |  | 19.2.1 Displaying SWISS-PROT identifiers | 417 |
|  | 19.3 | Creating High-quality Graphics with GD | 422 |
|  |  | 19.3.1 Using the GD module | 424 |
|  |  | 19.3.2 Displaying genes in EMBL entries | 426 |
|  |  | 19.3.3 Introducing mogrify | 429 |

|       |       |                                                                 |     |
| ----- | ----- | --------------------------------------------------------------- | --- |
|       | 19.4  | Plotting Graphs                                                 | 431 |
|       |       | 19.4.1 Graph-plotting using the `GD::Graph` modules             | 432 |
|       |       | 19.4.2 Graph-plotting using Grace                               | 433 |
|       |       | Where to from Here                                              | 439 |
|       |       | The Maxims Repeated                                             | 439 |

## 20 Introducing Bioperl — 441

|       |       |                                                         |     |
| ----- | ----- | ------------------------------------------------------- | --- |
|       | 20.1  | What is Bioperl?                                        | 441 |
|       | 20.2  | Bioperl's Relationship to Project Ensembl               | 442 |
|       | 20.3  | Installing Bioperl                                      | 442 |
|       | 20.4  | Using Bioperl: Fetching Sequences                       | 444 |
|       |       | 20.4.1 Fetching multiple sequences                      | 445 |
|       |       | 20.4.2 Extracting sub-sequences                         | 447 |
|       | 20.5  | Remote BLAST Searches                                   | 448 |
|       |       | 20.5.1 A quick aside: the `blastcl3 NetBlast` client    | 449 |
|       |       | 20.5.2 Parsing BLAST outputs                            | 450 |
|       |       | Where to from Here                                      | 451 |
|       |       | The Maxims Repeated                                     | 452 |

## A  Appendix A — 453

## B  Appendix B — 457

## C  Appendix C — 459

## D  Appendix D — 461

## E  Appendix E — 467

## F  Appendix F — 471

## Index — 475

# Preface

Welcome to *Bioinformatics, Biocomputing and Perl*, an introduction and guide to the computing skills and practices collectively known as *Bioinformatics*.

Bioinformatics is the application of computing techniques to the study of biology, and in particular biology research. Although the study of biology is hundreds of years old, the application of computing techniques to biology research is relatively new, with major advances occurring within the last decade. Consequently, the Bioinformatics field is evolving and maturing rapidly, and this has highlighted the need for a good, all-round introductory textbook. We believe that *Bioinformatics, Biocomputing and Perl* meets this need.

## What is in this Book?

After two introductory chapters, *Bioinformatics, Biocomputing and Perl* is divided into four main parts:

1. **Working with Perl.**
2. **Working with Data.**
3. **Working with the Web.**
4. **Working with Applications.**

Part I, *Working with Perl*, introduces programming to the student of Bioinformatics. Note that the intention is not to turn Bioinformaticians into software engineers. Rather, the emphasis is on providing Bioinformaticians with programming skills sufficient to enable them to produce bespoke programs when required in the course of their research.

The programming language of choice among Bioinformaticians, Perl, is used throughout Part I. Perl is popular because of its combination of excellent file-handling capabilities, native support for POSIX regular expressions and powerful scripting capabilities. If that sounds like *techno babble*, do not worry; the importance of these programming language features is explained in a less technical way later. Fortunately, Perl is not particularly difficult to learn. For instance, by the end of Chapter 3, the reader will know enough Perl to be able to produce simple, but useful, programs. This early material is then developed so that by the end of Part I, readers will be able to confidently create customised and customisable programs to solve diverse Bioinformatics problems.

In Part II, *Working with Data*, the emphasis shifts from creating bespoke Bioinformatics programs to exploring the tools and techniques used to organise, store, retrieve and process data. After explaining how to download datasets from the Internet, the Protein DataBank (PDB) is described in detail. A short chapter follows on the importance of non-redundant datasets, before discussion shifts to cover relational database management systems. How to create and use databases with the popular *MySQL* tool is described. In addition to using standard tools to interact with databases, the use of Perl programs to interrogate databases is also covered.

Part III, *Working with the Web*, covers a collection of web-based technologies that, once mastered, can be used to publish research -- both findings *and* data -- on the Internet. Electronic mechanisms allowing interaction with, and interrogation of, web-based data are explained. Perl again plays an important role in this part of the book, with HTML and CGI also covered.

Part IV, *Working with Applications*, describes a set of standard Bioinformatics tools and applications. Although it is often useful to be able to create a new tool from scratch, it can sometimes be more appropriate to take existing tools and control their execution and interaction. Scripting technologies, of which Perl is only one type, are particularly useful in this area. A discussion of "The Bioperl Project", and its importance, completes *Bioinformatics, Biocomputing and Perl*.

## Maxims, Commentaries, Exercises and Appendices

All but the first two chapters contain a collection of *maxims*. These are your authors' *snippets of wisdom*. At the end of each chapter, the maxims are repeated in list form. If, having worked through a chapter, the maxims are understood, it is an indication that the associated material has been understood. If, however, a maxim is not understood, it indicates that there is a need to review the material to which the particular maxim relates.

In addition to the maxims, chapters include *technical commentaries*. Unlike maxims, it is not necessary to fully understand the commentaries on first reading. If a technical commentary is not immediately understood, it is possible to safely continue to work through the text without too much difficulty.

The majority of chapters conclude with a set of exercises that are designed to expand upon the material introduced. It is highly recommended that these exercises are worked through, as it is only through practice and review that Bioinformatics computing skills are developed and honed.

A collection of appendices completes the book, providing information on, among other things, installing Perl on various platforms, the Perl on-line documentation and a list of Perl operators. An annotated list of references and suggestions for further reading are also presented as an appendix.

## Who Should Read this Book

This book targets three distinct readerships.

The main target is the student of biology, both under- and post-graduate. *Bioinformatics, Biocomputing and Perl* is designed to be *the* must-have, introductory Bioinformatics textbook. The biology student taking a Bioinformatics module will find this book to be a useful starting point and an essential desktop reference.

Another target is the qualified, professional or academic biologist who needs to understand more about Bioinformatics. The field of Bioinformatics is still relatively new and it is only now appearing as a feature within biology course outlines and syllabi. However, there are many qualified biologists "in the field" requiring a good primer. This book is designed to meet that need.

The final target is the computer scientist curious to understand how computing skills might be used within this growing field.

## What you Should know Already

It is assumed that some knowledge of computer use has already been acquired, including understanding the concept of a disk-file and knowing how to create one using an editor. On the Linux operating system, popular editors are `vi`, `pico` and `emacs`. On any of the Windows operating systems, `Notepad`, `WordPad` and `Word` are all editors, although the latter is a more sophisticated example. Macintosh users have `SimpleText` and `BBedit`. Any of these will suffice, so long as it allows for the creation and manipulation of plain text files. Later chapters (Parts III and IV) assume a working knowledge of HTML.

## Platform Notes

All of the examples in *Bioinformatics, Biocomputing and Perl* are designed to operate on the Linux operating system, in keeping with the current trend within the Bioinformatics community. There is no attempt to explain all that the reader needs to know about Linux, as the emphasis in this book is on explaining how to exploit the growing collection of tools that run *on top of* the Linux operating

## Accompanying Web-site

Details of the book's mailing list, its source code, any errata and other related material can be found on the book's web-site, located at:

    http://glasnost.itcarlow.ie/~biobook/index.html

## Your Comments are Welcome

The authors welcome all comments about *Bioinformatics, Biocomputing and Perl*. Send an e-mail to either of the following addresses:

    m.moorhouse@erasmusmc.nl

    paul.barry@itcarlow.ie

## Acknowledgements

Michael thanks his parents for their unwavering support, be it material, practical or emotional. Their endless hours of reading and re-reading the draft chapters and manuscript produced many points of very welcome constructive criticism. Although completing a PhD., moving country and starting a new job while writing a book is not something he'd recommend, Michael thanks those around him for helping when they could and for understanding why he was so busy. Also, thanks to all in the new Department of Bioinformatics, Erasmus MC, the Netherlands, who have offered their support and understanding.

   Paul thanks his father, Jim Barry, for taking the time to proofread the text (multiple times). As with Paul's first book, this one is better for his father's involvement. Thanks go to Karen Mosman (formerly with Wiley's Computing Division) for suggesting Paul when the Biology Division came looking for an author with Perl experience. The Institute of Technology, Carlow, was again supportive of Paul working on a textbook, and thanks are due to Dr Dave Dowling and Joe Kehoe for enthusiastically reviewing some of the early material. Paul's wife, Deirdre, held everything else together while the production of the manuscript consumed more and more of his time, while Joseph, Aaron and Aideen kept reminding Paul that there's more to life than computers and writing.

   Both authors thank the team at Wiley. Joan Marsh, this book's publishing editor, arranged for the authors to work together and never once complained when the draft manuscript went from being days late to weeks late to -- eventually -- six months late! This book's editorial assistant was Layla Paggetti, and both authors thank Layla for her prompt and efficient responses to their many queries. Robert Hambrook acted as production editor. As with Paul's first book, this one has benefited greatly from Robert's management of the production process.

A special word of thanks to those members of the computing and biology communities who produce such wonderfully useful software technologies and tools. There are many such individuals. Specific thanks to Richard Stallman, Linus Torvalds, Larry Wall, Tom Boutell, Andy Lester and Dr Lincoln D. Stein for sharing their software with the world and for providing the authors with technologies to write about. Paul also thanks Bill Joy (for vi) and Leslie Lamport (for LaTeX).

# 1

# Setting the Biological Scene

*Introducing DNA, RNA, polypeptides, proteins and sequence analysis.*

## 1.1 Introducing Biological Sequence Analysis

Among other things, this book describes a number of techniques used to analyse DNA, RNA and proteins.

To a molecular biologist, DNA is a very physical molecule: a polymer of nucleotides that are collectively called *deoxyribose nucleic acid*. It coils, bends, flexes and interacts with proteins, and is generally interesting. RNA is similar to DNA in structure, but for the fact that RNA contains the sugar *ribose* as opposed to *deoxyribose*. DNA has a hydrogen at the second carbon atom on the ring; RNA has a hydrogen linked through an oxygen atom.

In DNA and RNA, there are four *nucleotide bases*. Three of these bases are the same: *guanine* (G), *adenine* (A) and *cytosine* (C). The fourth base for DNA is *thymine* (T), whereas in RNA, the fourth base lacks a *methyl group* and is called *uracil* (U). Each base has two points at which it can join *covalently* to two other bases on either end, forming a *linear chain of monomers*. These chains can be quite long, with many millions of bases common in most organisms.

---

*Bioinformatics, Biocomputing and Perl.* Michael Moorhouse and Paul Barry
© 2004 John Wiley & Sons, Ltd   ISBN 0-470-85331-X

**2**   *Setting the Biological Scene*

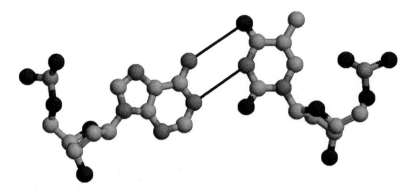

**Figure 1.1**   Adenine (A) and thymine (T) nucleotide bases (where the thin black lines indicate the three hydrogen bonds between the two bases).

Another interesting feature of nucleotide bases is that the four bases *hydrogen-bond* together in two exclusive pairs because of the position of the charged atoms along their *edges*, as shown in Figure 1.1 on page 2 and Figure 1.2 on page 3[1]. Three of these bonds form between C and G, whereas two form between A and T (or A and U in RNA).

These bonds, while considerably weaker than the covalent bonds between atoms, are enough to stabilise structures such as the famous *double helix*, in which the bases line up nearly perpendicular to the axis of the helix, as shown in Figure 1.3 on page 4. There are several important consequences of the *double helix*:

- Where there is a G in one chain, there is a C in the corresponding location in the other, and the two chains are said to be *complementary* to each other. The chains are often referred to as *strands*.

- This complementarity means that there is 50% redundancy in the information stored in both chains; consequently, only one chain is needed to store all the information for both (as one can be deduced from the other)[2].

- Because of the structure of the nucleotide bases, DNA molecules have *direction*. This is a subtle, but important, point. The phosphate backbones *attach* to the sugar rings at different locations: the 3' and 5' hydroxyl groups.

---

[1] These diagrams were produced with *Open Rasmol* on the basis of protein structure 1D66.

[2] Of course, in an evolutionary world, where DNA can be damaged, keeping a spare copy is an evolutionary advantage as an organism can often reconstruct the damaged regions from any intact parts.

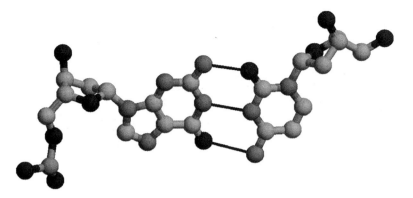

**Figure 1.2** Guanine (G) and cytosine (C) nucleotide bases (where the thin black lines indicate the three hydrogen bonds between the two bases).

When DNA is run in opposite directions, one end of the helix is the 3' end of one chain and the 5' end of the other. When the order of the nucleotide bases is written down, it is conventional to start at the nucleotides at the 5' (the 'left-most' nucleotide) end of the DNA molecule and work towards the 3' end at the right (the 'right-most' base). The importance of this directional feature will become clear later in this chapter, when *open reading frames* are described.

In general, RNA copies of DNA are made by a process known as *transcription*. For most purposes, RNA can be regarded as a *working copy* of the DNA *master template*. There is usually one or a very small number of examples of DNA in the cell, whereas there are multiple copies of the transcribed RNA.

A common term related to the number of nucleotide bases in a particular sequence is a reference to *base pairs*[3], for example "400 base pairs". This term is a generic term that can literally mean "400 paired bases". More often, though, it is used to acknowledge that while there are 400 nucleotides in a particular sequence being actively considered, there are another 400 nucleotides on the complementary strand running in the other direction. In this context, the use of base pairs is a tacit acknowledgement of their existence that may be of great importance, as the *feature* under investigation may be on *the other strand*. In nearly all cases, both strands should be considered.

There are many interesting features of DNA. As this discussion is an overview, a description of some of these features (such as *promoters, splice sites, intron/exon boundaries* and *genes*) is deferred until later chapters.

---

[3] Or "bp", for short.

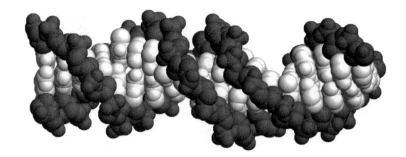

**Figure 1.3** The DNA "double helix" (where the backbones, in black, run in opposite directions).

## 1.2 Protein and Polypeptides

DNA is the *nobility* of the cellular world. Proteins are the *worker-serfs*.

To a biochemist, proteins are the *functioning units of cellular life*. Proteins do physically useful things such as catalysing reactions, processing energy rich molecules, pumping other molecules across cellular barriers and forming connective and motility structures. Proteins do just about anything else in the cell that can be considered "real work".

In molecular terms, proteins are chains technically termed *polypeptides* and formed from 20 different types of *amino acids*. These may be modified in different ways to alter their properties, the structure that is formed and the final function of the molecule. For example, certain amino acids can be *glycosylated*[4], which can be used as recognition *tags*, while other proteins associate with small molecules called *ligands* that have special properties useful in the catalysis of reactions.

The structure of a protein is generally more variable than DNA. It is at the level of proteins that the variety of the information contained in the order of DNA bases is used. The result is that the amino acid chain produced *fold* into structures that are closely linked to that particular protein's functional role within the cell (and these can vary enormously). This folding has another important consequence in that parts of a protein (i.e. its amino acids) can be physically close together in space, but distant in terms of their location in the sequence of the amino acids.

Consider, as an example, the well-studied *catalytic triad of chymotrypsin*. The critical parts of the protein for its function (which is to degrade other proteins) are the amino acids *asparate* at position 102 in the polypeptide chain, *histidine* at 57 and *serine* at 195. The triad is presented in Figure 1.4 on page 5. The right-hand side of the image shows the catalytic site in close-up, with the three critical amino acids located closely in physical space, but distant in sequence. The inset (left-hand image) shows the general structure of the protein demonstrating how the complex folding of the chain brings these residues together.

---

[4] Have sugars added.

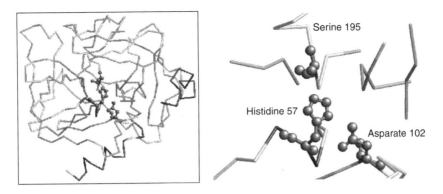

**Figure 1.4** The catalytic triad of chymotrypsin (PDB ID: 1AFQ).

## 1.3 Generalised Models and their Use

The relationships between DNA, RNA, protein, structure and function follow a generalised model. Unfortunately, like most generalisations, it is oversimplistic for many situations. If this is the case, why use it? There are two good reasons:

1. The model is a "good enough" description of what happens *most of the time*. Certainly, there are important exceptions. There are non-standard amino acids included in proteins via some other mechanism (which are ignored in this book). Possibilities such as the section of DNA coding for single protein being *discontinuous* are additional complexities that are considered later. However, overall, the model is a valuable approximation to reality that has useful predictive power when working with new systems.

2. The model is a "lie-to-children"[5]: it allows the basic features to be understood without confusing things by considering exceptions and enhancements. Once such a simple system is understood, it can be extended to cover more complex aspects and specific examples. In short, a start has to be made somewhere, and the generalised model is as good a place to start as anywhere.

Before considering the mechanisms by which information is conserved and converted *along the pathway*, let's consider another important point about the abstract nature of the data to be used.

Bioinformaticians are generally concerned with information at an abstract level: DNA, RNA and amino acid sequences are "just" strings of letters. It is sometimes easy to forget that these are actual representations of molecules that exist in the cellular world and, consequently, must interact with the physical

---

[5] *Jack Cohen, Ian Stewart* and *Terry Pratchett* discuss this concept and some general theories of science in their *Science of the Discworld* books. These are well worth a read if you fancy a laugh while pretending to work.

universe *in general*, let alone existing within a cellular environment. How much a Bioinformatician needs to know about the real-world context of the data being analysed depends on the analysis that is performed[6]. In some cases, quite superficial knowledge suffices, while others require a deeper understanding of the fundamental physical and biological processes at work.

Only through experience can the Bioinformatician hone the skill and professional judgement necessary to decide how much understanding of the underlying biological system is needed for any particular analysis. The idealistic response is "the more the better", which is like all ideals: something to aim at but rarely achieved in practice. Time is often a factor for the Bioinformatician. If too long is spent becoming versed in the biological background, the risk of not completing an analysis within a useful timescale will increase. Conversely, there is also the risk of an analysis being compromised because too little is known about the system under study. This is where the balance between the two extremes comes in. This book attempts to guide the reader in this regard through the examples presented and provide useful pointers beyond. However, in the end, it all comes down to *experience* and *professional judgement*.

## 1.4 The Central Dogma of Molecular Biology

The DNA to Functional Protein Structure Model discussed above is often referred to as the "Central Dogma of Molecular Biology". It is summarised in a slightly extended form in Figure 1.5 on page 6. The arrows represent *information flow* from that stored in the order of the DNA bases through the folding of the polypeptide chain to a fully functional protein.

### 1.4.1 Transcription

*Transcription* is the conversion of information from DNA to RNA, and is straightforward because of the direct correspondence between the four nucleotide bases of DNA and those of RNA.

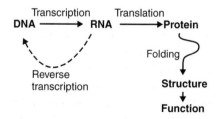

**Figure 1.5** The central dogma of molecular biology.

---

[6]This is so obvious that it is often forgotten.

There is an interesting exception in *RNA Retroviruses*, the most famous example being HIV (the Human Immunodeficiency Virus) that causes AIDS. In retroviruses, RNA is used as the information storage material. This is then copied (badly in the case of HIV) into DNA, which then integrates into the nucleic acid material of the cell under attack. This "trick" allows the virus (and its information) to lie dormant for long periods *in relative safety*, whereas the original RNA material is more likely to be actively degraded by cellular enzymes.

This RNA to DNA conversion ability is also useful for molecular biologists, as DNA can be more easily stored or manipulated using standard techniques. This has important implications, which are discussed later.

## 1.4.2 Translation

In a protein-coding region of DNA, three successive nucleotide bases, called *triplets* or *codons*, are used to code for each individual amino acid. Three bases are needed because there are 20 amino acids but only four nucleotide bases: with one base there are four possible combinations; with two bases, 16 ($4^2$); with three, 64 ($4^3$), which is more than the number of amino acids.

The *RNA transcript* is used by a complex molecular machine called the *ribosome* to translate the order of successive codons into the corresponding order of amino acids. Special *stop codons*, such as UAA, UAG and UGA, induce the *ribosome* to terminate the elongation of the polypeptide chain at a particular point. Similarly, the codon for the amino acid *methionine* (AUG in RNA) is often used as the *start signal* for translation.

The section of DNA between the *start* and *stop* codons is called an *open reading frame*. There is a complication in that the codons found depend on how the sequence of nucleotide bases is divided. This is dependent on where the count starts. There is no biological reason why the first nucleotide base reported in a DNA sequence should be related to the protein coding regions.

A common solution is to calculate the codons produced from all possible open reading frames and select the most plausible on the basis of the results. The *correct* open reading frame for a particular region of DNA is generally that which has the longest distance between any start and stop codons. Though there are exceptions, especially in some viruses and bacteria, each nucleotide is involved in coding for only one amino acid and, hence, only one open reading frame is correct. The incorrect reading frames are generally short and as a consequence, do not resemble recognisable proteins.

With three nucleotide bases in each codon, it is reasonable to assume that there are reading frames starting at the first, second and third nucleotide bases relative to a particular nucleotide. This is due to the fact that all subsequent reading frames are repeated and could start to occur anywhere else in the sequence. Consequently, it is easiest to start at the beginning. It is also important to consider the other DNA chain that *base-pairs* with the one that you have as an example, as this has *another* three reading frames. By convention, the reading on

**8**  *Setting the Biological Scene*

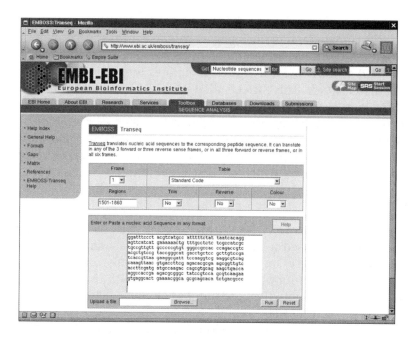

**Figure 1.6**  The EMBOSS/Transeq page at the EBI.

the sequence under study are referred to as +1, +2 and +3, while those on the complement strand are −1, −2 and −3.

The effects of choosing the correct and incorrect reading frames can be investigated using the *Transeq* tool contained in the *EMBOSS* suite of programs. As these tools are discussed later in this book, a number of the details are glossed over here in favour of illustrating the point at hand. Figure 1.6 on page 8 shows the *Transeq* interface provided by the EBI at the following Internet address:

```
http://www.ebi.ac.uk/emboss/transeq/
```

For this example, consider bases Bases 1501 through 1800 from EMBL entry M245940. This sequence is chosen because it contains the MerP protein. These particular bases are easy to extract from a disk-file using any text editor. From the entry, the six lines of DNA bases (near the end of the EMBL data-file) can be copied. The line numbers at the end of each line can be removed and then the resulting data can be pasted into the box on *EMBOSS/Transeq* WWW form (refer to Figure 1.6). Here's what the data looks like *before* the editing takes place:

```
     ggatttccct acgtcatgcc atttttctat taatcacagg agttcatcat gaaaaaactg     1560
     tttgcctctc tcgccatcgc tgccgttgtt gccccgtgt gggccgccac ccagaccgtc      1620
```

```
acgctgtccg taccgggcat gacctgctcc gcttgtccga tcaccgttaa gaaggcgatt    1680
tccaaggtcg aaggcgtcag caaagttaac gtgaccttcg agacacgcga agcggttgtc    1740
accttcgatg atgccaagac cagcgtgcag aagctgacca aggccaccga agacgcgggc    1800
tatccgtcca gcgtcaagaa gtgaggcact gaaaacggca gcgcagcaca tctgacgccc    1860
```

If desired, the space between each group of ten letters can be removed using any editor's *search-and-replace* function. However, in the *raw* sequence, space characters and newlines are ignored, so it is OK to leave them as-is when pasting the data into the form.

The stand-alone, command-line version of *Transeq* has a parameter, called -regions, that restricts translation to a specified range of bases. To use this feature on the WWW form, insert "1501-1860" into the "Regions" box.

> **Technical Commentary:** Note that the line numbers on the right-hand side of the above extracted data are actually the index of the last base on the line. This means that 1501 is the first base on the line that ends with 1560, as the bases are arranged in six blocks of ten per line.

The results of this web-run are not shown. Here is the *correct* result, which is reading frame +1 relative to the start point of the sequence just selected:

```
GFPYVMPFFY*SQEFIMKKLFASLAIAAVVAPVWAATQTVTLSVPGMTCSACPITVKKAI
SKVEGVSKVNVTFETREAVVTFDDAKTSVQKLTKATEDAGYPSSVKK*GTENGSAAHLTP
```

The underlined section is the MerP protein sequence. It starts with a Methionine (M) *start* signal codon, which is ATG, as this is the DNA representation, not RNA. It ends with * *stop* codon (which is TGA in DNA). The *start* and *stop* codons are underlined in the original sequence block above. The rest of the triplet of bases (the other codons) are translated by looking them up in standard codon translation tables. These vary very little between organisms.

This translation of the DNA for the MerP protein is also documented in the EMBL disk-file in annotation included with the original M15049 EMBL entry's FT annotation (where "F" and "T" are taken from "feature"):

```
FT      CDS       1549..1824
FT                /codon_start=1
FT                /db_xref="GOA:P13113"
FT                /db_xref="SWISS-PROT:P13113"
FT                /transl_table=11
FT                /gene="merP"
FT                /product="mercury resistance protein"
FT                /protein_id="AAA98223.1"
FT                /translation="MKKLFASLAIAAVVAPVWAATQTVTLSVPGMTCSACPITVKKAIS
FT                KVEGVSKVNVTFETREAVVTFDDAKTSVQKLTKATEDAGYPSSVKK"
```

Note that all of the hard work is already done, including a cross reference to the SWISS-PROT database (the "/db_xref=SWISS-PROT:P13113" bit) as well as the official translation of the DNA sequence[7].

---

[7] We will have more to say about SWISS-PROT and EMBL in later chapters.

This introduction is purposefully straightforward. Things become more difficult when all that's at hand is a small piece of DNA, the order of the bases and, maybe, the name of the organism. Using these data to identify a protein is returned to later in *Bioinformatics, Biocomputing and Perl*.

Once produced, the polypeptide chain must by folded in order to become an active protein in the functional form. A common assertion is that all the information needed to produce the defined structure of the fully functional protein is contained in the amino acid sequence. In a very general sense, this is true. However, it is only correct when the environment within which the polypeptide exists is taken into account.

## 1.5 Genome Sequencing

The sequencing of an entire genome – the DNA content of a particular organism – is now relatively routine. Originally, it was performed in a very "cottage industry" way, with small groups of researchers working away, in relative isolation, at sequencing small sections of the complete genome.

Today, genome sequencing is "big science", and there are numerous specialised genome sequencing centres around the world, such as The Welcome *Trust Sanger Institute* in the United Kingdom and The Center *for Genome Research* in the United States. A number of commercial organisations sequence genomes on a for-profit basis, with *Celera Genomics* the most famous – some would say "infamous" – because of the company's efforts to beat the publicly funded *Human Genome Project* in being first to publish the draft human genome sequence. This was in an effort to copyright and/or patent the information and, consequently, charge money for the usage rights[8].

In *Bioinformatics, Biocomputing and Perl*, the emphasis is on analysing the DNA and protein sequences rather than understanding the technical details of the methods by which the sequences are produced. However, it is important to have (at least) a rudimentary knowledge of the technologies used to produce the sequences. This allows the reader to better understand both the successes and the problems associated with the processes, as well as how they influence the data analysed. This description is very brief and intended to summarise the more thorough treatments found in any general biochemistry or molecular biology textbook.

Nowadays, most DNA is sequenced using the *Dideoxynucleotide (Chain Termination) Method* developed by *Fredrick Sanger* and his colleagues. This method uses a modified *DNA polymerase enzyme* to make copies of the DNA present in an original sample. As well as the normal DNA nucleotide bases present in the reaction mixture, special *di-deoxy versions* are also included. These have hydrogen atoms instead of hydroxyl groups in the ribose sugar at two positions: the

---

[8] The scoundrels! Jeez ... why didn't *we* think of that?

2' (as per normal DNA bases) and also at the 3' position. This means that when the DNA polymerase adds a di-deoxy base to the elongating DNA chain, no more bases can be added to that chain. This is because the hydrogen at the 3' position is non-reactive compared to the hydroxyl group normally present. The result is that the chain is essentially *blocked* from further extension at this length. As all four di-deoxy nucleotides are added to the reaction mixture, there will be *blocked* examples of the DNA molecules that terminate at every base.

These molecules can be separated from each other by the use of a *polyacrylamide gel lattice*, as shorter DNA molecules pass through it quickly, while longer ones take more time. Each di-dedoxy nucleotide is labelled with a different fluorescent marker corresponding to the base type: A, T, G and C. This tag can be excited by a laser scanning at a particular location and the base passing that point at a particular time can be *read off*. The length of this "read" is typically about 500 bases before the separation between the molecules becomes too poor to determine which molecule is passing under the laser excitation position. Actually, longer reads are possible but can result in reduced accuracy if special techniques are not employed. For the purposes of this book, 500 bases is assumed to be enough. Even if this were 250 or 1000, it would not algorithmically affect the next step, which is sequence assembly. All that's required is to do more or less depending on the actual value chosen.

## 1.5.1 Sequence assembly

500 bp (base pairs nucleotides) is a short piece of DNA compared to the total found in organisms. This can code for a protein of slightly over 165 aminos[9], which is a "none-too-large protein". Yet even viruses that are not self sufficient have many kilobases of DNA that have been sequenced. The general technique is to sequence many 500 bp regions and then *stitch them back together*. This has allowed the DNA sequence for a particular organism, commonly referred to as "The Genome", to be found. Nowadays, sequencing the genome is one of the standard stages in the analysis of any sufficiently interesting organism, and the *threshold of interest* that must be reached before resources are committed to such a project continues to fall. The process is as follows:

- An individual organism (or a range of individual organisms) is selected as a representative sample.
- The DNA of the organism is extracted.
- The DNA is fragmented and stored in *biological vector molecules*. Typically, a series is used from those such as *bacterial artificial chromosomes (BAC)* to store large amounts of DNA (up to many hundred of thousands of bases) to cosmids containing up to 40,000 bases.

---

[9] 500/3 = 166.67, recalling that there are three bases in each codon.

- The DNA stored in these vectors are sequenced in sections of around 500 bases at a time and then re-assembled. This is accomplished by the use of the *di-deoxy chain termination sequencing method*, as described above.

There are differences in the methods employed here, particularly the type and size of vectors used and the strategy used for their selection. All these factors influence the re-assembly process and the *coverage* of the resultant sequence, which may contain large "gaps" that need filling. Determining the first example genome for an organism is the hard part. After that, it is relatively easy to re-sequence the parts of the organism that different research projects find interesting, even if these "interesting parts" tend to be a tiny fraction of the whole genome. So, a genome is the complete DNA content of a cell that codes from an organism. As an indication of the relative sizes involved in sequencing a protein, consider that a human cell contains about two billion bases, while the *Escherichia coli* bacterium has approximately four million. Viruses tend to have a few tens of thousands.

## 1.6 The Example DNA-gene-protein System We Will Use

Throughout *Bioinformatics, Biocomputing and Perl*, a relatively "nice" example of DNA and protein sequences is used to explain the basic concepts of sequence analysis. The DNA-gene-protein system we will use is the *Mer Operon*. This is a set of genes often found in bacteria that are important for the detoxification of mercury by the conversion of $Hg^{2+}$ ions to the less toxic $Hg$ metal.

The system has been well characterised and the following genes have been identified in it (refer to Figure 1.7 on page 13):

- MerA is *mercury reductase* (Enzyme Classification Number: 1.16.1.1). This is the protein that uses NADPH to reduce $Hg^{2+}$ (mercury) ions.
- MerR is the regulator protein that represses the production of the Mer proteins. When $Hg^{2+}$ ion binds to this protein, the transcription of the other Mer genes is stimulated.
- MerP, MerT and MerC are membrane-associated proteins that sequester free $Hg^{2+}$ ions until they can be detoxified by MerA.
- MerB is the protein *organomercurial lyase* (Enzyme Classification Number: 4.99.1.2). This cleaves the carbon–mercury bond formed in other structures releasing $Hg^{2+}$ ions for detoxification.

The specific examples used are from the bacteria *Serratia Marcescens*, and their DNA sequences span the two EMBL database entries, M15049 and M24940. Although these entries contain most of the genes that have been identified in the

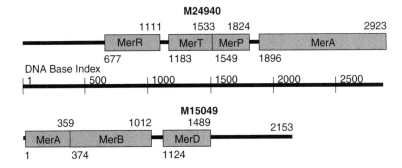

**Figure 1.7** The Mer Operon example DNA-gene-protein.

*Mer Operon*, some are still absent. However, the MerA and MerT genes that form the "core" of the system are always present. Refer to the following web-site for more information on Mer Operon:

http://www.uga.edu/cms/FacAOS.html

As stated earlier, an advantage to studying this system is that it has been so well characterised. So, after a particular analysis is complete, it is possible to look up the "right answer" and compare it with what was found as the result of the analysis. If the two results are similar, then the assumption is that the analysis worked.

## Where to from Here

This chapter sets the scene for this book from a *biological perspective*. In the next chapter, the scene is set again, this time from a *technological perspective*.

# 2
# Setting the Technological Scene

*Perl's relationship to operating systems and applications.*

## 2.1 The Layers of Technology

An objective of this book is to enable the reader to acquire an understanding of, and ability in, the Perl programming language as the main enabler in the development of bespoke computer programs for use in the area of Bioinformatics. As a prelude, let's set the technology scene.

Modern computers are organised around two main components: *hardware* and *software*. The hardware is the stuff that can be seen and touched: screens, keyboards, printers, mice, and so on. Hardware also includes network connections, hard disks and ZIP drives. In order to use hardware, technology is required to *drive* it. This is the role of software. Without software, hardware is all but useless.

Software is typically categorised *by type*. It is useful to think of the types of software as being organised into technology layers (see Figure 2.1 on page 16).

The category of software that is closest to the hardware is the *operating system*. This interacts directly with the hardware and is responsible for ensuring the efficient and equitable use of all hardware resources available. Example operating systems, of which there are many, include Linux, UNIX, Windows, Mac

---

*Bioinformatics, Biocomputing and Perl.* Michael Moorhouse and Paul Barry
© 2004 John Wiley & Sons, Ltd   ISBN 0-470-85331-X

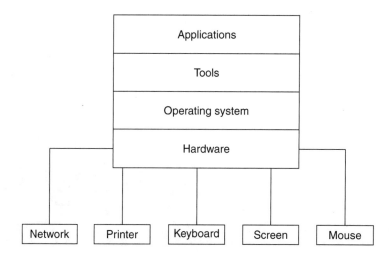

**Figure 2.1** The layers of technology.

OS X, MS-DOS and VMS. Like hardware, operating systems on their own are not very useful.

Another category of software, known as *tools*, takes advantage of what the operating system has to offer, enabling a set of services to be made available to application builders, that is, programmers. The tools category includes programming languages, databases, editors and interface builders. So Perl is, first and foremost, a software tool. Tools provide an environment within which *applications* can be created and deployed.

Applications are, by far, the most useful category of software. The application layer also has the largest diversity, and includes software such as web browsers, e-mail clients, web servers, word processors, spreadsheets and so on. It is this layer that users interact with to get their work done.

The overall process is that applications are built with tools that use the services provided by the operating system, which in turn interacts with the hardware.

## 2.1.1 From passive user to active developer

Since it is often the case that pre-existing applications do not provide a sufficiently specific solution to a user's needs, there continues to be a need to develop bespoke computer programs tailored to meet the particular, and sometimes unique, requirements identified in the user environment. The emphasis in this book is on acquiring an understanding of, and ability in, the Perl programming language.

By the end of *Bioinformatics, Biocomputing and Perl*, the reader will no longer be a *passive user* who simply clicks web-page links and selects an option from a menu, but will instead be an *active developer*, capable of building web-pages and bespoke computer programs.

## 2.2 Finding `perl`

As mentioned in the Preface, this book assumes that the Linux operating system is being used. If so, the Perl programming language and its environment should already be installed. A method of confirming this is detailed below. If Linux is *not* running, don't worry: the vast majority of the program code in this book should work on any version of `perl`, regardless of the operating system used. Please refer to the *Installing Perl* appendix on page 453 for instructions on installing Perl onto any one of a variety of operating systems.

### 2.2.1 Checking for `perl`

On Linux, check if something is installed by using the `whereis` command. Take care to use the correct case since Linux operating systems are case-sensitive (generally system tool names such as `whereis` are all lower case, as here, but not always):

```
whereis  perl
```

When the above command is executed on Paul's computer (which is running a recent version of *RedHat Linux*)[1], the results are:

```
perl: /usr/bin/perl /usr/share/man/man1/perl.1.gz
```

This confirms that "perl" is in the `/usr/bin/` directory location, and there is also "`perl.1.gz`" in the `/usr/share/man/man1/` directory location. The former is the actual `perl` program, the latter is part of the Perl documentation[2].

Another Linux command, `which`, reports on the version of `perl` that executes when the `perl` program is invoked. Again, using Paul's computer, this command:

```
which  perl
```

produces this result:

```
/usr/bin/perl
```

---

[1] Michael's computer, which is running *SuSE Linux*, also reports this directory location for `perl`. Other computers may report `/usr/bin/perl5.00503` as the location for `perl`, which looks a little strange. This is an older version of `perl`, which will run most of the Perl in this book, except for those programs that require the installation of some very specific modules.

[2] This sentence serves to illustrate a convention in the Perl programming community: when referring to the tool that executes a Perl program, we refer to it as "`perl`", whereas the programming language itself is referred to as "Perl".

The actual location of the `perl` program is confirmed to be the `/usr/bin/` directory location. Note that it is possible to have more than one `perl` installed on a computer, so the `whereis` command may report more than one directory location. The `which` command confirms which of the alternatives is actually executed. Note that another very popular directory location for `perl` is:

```
/usr/local/bin/perl
```

Now, make a note of the `perl` directory location reported by your computer, as this information is needed in the next chapter.

## Where to from Here

Having lulled the reader into a rather comforting, but false, sense of security with this less-than-demanding technical chapter, the next chapter introduces the more taxing subject of the basics of programming, Perl style. It is time to get your hands dirty.

# Part I
# Working with Perl

# 3
# The Basics

*Getting started with Perl for Bioinformatics programming.*

## 3.1  Let's Get Started!

There is no substitute for practical experience when first learning how to program. So, here is our first Perl program, called `welcome`:

```
print "Welcome to the Wonderful World of Bioinformatics!\n";
```

When executed by `perl`[1], this small program displays the following, perhaps rather not unexpected, message on screen:

```
Welcome to the Wonderful World of Bioinformatics!
```

This program could not be easier. A single Perl command, `print` in this program, tells `perl` to display on screen the phrase found within the double-quotes. Use any text editor to create the `welcome` disk-file on a computer (it is required in the next section). Now, let's look at another way to write `welcome`:

```
print "Welcome ";
print "to ";
print "the ";
print "Wonderful ";
print "World ";
print "of ";
print "Bioinformatics!";
print "\n";
```

[1] We will learn how to do this is in just a moment.

This considerably longer program, called `welcome2`, displays *exactly the same message* as our first Perl program. Rather than displaying the phrase as a whole, as was the case with `welcome`, this program displays each word from the phrase individually, that is, with its own `print` command. Of note is the last `print` command, which displays \n. Just what exactly is \n? It's how to tell `perl` to display, or take, a new line.

These two programs serve to illustrate and highlight our first programming maxim[2].

> **Maxim 3.1** *Programs execute in sequential order.*

The `welcome2` program displays the word "Wonderful" *before* displaying the word "World". That is, the `print` commands are executed *in sequence*, one after the other.

> **Technical Commentary:** Within Perl, and almost all other programming languages, each line in a program is referred to as a "statement". Perl statements end with, and are separated from any other statements by, a semicolon, that is, the ";" character.

Here's another programming maxim highlighted by these two programs.

> **Maxim 3.2** *Less is better.*

As far as these programs are concerned, the *smaller* of the two, `welcome`, is the *better* of the two. By giving each word in the phrase its own `print` command, the `welcome2` program is more complex than it needs to be. It is also harder to understand. This is in spite of the fact that it is *functionally identical* to `welcome`. Adding complexity to programs for no benefit is a practice to be avoided. Put another way, the second maxim could be rewritten as follows.

> **Maxim 3.3** *If you can say something with fewer words, then do so.*

## 3.1.1 Running Perl programs

Prior to actually running a program, it is prudent to first check Perl programs for obvious errors. To do this for `welcome`, type the following at the Linux command-line (where the -c stands for "check"):

```
perl -c welcome
```

If all is well, `perl` responds with:

```
welcome syntax OK
```

---

[2] A *maxim* is a general truth or a rule of thumb expressed as a sentence.

Let's assume that the welcome program contains an error, specifically that the word "print" is entered as "pint". When the syntax-checking command-line is entered, the following messages appear:

```
String found where operator expected at welcome line 3,
    near "pint "Welcome to the Wonderful World of Bioinformatics!\n""
        (Do you need to predeclare pint?)
syntax error at welcome line 3,
    near "pint "Welcome to the Wonderful World of Bioinformatics!\n""
welcome had compilation errors.
```

When messages such as this appear, don't panic! This is perl's way of indicating that there is something wrong with the program. Look at the messages and the program again and check the spaces, as quotation marks and semicolons are most likely to get left out or misplaced. Commands can also be misspelt, as is the case here, resulting in a syntax error. Now, just what exactly is "syntax", and why is it OK or in error?

In any written language, syntax refers to the way words are arranged to form phrases and sentences. When referring to computer programs in any programming language, syntax refers to the arrangement of program statements. Specifically, the arrangement of statements as defined by the programming language's rules and regulations is known as its syntax.

So, the perl program is happy that the welcome program contains only legitimate Perl statements, and that no syntax rules have been violated.

## 3.1.2 Syntax and semantics

It is important to understand that a Perl program may be *syntactically correct*, but *semantically wrong*. Semantics has to do with the *meaning of language*. For a Perl program to be syntactically correct but semantically wrong means that the program satisfies the rules and regulations of the programming language, but does *not* do what you expected it to do.

For example, here is a syntactically correct but semantically wrong Perl program, called whoops:

```
print ; "Welcome to the Wonderful World of Bioinformatics!\n";
```

When "perl -c whoops" is executed, the familiar "whoops syntax OK" message appears. So syntactically, everything is OK. However, try executing this command-line, which actually runs the program (note: the -c is missing):

```
perl whoops
```

And nothing appears on screen. Oh dear.

The whoops program is semantically incorrect, in that it does not do what we were expecting it to do. In fact, it does nothing. The problem is that the `print` command has been terminated too early. Look at that ";" character right after the word `print` in the program. What that tells `perl` is that the `print` command has finished printing. As `print` has nothing to print, nothing displays on screen! And as the program has not told `perl` what to do with the friendly message, `perl` does nothing with it. Which is probably the *safest* thing for the program to do.

Surely, `perl` should spot that something is not quite right here? The fact that `perl` sees the message and then decides not to do anything with it should mean something and – if nothing else – should be reported to the programmer.

You are right, it should. But, as programs go, `perl` is the strong, silent type. The problem is that `perl` has not been asked to highlight anything out of the ordinary. All that was required was a syntax check. In contrast, this next command-line instructs `perl` to report potential problems (where -w stands for "warnings"):

```
perl -c -w whoops
```

Now, in addition to performing a syntax check (with -c), we have asked `perl` to look for and report on anything else that might be strange. Here's what `perl` has to say about `whoops` now:

```
Useless use of a constant in void context at whoops line 1.
whoops syntax OK
```

The `perl` program informs the programmer that a "useless use" of something has occurred. In this case, it is the friendly message that is of no use. Note that the syntax is still OK, but the warning message is a clue to look at the program for possible semantic errors.

> **Technical Commentary:** Programmers often refer to semantic errors by another name: *logic errors*.

In learning about syntax and semantics, we rather sneakily demonstrated just how easy it is to execute any Perl program: simply invoke `perl` without the -c switch, as follows:

```
perl welcome
```

This command-line produces this rather triumphant output, which is repeated from the start of this chapter:

```
Welcome to the Wonderful World of Bioinformatics!
```

### 3.1.3 Program: run thyself!

It is possible, on computers running Linux and other UNIX-like operating systems, to arrange for a program to automatically invoke `perl` when necessary. Look at this command-line[3], which is *executed against* the soon-to-be-discussed welcome3:

```
chmod   u+x   welcome3
```

This `chmod` command tells Linux that the welcome3 program can be executed, and it assumes that the following line[4] appears as the first line of welcome3:

```
#! /usr/bin/perl -w
```

The welcome3 program can now be invoked like this, in which the leading ./ tells the Linux operating system to find the welcome3 program in the current directory:

```
./welcome3
```

But it is perfectly OK to still invoke the program like this:

```
perl   welcome3
```

Both techniques are valid and either may be used. Try them for yourself. This leads in rather nicely to the next maxim.

> **Maxim 3.4** *There's more than one way to do it.*

This is also the Perl programming language's *motto*. It is actually more of a philosophy. The central idea being that whatever works for the Perl programmer works for Perl, assuming – of course – it is legitimate Perl. There are many references to this maxim throughout *Bioinformatics, Biocomputing and Perl*

> **Technical Commentary:** The Linux `chmod` command changes the *mode* of a disk-file. Typically, a disk-file is not created as a program, but rather as an *ordinary* disk-file that can be read from or written to. When the mode of the disk-file is changed to *executable*, the disk-file is turned into something that can be executed from the command-line. That is the purpose of "`chmod u+x`". The "u" refers to the user (or owner) of the disk-file, and the "+x" turns on the disk-file's ability to execute.

---

[3] As the welcome2 program is essentially the same program as welcome, we have nothing further to do with it at this stage. That said, it does make a short comeback later in this chapter when used with another example program.

[4] As discussed at the end of the previous chapter, this *may* not be where your `perl` is, so be sure to substitute the correct location here.

## 3.2 Iteration

In the previous section, in addition to learning how to syntax check and execute programs, the concept that programs are a sequence of statements was also introduced[5]. If all that could be accomplished by a program was to execute a simple sequence of statements, the vast majority of programs would not be very useful. So, programming languages support additional mechanisms, known as *programming constructs*, to do more interesting things. One such mechanism is called *iteration*, which is just another word for *repetition*. Here is an example of an iteration from the non-programming world:

*Heat the pie in the oven until the sugar glazes.*

We do something, that is, heat the pie, until something is true, that is, the sugar glazes. Another way of expressing this iteration is:

*While the sugar is still sugar, heat the pie in the oven.*

or:

*While the sugar is not glazed, heat the pie in the oven.*

These latter iterations are less intuitive when compared to the first, mainly because the *test* to see if something is true occurs first, that is, check the state of the sugar, *before* the something to do, that is, heat the pie. The second "while" iteration is the least intuitive, as the check is for a negative, that is, the sugar is *not* glazed and, as a result of this check being true, that is, a positive, the pie continues to heat.

Compared to the original iteration, which used "until", the two "while" iterations seem to have things the wrong way around. It is more natural to say "I'll stand by the fire until I warm up", as opposed to "While I'm cold, I'll stand by the fire", or the truly awful "While I'm not hot, I'll stand by the fire".

Unfortunately, programming languages favour the use of iterations based on the use of "while". Although it is possible to write iterations using "until", such usage tends to be less common in practice.

### 3.2.1 Using the Perl `while` construct

A quick example illustrates the use of the `while` construct in Perl. This next program is called `forever`:

---

[5] Note that this *sequence* is a very different sequence to the *Bioinformatics sequences* we encounter later in this book. Here, "sequence" simply means "one after the other".

## Iteration

```perl
#! /usr/bin/perl -w

# The 'forever' program - a (Perl) program,
# which does not stop until someone presses Ctrl-C.

use constant TRUE    => 1;
use constant FALSE   => 0;

while ( TRUE )
{
    print "Welcome to the Wonderful World of Bioinformatics!\n";
    sleep 1;
}
```

Using the chmod command from earlier, make forever executable and execute it as follows:

```
chmod   u+x   forever
./forever
```

The screen should start to fill with copies of the message, and the program keeps printing the message once every second until Ctrl-C is pressed. Think of Ctrl-C as meaning "cancel". Press and hold down the Ctrl key on your keyboard, then tap the C key to stop the program:

```
Welcome to the Wonderful World of Bioinformatics!
Welcome to the Wonderful World of Bioinformatics!
Welcome to the Wonderful World of Bioinformatics!
Welcome to the Wonderful World of Bioinformatics!
Welcome to the Wonderful World of Bioinformatics!
Welcome to the Wonderful World of Bioinformatics!
Welcome to the Wonderful World of Bioinformatics!
Welcome to the Wonderful World of Bioinformatics!
Welcome to the Wonderful World of Bioinformatics!
    .
    .
    .
```

The forever program repeatedly prints the message on screen, and it continues to print the message while TRUE is true. So the program runs *forever*. The Ctrl-C key combination must be pressed to stop the program executing.

**Technical Commentary:** Rather than use the word *iteration* or *repetition* to refer to this mechanism, many programmers favour the use of the word *loop*. In this context, *loop* is both a *noun* and a *verb*. Typical programmer utterances might be "the loop prints the message five times" or "this program loops forever". Programs that loop forever, just like the forever program in this section, are referred to as an *infinite loop*. The infinite loop is generally regarded as a *very bad thing*, and programmers are encouraged not to introduce such loops into programs.

There is a lot going on in the `forever` program and we are introducing a number of new concepts, so let's look at `forever` in more detail.

The first three lines of the program all start with the # character. The first line is the *run thyself!* line from the last section. The other two require further explanation:

```
#! /usr/bin/perl -w

# The 'forever' program - a (Perl) program,
# which does not stop until someone presses Ctrl-C.
```

In Perl, the # character denotes the start of a comment that runs from the # character to the end of the current line. A *comment* is targeted at the person reading the program, not `perl`. This means that `perl` ignores lines that start with #.

> **Technical Commentary:** Actually, the # character can appear anywhere on the line, not just at the start. When `perl` encounters the # character, everything from the # character to the end of the current line is ignored by `perl`.

Good programmers always add comments to their programs. In effect, comments help document the program and, in doing so, facilitate the application of future changes to a program, especially when the programmer making the changes is not the original author. Time for another maxim.

> **Maxim 3.5** *Add comments to make future maintenance of a program easier for other programmers and for you.*

Having said that, the program code that appears throughout *Bioinformatics, Biocomputing and Perl* is notable in that it is devoid of comments. This is deliberate, as the book is itself a comment on the program code. Two *constant definitions* come after the comment lines:

```
use constant TRUE    => 1;
use constant FALSE   => 0;
```

Unlike other programming languages, Perl has a rather strange notion of what is true and what is false. To keep things simple, for now, note that Perl treats a value of 1 as true and a value of 0 as false. Since it is not desirable for the values of true and false to change within a program, it is prudent to take advantage of Perl's mechanism to define these two values as *constants*. A constant is a value that cannot change while a program is running. If we try to change a constant value by, for instance, trying to add 1 to the value chosen to represent false, the `perl` program complains loudly and refuses to process the program further.

Rather than using 1 and 0 as true and false, we give the constants nice, human-friendly names: TRUE and FALSE. It is a convention to give constants all UPPERCASE names, although the use of UPPERCASE is not required. However, when programming, it is always advisable to follow existing conventions.

> **Technical Commentary:** Perl is case-sensitive. This means that when naming variables in Perl, case is *significant*. So, "TRUE" is a different symbol to "`true`". Other programming languages – notably Pascal – are not as fussy.

Having declared the truth values to be constants, the program uses the value for true in the very next line:

```
while ( TRUE )
```

Which is much easier to read and understand than this:

```
while ( 1 )
```

Isn't it? Even though we are yet to describe this line in detail, "`while ( TRUE )`" should mean more than "`while ( 1 )`". Could we possibly use this program to demonstrate *another* maxim? Yes, we can.

> **Maxim 3.6** *When using constant values, refer to them with a nice, human-friendly name as opposed to the actual value.*

This practice has a very important implication. To understand the implication, imagine writing a program that is 3000 lines long. Throughout the program TRUE has been used extensively, appearing in 42 different places in the program. Next, imagine that throughout the program, TRUE has been used, in error, for FALSE. Changing all 42 occurrences of TRUE is as easy as instructing an editor to execute a global search-and-replace, changing all occurrences of TRUE to FALSE.

Now imagine that 1 has been used as the value for true. Now there is a problem. A global search-and-replace cannot be used to change all the occurrences of 1 to 0, since 1 may not have been used to mean true at every point in the program that 1 appears. For instance, 1 might be added to some other value in the program to *increment* its value, and if this occurrence of 1 is changed to 0, the increment will no longer work, since adding zero to some value does *not* change the value. The program will be syntactically correct but semantically wrong, and all because constants were not used when it was appropriate to do so. The conclusion is clear: *use constants!*

To return to the "`while ( TRUE )`" line from `forever`, an iteration is started that will run for as long as TRUE is true. In other words, the iteration runs forever.

In Perl, the `while` construct encloses a collection of lines to repeat within a *block*. A block in Perl is any collection of program statements enclosed in *curly braces*, which are also known as *squigglies*. The start of the block is marked by the "{" character and the end of the block is marked by the "}" character. Here's the block associated with the `while` statement from `forever`:

```
{
    print "Welcome to the Wonderful World of Bioinformatics!\n";
    sleep 1;
}
```

This block contains two program statements; one is the familiar `print` command that displays the message, and the other is an invocation of Perl's `sleep` command, which pauses the program for the indicated number of seconds, which, in this case, is 1. So, the `forever` program prints the message every second, and keeps printing the message every second until `Ctrl-C` is pressed.

Think of blocks as a way of grouping program statements together, allowing them to be treated (and repeated) as a single entity.

> **Maxim 3.7** *Use blocks to group program statements together.*

To hammer home the point recommending the use of constants for the truth values, note that the length of time to sleep for has the same value as TRUE, that is, 1. However, its meaning is very different. We are asking `perl` to "sleep for one second", not to "sleep true", the latter being another example of syntactic correctness, but incorrect semantics.

## 3.3 More Iterations

Iterating forever is occasionally useful. However, it is more common for iterations to execute for a specific number of occurrences. The loop in the `forever` program kept going while the value of TRUE remained true, which meant that `forever` was designed never to stop[6]. The *thing to check* at the top of the loop has a generic name: *condition*. The general form of a *while loop* in Perl is:

*while ( some condition is true )*
*{*
  *do something*
*}*

Note that as this isn't actual Perl code, it is not shown in the usual `program` font.

We enclose the *something to do* in a block enclosed in curly braces. Note that the condition is itself enclosed in parentheses, the "(" and ")" characters. Just

---
[6] Of its own accord, that is. Remember: we were able to stop it by pressing `Ctrl-C`.

as curly braces are required around a block, parentheses are required around a condition. But just what is a condition?

The short answer is anything that can result in a value of true or false. This is an answer worthy of another maxim.

> **Maxim 3.8** *A condition can result in a value of true or false.*

The longer answer is a little more complicated. It is complicated by the fact that conditions can themselves be complicated, of which more later. However, before starting to use and learn about more complex conditions, we first need to introduce variable containers.

## 3.3.1 Introducing variable containers

Earlier in this chapter, constants were described. To recap, a constant is a container within a program whose value cannot be changed under any circumstance. The opposite of a constant is a variable container, or *variable* for short. A variable's value can change over the lifetime of the program. In other words, a variable's value can *vary*.

> **Maxim 3.9** *When you need to change the value of an item,*
> *use a variable container.*

For instance, it might be a requirement to repeat a loop ten times. A count is kept of the number of iterations. When the count reaches ten, the loops ends and stops iterating. To do this, use a variable.

Perl, probably more than any other programming language, has excellent support for all types of variable containers. The simplest type of variable container is the *scalar*. In Perl, scalars can hold, for example, a number, a word, a sentence or a disk-file. Within Perl programs, scalars are given a name prefixed with a dollar sign ($). Here are some example scalar names:

```
$name
$_address
$programming_101
$z
$abc
$count
```

Scalar variable container names always start with the dollar sign ($) followed by at least one other character, as long as that character is a lowercase letter (a - z), an uppercase letter (A - Z) or an underscore (_).

Look at the list of six example scalar names. They all are *correctly formed*. However, they are not all equally descriptive of the contents of the variable. Simply by looking at them, a reasonably good idea of what $name, $_address and $count will be used for can be formed. Equally, an assumption can be made about what the $programming_101 scalar is being used for. But not so with $z and $abc.

Without examining the program within which $z appears, it is impossible to determine what the variable is being used for, and reading the program within which $z appears may not help at all. Using a single letter for a variable name is rarely justified, as a single letter is not enough to convey any sort of meaning. Even though more than a single letter has been used to name $abc, this is also a poor choice of name. Both $z and $abc highlight a key maxim.

> **Maxim 3.10** *Don't be lazy: use good, descriptive names for variables.*

### 3.3.2 Variable containers and loops

To demonstrate the use of variable containers within loops, a version of `forever` that displays ten messages and then stops can be created. This new program is called `tentimes`, and here it is:

```perl
#! /usr/bin/perl -w

# The 'tentimes' program - a (Perl) program,
# which stops after ten iterations.

use constant HOWMANY    => 10;

$count = 0;

while ( $count < HOWMANY )
{
    print "Welcome to the Wonderful World of Bioinformatics!\n";
    $count++;
}
```

To run this program, use the now familiar `chmod` command to make it executable, then invoke it as follows:

```
chmod u+x tentimes
./tentimes
```

This program displays the following on screen:

```
Welcome to the Wonderful World of Bioinformatics!
Welcome to the Wonderful World of Bioinformatics!
Welcome to the Wonderful World of Bioinformatics!
Welcome to the Wonderful World of Bioinformatics!
Welcome to the Wonderful World of Bioinformatics!
Welcome to the Wonderful World of Bioinformatics!
Welcome to the Wonderful World of Bioinformatics!
Welcome to the Wonderful World of Bioinformatics!
```

```
Welcome to the Wonderful World of Bioinformatics!
Welcome to the Wonderful World of Bioinformatics!
```

The `tentimes` program is not very different from the `forever` program. The usual first line is followed by two lines of comment. A constant is then defined:

```
use constant HOWMANY    => 10;
```

The `HOWMANY` constant is used to control the number of times the loop iterates. Once the program has displayed the message `HOWMANY` times, it stops. A scalar variable container called `$count` is used to count how many times the loop iterates. Before starting the loop, the `$count` scalar is given a value of zero:

```
$count = 0;
```

Read this line as "set the `$count` scalar to equal zero" or "`$count` becomes equal to zero". The = symbol is referred to as the *assignment operator*.

> **Technical Commentary:** In Perl, it is not necessary to set the value of a variable container before it's used. Perl has a number of rules that are applied to the first usage of a variable container, and `perl` sets a scalar to zero if it is first used within a *numeric context*. This feature can be very convenient. However, it is always a good idea to give variable containers an explicit starting value, as it indicates precisely what the intentions for the variable are. Any programmer reading the `tentimes` program should be in no doubt that the `$count` scalar is to be used within a numeric context (not that the use of the word "count" wasn't a big enough clue already).

With the `$count` scalar set and the `HOWMANY` constant defined, the condition for the loop can now be written:

```
while ( $count < HOWMANY )
```

The loop continues to iterate while the value of `$count` is less than the value of `HOWMANY`. Note the use of the standard symbol (borrowed from Mathematics) for less than, namely, <.

> **Technical Commentary:** The use of such symbols is common to all programming languages, not just Perl. Rather than refer to them as symbols, programming languages use the word *operator*. Throughout this part of *Bioinformatics, Biocomputing and Perl*, a number of operators are used accompanied by appropriate explanation. For a review of the list of operators supported by Perl, see *Appendix A: Perl Operators* on page 457.

The block executed by the loop comes next:

```
{
    print "Welcome to the Wonderful World of Bioinformatics!\n";
    $count++;
}
```

The block contains two program statements. The first displays the usual message. The other applies the ++ operator to the $count scalar. This operator is the *increment operator*, and when used, adds 1 to the value of a numeric scalar. As $count started out with a value of zero, this statement sets its value to 1. The next time the loop iterates, $count will have the value 2, and so on.

So, as the loop iterates, the message displays and the value for $count increases until it reaches the value of 10. At this point, the value of $count is no longer less than the value of HOWMANY, and the loop then ends, as does the program.

By employing the services of a simple scalar variable container and two operators (< and ++), a loop has been written that iterates a specified number of times.

## 3.4 Selection

The third basic building block of programming is *selection*. The use of a selection mechanism allows a program to choose one of a number of possible courses of action. Here's a simple selection from the real world:

*I'll eat if I'm hungry, otherwise, if I'm not hungry, I'll sleep.*

Another way of saying this would be:

*If I'm hungry, I'll eat, otherwise I'll sleep.*

And here's the general form of the selection statement in Perl:

```
if ( some condition is true )
{
    do something
}
else
{
    do something else
}
```

The first point to note is that the `else` part of the selection is *entirely optional*. As expected, the `else` block is only executed if the condition fails. The `if` block is executed if the condition is true. Note that, as with loops, blocks of program statements are associated with each part of the selection.

The use of a condition is central to the workings of the selection mechanism, just as it is with loops. If the condition is true, the first block of program statements is executed. If the condition is false, the second block of program statements executes.

### 3.4.1 Using the Perl `if` construct

Here is another variation on the `forever` program that prints the message five times. This program is called `fivetimes`:

```
#! /usr/bin/perl -w

# The 'fivetimes' program - a (Perl) program,
# which stops after five iterations.

use constant TRUE     => 1;
use constant FALSE    => 0;

use constant HOWMANY  => 5;

$count = 0;

while ( TRUE )
{
    $count++;
    print "Welcome to the Wonderful World of Bioinformatics!\n";
    if ( $count == HOWMANY )
    {
        last;
    }
}
```

The first three lines are as expected: the standard first line followed by two comment lines. Then come three constant definitions that require no explanation as they have been seen before. As in the `tentimes` program, the `$count` scalar is initially set to zero. The loop then begins:

```
while ( TRUE )
{
    $count++;
    print "Welcome to the Wonderful World of Bioinformatics!\n";
```

Note that the loop condition is simply TRUE, which is defined as a constant value of 1. The value of 1 represents a true value in Perl, which results in this block looping forever. As in the `tentimes` program, the first two program statements in the block increment `$count` and display the message. Then comes the selection:

```
    if ( $count == HOWMANY )
    {
        last;
    }
}
```

## 36  The Basics

The `if` selection statement uses the *numeric equality operator* (==) to check if the value of $count is equal to the value of the HOWMANY constant. If it is not equal, that is, if the condition fails, the block associated with the `if` statement is not executed, which results in another iteration beginning. If $count is equal to HOWMANY, the block associated with the `if` statement is executed. This block contains a single Perl command, `last`, which forces `perl` to exit from the current loop regardless of whether or not the loops condition is true or false. In effect, the `last` command is a loop short-circuit, and its use here ensures that the `fivetimes` program stops after five iterations.

If you are wondering which "stop the loop" technique is the best to use, either (a), the iteration condition test from `tentimes` or (b), the selection condition test combined with the `last` command from `fivetimes`, the maxim on page 25 provides the answer.

## 3.5 There Really is MTOWTDI

Where MTOWTDI stands for "more than one way to do it". This philosophy is one of the great strengths of Perl, but care is needed. Let's illustrate the good and the bad of this philosophy, by example, starting with a couple of not so good examples followed by a couple of much improved ones. Here's another example program, called oddeven:

```perl
#! /usr/bin/perl -w

# The 'oddeven' program - a (Perl) program,
# which iterates four times, printing 'odd' when $count
# is an odd number, and 'even' when $count is an even
# number.

use constant HOWMANY    => 4;

$count = 0;

while ( $count < HOWMANY )
{
    $count++;
    if ( $count == 1 )
    {
        print "odd\n";
    }
    elsif ( $count == 2 )
    {
        print "even\n";
    }
    elsif ( $count == 3 )
    {
        print "odd\n";
```

```
        }
        else    # at this point $count is four.
        {
            print "even\n";
        }
    }
```

The comments at the top of the program explain its purpose. The program iterates, and as it iterates it examines the value of the $count scalar. When $count is an odd number, the word "odd" displays on screen. When $count is an even number, the word "even" displays on screen.

Of note is that the if selection statement is a *multi-way* selection. It has four blocks, with one of the blocks executing when the value of $count is 1, 2, 3 or 4. When $count has a value of 2 or 3, the blocks associated with Perl's elsif are executed[7]. The trailing else block does not have a condition associated with it, as it assumes that $count is not 1, 2 or 3, so it must be 4. Note the use of a comment to document this assumption.

This long if statement also highlights another property of this selection mechanism, that only one block is executed on each iteration. The value of the $count scalar *controls* which block is executed.

Use chmod to turn the oddeven program into a file that can be executed, then invoke it with the "./oddeven" command-line. As expected, the program displays the following on screen:

```
odd
even
odd
even
```

Whew! We can all sleep tonight: the oddeven program has confirmed that 1 and 3 are odd numbers, and 2 and 4 are even numbers.

Now, let's look at another program that does exactly the same thing as oddeven. This program is called terrible:

```
#! /usr/bin/perl -w
# The 'terrible' program - a poorly formatted 'oddeven'.
use constant HOWMANY => 4; $count = 0;
while ( $count < HOWMANY ) { $count++;
if ( $count == 1 ) { print "odd\n"; } elsif ( $count == 2 )
{ print "even\n"; } elsif ( $count == 3 ) { print "odd\n"; }
else    # at this point $count is four.
{ print "even\n"; } }
```

---

[7] Note the strange spelling. Other programming languages use elseif or else if, which are both illegal syntax as far as Perl is concerned.

Yikes! What a mess. Look closely. Notice that the program statements that make up the `terrible` program are *exactly the same* as those that make up the `oddeven` program. The results produced from `terrible` are *exactly the same* as those produced by `oddeven`, namely:

```
odd
even
odd
even
```

The difference between the two programs has to do with how they are laid out, or *formatted*. The `oddeven` program uses plenty of whitespace, blank lines and indentation to present the program statements in such a way that they are easy for another programmer, and you, to read. The use of indentation helps the reader of `oddeven` see which blocks of code are associated with which condition tests. The `terrible` program, on the other hand, squeezes as much as possible onto as few lines as possible. It is just about readable, but it is all but impossible to see which blocks are associated with which condition tests, let alone work out what the program actually does.

> **Technical Commentary:** Like a lot of modern programming languages, Perl is classified as *free format*. This means that you can write a program using whatever formatting you prefer, as `perl` can just as easily process a well-formatted program, such as `oddeven`, as it can a poorly formatted program, such as `terrible`. Do yourself and everyone else a favour, and be sure to format your programs to be as *readable* as possible.

It is time for a new maxim, as it has been quite a while since the last one.

> **Maxim 3.11** *Use plenty of whitespace, blank lines and indentation to make your programs easier to read.*

Do not be misled into thinking that comments are central to the future understanding of a program, as suggested by an earlier maxim. Certainly, comments are important, but all the comments in the world will not make up for improperly formatted and poorly laid out code. Take another quick look at the `terrible` program. It really is a *mess*.

Now for the good. Here is another version of `oddeven`. This program is called `oddeven2` and it produces *exactly the same* output as both `oddeven` and `terrible`:

```perl
#! /usr/bin/perl -w

# The 'oddeven2' program - another version of 'oddeven'.

use constant HOWMANY    => 4;

$count = 0;
```

```
    while ( $count < HOWMANY )
    {
        $count++;
        if ( $count % 2 == 0 )
        {
            print "even\n";
        }
        else    # $count % 2 is not zero.
        {
            print "odd\n";
        }
    }
```

The key program statements are these:

```
    if ( $count % 2 == 0 )
    {
        print "even\n";
    }
    else    # $count % 2 is not zero.
    {
        print "odd\n";
    }
```

That percentage sign (%) is another Perl operator, the *modulus, %* operator. Given two numbers, "A % B" returns the *remainder* after A has been divided by B, assuming both are positive numbers.

> **Technical Commentary:** The use of the word "return" in the last paragraph requires further explanation. When programmers state that some program statement "returns" some value, they mean that the program statement produces some result that then becomes available to the program. That is, the program statement actually has a *value associated with it*. In the examples that follow, each statement results in the remainder being "returned" to the program, which then prints the resulting value, that is, the value of the remainder after the modulus operator has executed.

Here are some examples of the modulus operator in action:

```
    print 5 % 2, "\n";      # prints a '1' on a line.
    print 4 % 2, "\n";      # prints a '0' on a line.
    print 7 % 4, "\n";      # prints a '3' on a line.
```

The key point is that an odd number divided by 2 yields a remainder of 1, whereas an even number divided by 2 yields a remainder of 0.

The `oddeven2` program exploits this mathematical property, thanks to Perl's modulus operator and the `$count` scalar. When "`$count % 2`" yields zero, the word "even" is printed, and when "`$count % 2`" yields anything other than zero, the word "odd" is printed.

The oddeven2 program is shorter than oddeven. But is this enough to make it *better*? On its own, it is not. However, the oddeven2 program is better because it is *easier to extend*. Specifically, if the program is changed to iterate 20 times, we need only make one change: the value for HOWMANY becomes 20 instead of 4. Contrast this to the changes required to the oddeven program: adding 16 new elsif blocks for each of the new values of $count, 4 through 19. This is probably too much work to be worth the effort. However, the change to oddeven2 is trivial, which is an excellent example of taking the maxims to heart. And we can do even better. Take a look at this program, called oddeven3:

```
#! /usr/bin/perl -w

# The 'oddeven3' program - yet another version of 'oddeven'.

use constant HOWMANY    => 4;

$count = 0;

while ( $count < HOWMANY )
{
    $count++;
    print "even\n" if ( $count % 2 == 0 );
    print "odd\n"  if ( $count % 2 != 0 );
}
```

Let's take a look at the print commands more closely. Here's the first:

```
print "even\n" if ( $count % 2 == 0 );
```

Read this program statement as "print the word 'even' if and only if the value of $count modulus 2 is equal to zero". Unlike the if statements that appeared in oddeven and oddeven2, in which the condition test comes *before* the block, in oddeven3, the condition test comes *after* the program statement. Such an arrangement is known as a *statement qualifier* in Perl. The second print statement in the oddeven3 program is this:

```
print "odd\n"  if ( $count % 2 != 0 );
```

This program statement introduces a new operator, !=, which means "not equal to". This statement can be read as "print the word 'odd' if and only if the value of $count modulus 2 is *not* equal to zero".

When the block of program statements associated with a particular if condition test is small (as is the case with oddeven), it is often more natural to use a statement qualifier, specifying "print if" as opposed to "if print". Both work, of course.

## 3.6 Processing Data Files

The programs developed thus far have served their purpose in demonstrating the basic programming mechanisms of sequence, iteration and selection. However, although academically interesting, these programs have not really performed any useful function. It is only when data from outside a program comes into the picture that this changes. Getting data into a Perl program is *surprisingly easy*.

In order to demonstrate just how easy, we need to introduce another, rather special, Perl operator. The *input operator* looks like this:

```
<>
```

When perl encounters this operator within a program, it looks for and returns a line of input from *standard input*. This is the name given to the mechanism that is currently providing input data to the program. Unless perl is told otherwise, the default input mechanism is the keyboard. This means that a program takes *a line of data* from the keyboard whenever the input operator is used.

It is useful to think of a program's input as its *data*. Conversely, think of a program's output as its *results*. Consider this program statement:

```
$line = <>;
```

The scalar variable container, called $line, is assigned its value from the input operator. In other words, a line is read from the keyboard and put into the $line scalar. Here is a small program called getlines that exploits the above program statement:

```
#! /usr/bin/perl -w

# The 'getlines' program which processes lines.

while ( $line = <> )
{
    print $line;
}
```

The first line of this program is the usual *run thyself!* line. A one-line comment follows, then the remainder of the program is a loop. What is strange about this loop is that the condition part does not result in a numeric value, unlike the other loops seen so far in this chapter. Instead, the getlines program has a condition part that uses <> to look for and return a line from standard input. The line, when available, is assigned to the $line scalar, which is then checked for *trueness*. But how can a line be checked for *trueness*?

Earlier in this chapter, we alluded to the fact that the values for true and false in Perl are a little strange. It turns out that, in addition to using numerics to represent true and false, strings[8] also have a truth value. The rule is simple: A string with no characters is false, otherwise it is true.

Returning to the `getlines` program, note that the `<>` operator returns a line from standard input and assigns it to `$line`. The *trueness* of the `$line` scalar is then tested. If `$line` contains one or more characters it is considered to be true, otherwise it is considered to be false. That is, if it contains no characters, it is an *empty string*, and "empty" implies false. Obviously! Remember the earlier warning in this chapter that Perl had its own unique *notion* of truth.

So, the loop in `getlines` keeps iterating while there are lines of input arriving from standard input, that is, the keyboard. The single program statement within the loop simply displays the line on screen using Perl's `print` command.

> **Technical Commentary:** In addition to *standard input*, Perl has *standard output*, the default place to display normal messages, and *standard error*, the default place to display error messages. Unless told otherwise, `perl` uses the screen as the default for both standard output and standard error. To make things convenient, standard input, standard output and standard error go by the shorthand names of `STDIN`, `STDOUT` and `STDERR` respectively.

The now familiar command makes `getlines` executable:

```
chmod   u+x   getlines
```

Run the `getlines` program as follows:

```
./getlines
```

The program takes a new line, then nothing appears to happen. What is actually happening is that `getlines` is waiting for some input to arrive from standard input. Go ahead and type something at the keyboard, remembering to press the `Enter` key at the end of each line typed. Immediately upon pressing `Enter`, the `getlines` program displays what is typed on screen. It iterates for as long as lines are typed at the keyboard, as the `<>` operator takes what was typed and assigns it to the `$line` scalar, which is then checked for *trueness*. As long as there is something typed, `$line` results in a true value.

To signal to the `getlines` program that typing is finished, press `Ctrl-D`. This sends an *end-of-file message* to the current program. Think of `Ctrl-D` as signalling "Done".

---

[8] Sequences of zero or more characters.

## 3.6.1 Asking `getlines` to do more

Type the following command-line:

```
./getlines terrible
```

And, as awful as it is, the `terrible` program appears on screen. Try this command-line:

```
./getlines terrible welcome3
```

This time, not only does the `terrible` program appear on screen but it is also immediately followed by the `welcome3` program. In fact, it is the *contents* of the disk-files that appear. The fact that these two disk-files are Perl programs is of no consequence to the `getlines` program, it just sees them as a collection of lines to display. Now, no changes have been made to the `getlines` program, so how is this display of disk-files occurring? What are we not telling you?

Nothing, if truth be told. The `getlines` program is still reading lines from standard input until there are no more lines to read. However, with the above command-lines, rather than looking to the keyboard for data, the `getlines` program looks to the disk-files for data. What `perl` does is open the first disk-file (`terrible`) and read each line from the disk-file, passing the lines one at a time to the `<>` operator within the `getlines` program. When the data (the lines) within `terrible` are exhausted, `perl` closes the disk-file and then opens the `welcome3` disk-file and reads its data one line at a time. When there are no more lines or disk-files to process, the `Ctrl-D` end-of-file message is passed to `getlines`.

This is really cool. As the `getlines` program uses standard input, it uses the keyboard by default. When used in association with a named disk-file, it uses the contents of the disk-file as input, and there can be more than one named disk file.

> **Technical Commentary:** Programmers refer to the list of "things" on the command-line that follow a program name as its command-line *arguments* or *parameters*. The last invocation of `getlines` has two command-line arguments, the word "`terrible`" and the word "`welcome3`".

As stated at the start of this section, getting data into `perl` is not difficult. This is one of Perl's main strengths – processing disk-files that contain textual data – and it goes a long way to explaining Perl's popularity as the programming language of choice within the Bioinformatics community. This is no accident. Perl is a very powerful text processor. The *icing on the cake* is a technology called *regular expressions*, which is introduced in the next section.

## 3.7 Introducing Patterns

As strange as this may sound, Perl has another programming language built into it. This language within a language makes extensive use of Perl's *regular expression, pattern-matching* technology.

The Perl on-line documentation[9] defines a regular expression to be "simply a string that describes a pattern". The *pattern* identifies *what* it is hoped to match. The actual *how* of finding the pattern is taken care of by the `perl` program.

> **Technical Commentary:** A programming language that allows the programmer to specify what is required is often referred to as a *declarative* language. The programmer "declares" what's required, and the technology works out the details. On the other hand, a programming language that allows the programmer to specify exactly how a result is to be arrived at is often referred to as a *procedural* language. The programmer defines the "procedure" to be followed, and the technology blindly follows the instructions. Most programming languages can be classified as one or the other, either declarative or procedural. Remarkably, Perl can be one or the other, or *both*.

The definition of regular expression patterns is a complex topic, and an entire chapter is devoted to the details later in *Bioinformatics, Biocomputing and Perl*. For now, and by way of introducing regular expressions, a very simple pattern will be used to demonstrate the potential of this programming mechanism. Take a look at the next program, called `patterns`:

```
#! /usr/bin/perl -w

# The 'patterns' program - introducing regular expressions.

while ( $line = <> )
{
    print $line if $line =~ /even/;
}
```

This program is very similar to the `getlines` program from the last section. Changes were made to the comment, of course, and to the `print` command within the loop's block. Let's look at the changed `print` command in more detail:

```
print $line if $line =~ /even/;
```

Before describing this program statement in detail, here's the English language equivalent: display the contents of the scalar called `$line` if and only if the scalar called `$line` *contains* the pattern "even".

Another new operator is introduced here. It is called the *binding operator*, and it looks like this: `=~`. This operator compares something (usually a scalar variable

---

[9] See the `perlretut` manual page.

container) against a pattern[10]. For now, a pattern is defined as any sequence of characters surrounded by the forward-leaning slash character (i.e., "/"). In the example above, the pattern is the word "even". Specifically, it is the letter "e", followed by the letter "v", followed by the letter "e", followed by the letter "n". If the contents of $line contains the pattern *anywhere in the line*, it is said to match.

> **Technical Commentary:** When programmers refer to a character that surrounds something of interest, such as the forward-leaning slash surrounding the patterns in this section, they call that character a *delimiter*. The character delimits the something of interest. The "/" character is the default delimiter for regular expression patterns in Perl.

A few examples will illustrate what's going on. Try this command-line:

```
./patterns    terrible
```

The `patterns` program reads the contents of the disk-file called `terrible` one line at a time looking for a match on the pattern. When the pattern is found, `patterns` displays the matching line. It finds matching lines as follows:

```
# The 'terrible' program - a poorly formatted 'oddeven'.
{ print "even\n"; } elsif ( $count == 3 ) { print "odd\n"; }
{ print "even\n"; } }
```

Here's another invocation of `patterns`, this time against the `oddeven` disk-file:

```
./patterns    oddeven
```

Again, the `patterns` program reads the contents of the disk-file called `oddeven` one line at a time looking for a match on the pattern. When it is found, it displays the matching line. As with the `terrible` disk-file, the program finds matching lines:

```
# The 'oddeven' program - a (Perl) program,
# is an odd number, and 'even' when $count is an even
        print "even\n";
        print "even\n";
```

Note that as the `oddeven` program is formatted correctly, it is easier to spot the pattern on the displayed lines. Here is one final invocation of the `patterns` program:

```
./patterns    welcome2
```

---

[10] If you are wondering why this operator is called "bind" and not "compare", wonder no longer. The word "compare" was already taken, so "bind" was chosen instead. So, we refer to a scalar *binding* to a pattern, as opposed to *being compared* to a pattern. Conceptually here, "bind" and "compare" both mean the same thing.

This invocation produces no output. This is perfectly OK, as the `welcome2` program does not contain the pattern "`even`".

To finish off this quickie introduction to Perl's regular expression, pattern-matching technology, let's conclude with another maxim.

> **Maxim 3.12** *Patterns tell `perl` what to look for, not how to find it.*

## Where to from Here

In this chapter, sequence, iteration and selection and the basic building blocks of programming, were discussed. The *three C's*: constants, comments and conditions were introduced. The use of simple variable containers helped to keep things interesting, as did the use of some Perl operators and its pattern-matching technology.

In the next chapter, additional variable containers are described, and additional Perl programming constructs are introduced.

## The Maxims Repeated

Here's a list of the maxims introduced in this chapter.

- *Programs execute in sequential order.*
- *Less is better.*
- *If you can say something with fewer words, then do so.*
- *There's more than one way to do it.*
- *Add comments to make future maintenance of a program easier for other programmers and for you.*
- *When using constant values, refer to them with a nice, human-friendly name as opposed to the actual value.*
- *Use blocks to group program statements together.*
- *A condition can result in a value of true or false.*
- *When you need to change the value of an item, use a variable container.*
- *Don't be lazy: use good, descriptive names for variables.*
- *Use plenty of whitespace, blank lines and indentation to make your programs easier to read.*
- *Patterns tell `perl` what to look for, not how to find it.*

# Exercises

1. Write a program that displays the message "Hey, look Ma – I can program!" six times, sleeping for three seconds between each iteration.

2. Adapt the program from the last exercise to iterate 30 times. Arrange for the message to appear only when the iteration count is evenly divisible by three. To speed things up a little, remove the three-second sleep.

3. Write a program that initialises the $count scalar to ten, then iterates, displaying a message of your choosing, until such time as the value of $count is zero. [Hint: Review the list of Perl operators in Appendix A on page 457].

4. Write a program that searches through the terrible program looking for the word "count". How many times does the word "count" appear in terrible?

# 4
# *Places to Put Things*

*Exploring Perl's built-in variable containers: arrays and hashes.*

## 4.1 Beyond Scalars

Chapter 3, *The Basics*, introduced the scalar variable container: a place to put *one of something*. Perl provides a rich collection of places to put things. In this chapter, two of these other places, arrays and hashes, are explored.

## 4.2 Arrays: Associating Data with Numbers

It is often convenient to take a number of scalar values and treat them as one unit. Perl supports this idea with *arrays*. Whereas a scalar contains a single value, an array contains a *collection of scalar values*.

Arrays are named in a similar way to scalars, with the exception that the "$" that prefixes the scalar name is replaced with "@". Keeping the *Don't be lazy: use good, descriptive names for variables* maxim in mind, here are some good array names:

```
@list_of_sequences
@totals
@protein_structures
```

## 50   *Places to Put Things*

An array is typically populated with a *list*. Lists in Perl are a collection of scalar values separated by commas and enclosed in parentheses. Here is a small list of three short DNA sequences:

```
( 'TTATTATGTT', 'GCTCAGTTCT', 'GACCTCTTAA' )
```

Note that these short DNA sequences are strings of letters, so it is necessary to enclose them in single quotes (') to have them treated as a scalars. The reader needs to develop a good understanding of lists, as they are extensively used in Perl.

**Maxim 4.1** *Lists in Perl are comma-separated collections of scalars.*

To put the list of DNA sequences into an array, *assign* the list to the array as follows:

```
@list_of_sequences = ( 'TTATTATGTT', 'GCTCAGTTCT', 'GACCTCTTAA' );
```

Note the use of the assignment operator (=) and the semicolon at the end of the line. It is now possible to refer to this entire list of DNA sequences with one name, namely, @list_of_sequences.

Figure 4.1 shows pictorially the current state of the @list_of_sequences array. Not only does perl arrange to put the three short DNA sequences into the array but perl also numbers them. Starting at zero, each value in the array has a unique number associated with it, as indicated in the figure. This number is referred to as the value's *array index*. Consequently, each scalar value in the array is now *associated* with a unique number.

The array index is used to refer to an individual value of the named array. Here's a very important maxim.

**Maxim 4.2** *Perl starts counting from zero, not one.*

Here's how to display "GCTCAGTTCT" on the screen:

```
print "$list_of_sequences[1]\n";
```

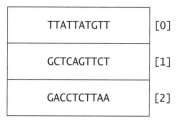

**Figure 4.1**   The @list_of_sequences array.

The scalar value is accessed *via* its array index. Take another look at this line of Perl code: notice how list_of_sequences is prefixed with "$" and not "@". What is going on? Surely, arrays need to be prefixed with "@"? That's correct, they do. When referring to an *entire* array, it is necessary to prefix the array name with "@". However, when referring to an individual value stored in an array, commonly referred to as an array *element*, the value being referred to is a scalar value and, in Perl, scalar values are prefixed with "$".

## 4.2.1 Working with array elements

In addition to accessing an individual element within an array, it is also possible to assign a new value to an array element. Consider these two Perl statements:

```
$list_of_sequences[1] = 'CTATGCGGTA';
$list_of_sequences[3] = 'GGTCCATGAA';
```

The first statement *changes* the value associated with array index 1 to the value indicated, another short, but different, DNA sequence. The previous value of $list_of_sequences[1], which was "GCTCAGTTCT", is overwritten by this assignment statement. Recall that it is an individual element of the array that is being accessed, so it is necessary once again to prefix the array name with "$".

The second Perl statement is interesting. Until this statement was executed, the array contained three scalar values. By referencing a new array index, perl arranges to *dynamically grow* the size of the @list_of_sequences array as needed. After these two statements execute, the original array has changed and grown to look like Figure 4.2.

## 4.2.2 How big is the array?

It is often useful to determine the size of an array, where "size" refers to the number of elements currently in the array. In true Perl style, and remembering

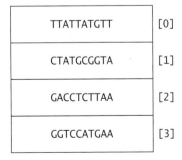

**Figure 4.2** The grown @list_of_sequences array.

the *There's more than one way to do it* maxim, determining the size of an array can be accomplished in a number of ways, as follows:

```
print "The array size is: ", $#list_of_sequences+1, ".\n";
print "The array size is: ", scalar @list_of_sequences, ".\n";
```

When executed by `perl`, both statements display the following message on screen:

```
The array size is: 4.
```

When the name of the array is prefixed with "$#", the value returned by `perl` is equal to the largest array index associated with the named array. The largest array index for the @list_of_sequences array is 3. Array indexes are numbered from 0, so 1 is added to the $# value to calculate the number of elements in the array, which is 4 in this case.

Typically, a list is always assumed to be operating within what is known as *list context*. In other words, a list is treated just like, well, a list. That said, it can sometimes make sense to treat a list as something other than a list, specifically, to treat it is a scalar, using it in what's known as *scalar context*. Here's a new maxim to highlight the importance of *context* in Perl.

**Maxim 4.3** *There are three main contexts in Perl: numeric, list and scalar.*

Perl's `scalar` subroutine forces `perl` to treat a list as in *scalar context*. When used with an array, the `scalar` subroutine first takes the array and turns it into a list, then *evaluates the list as a scalar*. Doing this has to make some sense to `perl`, and the only thing that makes sense is for `perl` to look at the list, count the number of elements in the list, then return the resulting *scalar value*, which is 4 in this case, the number of elements in the @list_of_sequences array.

### 4.2.3 Adding elements to an array

Once an array has been created, adding elements to it is not difficult. As demonstrated in the last section, the introduction of a new array index adds an element into a named array. Another technique for adding single elements is as follows:

```
@sequences = ( 'TTATTATGTT', 'GCTCAGTTCT', 'GACCTCTTAA' );
@sequences = ( @sequences, 'CTATGCGGTA' );
```

After the first line of code, the @sequences array contains three short DNA sequences. After the second line of code, @sequences contains an additional array element, one which holds the scalar value "CTATGCGGTA". Unlike the previous technique, this method does not require the programmer to specify the next array index. The `perl` interpreter looks at the @sequences array and works out the value of the next available array index. Read the second line of code as "take the current elements of the @sequences array and add the element 'CTATGCGGTA', then assign all the elements back to the @sequences array, overwriting any elements that were there previously".

To confirm that the addition of the element has indeed occurred, display the entire array on screen with the following line of Perl code:

```
print "@sequences\n";
```

This line of code produces the expected outcome:

TTATTATGTT GCTCAGTTCT GACCTCTTAA CTATGCGGTA

Now, watch what happens if we forget to include @sequences on the right-hand side of the assignment operator:

```
@sequences = ( 'TTATTATGTT', 'GCTCAGTTCT', 'GACCTCTTAA' );
@sequences = ( 'CTATGCGGTA' );
print "@sequences\n";
```

The following is displayed:

CTATGCGGTA

Whoops! This code inadvertently deletes the original contents of the array. So be careful. Adding a list of elements to an existing array is accomplished like this:

```
@sequences = ( 'TTATTATGTT', 'GCTCAGTTCT', 'GACCTCTTAA' );
@sequences = ( @sequences, ( 'CTATGCGGTA', 'CTATTATGTC' ) );
print "@sequences\n";
```

Which, as expected, produces this output:

TTATTATGTT GCTCAGTTCT GACCTCTTAA CTATGCGGTA CTATTATGTC

Finally, two existing arrays are combined into a third array as follows:

```
@sequence_1 = ( 'TTATTATGTT', 'GCTCAGTTCT', 'GACCTCTTAA' );
@sequence_2 = ( 'GCTCAGTTCT', 'GACCTCTTAA' );
@combined_sequences = ( @sequence_1, @sequence_2 );
print "@combined_sequences\n";
```

This produces the following output:

TTATTATGTT GCTCAGTTCT GACCTCTTAA GCTCAGTTCT GACCTCTTAA

### 4.2.4 Removing elements from an array

Perl's `splice` subroutine removes any number of elements from an array and can, optionally, replace the removed elements with new ones. Interestingly, `splice` returns the removed elements.

The `splice` subroutine takes one mandatory parameter and three optional ones. The mandatory parameter[1] is the name of the array. The optional parameters indicate where in the array to start removing elements (the OFFSET), how many elements to remove (the LENGTH) and an array of elements (the LIST) with which to replace the removed ones. Let's see `splice` in action:

```perl
@sequences = ( 'TTATTATGTT', 'GCTCAGTTCT', 'GACCTCTTAA', 'TTATTATGTT' );
@removed_elements = splice @sequences, 1, 2;
print "@removed_elements\n";
print "@sequences\n";
```

The array `@removed_elements` contains the value "GCTCAGTTCT" and the value "GACCTCTTAA". The `@sequences` array is two elements shorter, as confirmed by the output generated:

```
GCTCAGTTCT GACCTCTTAA
TTATTATGTT TTATTATGTT
```

Look closely at the call to `splice`. The value of OFFSET is 1 and LENGTH is 2. Recalling that Perl starts counting from zero, an OFFSET of 1 refers to the second element in the array. The LENGTH value controls how many elements are removed from the array *starting at the element referred to by OFFSET*. In this case, LENGTH is 2 so this code removes two elements from the named array, returns the removed elements and assigns them to `@removed_element`.

Be careful with `splice`. If no value for LENGTH is provided, every array element from the OFFSET to the end of the array is removed. Similarly, if no value for OFFSET is provided, *every* array element is removed. In effect, the array is emptied of all its elements.

However, if emptying an array is the required action, assigning an empty list to the array also empties it. This method has the added advantage of being much faster than using `splice`:

```perl
@sequences = ();
```

### 4.2.5 Slicing arrays

To access a number of array elements but not remove them, use an *array slice*. Unlike `splice`, which is a special purpose subroutine built into `perl`, slicing is an extension to the array indexing mechanism. Rather than providing a contiguous string of array elements as does `splice`, a slice can refer to a *list* or *range* of array

---
[1] Mandatory parameters are also referred to as "required parameters".

indexes. Additionally, the @ prefix is used when referring to the named array, as a slice produces a list, not a scalar.

For example, to access the values at the second, fifth and tenth array index locations for an array called @dnas, specify the slice as follows:

```
@dnas[ 1, 4, 9 ]
```

Use Perl's *range operator*, .., to specify a sequential collection of array indexes. Here's how to access the second through tenth array elements of @dnas:

```
@dnas[ 1 .. 9 ]
```

Here's some code that highlights the differences between slicing and `splice`:

```
#! /usr/bin/perl -w

# The 'slices' program - slicing arrays.

@sequences = ( 'TTATTATGTT', 'GCTCAGTTCT', 'GACCTCTTAA',
               'CTATGCGGTA', 'ATCTGACCTC' );
print "@sequences\n";
@seq_slice = @sequences[ 1 .. 3 ];
print "@seq_slice\n";
print "@sequences\n";
@removed = splice @sequences, 1, 3;
print "@sequences\n";
print "@removed\n";
```

Which, when executed, produces the following results:

```
TTATTATGTT GCTCAGTTCT GACCTCTTAA CTATGCGGTA ATCTGACCTC
GCTCAGTTCT GACCTCTTAA CTATGCGGTA
TTATTATGTT GCTCAGTTCT GACCTCTTAA CTATGCGGTA ATCTGACCTC
TTATTATGTT ATCTGACCTC
GCTCAGTTCT GACCTCTTAA CTATGCGGTA
```

Let's go through this program in detail. After the standard first line and a short comment, a five element list is assigned to the @sequences array, and the entire array is displayed on screen:

```
@sequences = ( 'TTATTATGTT', 'GCTCAGTTCT', 'GACCTCTTAA',
               'CTATGCGGTA', 'ATCTGACCTC' );
print "@sequences\n";
```

Note that the list is on two lines. This is a perfectly acceptable practice, as Perl is a *free-format* language. Here, the list is written in this way to fit within the width of this page, but it could also have been written on a single line.

The next line takes a slice of the @sequences array, requesting the values at array index 1 through 3 (that is, the second through fourth values in

**56** *Places to Put Things*

the array). The list of values returned from the slice is assigned to an array called @seq_slice, then displayed on screen together with the @sequences array:

```
@seq_slice = @sequences[ 1 .. 3 ];
print "@seq_slice\n";
print "@sequences\n";
```

Refer back to the output from this program to confirm that the @sequences array has not been modified by the creation of this slice. The next line does indeed modify @sequences because of the use of the splice subroutine:

```
@removed = splice @sequences, 1, 3;
print "@sequences\n";
print "@removed\n";
```

As opposed to slicing, which requested a copy of the values stored in each of the array elements, the splice subroutine removes the array elements from @sequences. These are assigned to another array, this one called @removed, which is displayed on screen after displaying what's left of @sequences. Refer back to the output generated by this program to confirm this.

> **Maxim 4.4** *To access a list of values from an array, use a slice.*

> **Maxim 4.5** *To remove a list of values from an array, use* splice.

## 4.2.6 Pushing, popping, shifting and unshifting

Although splicing and slicing are useful, more often single values need to be added or removed either at the start or the end of the array. The techniques used in the previous subsections can be used to do this. However, as the requirement is so common, Perl provides four subroutines to make adding and/or removing from the start and/or end of an array convenient:

**shift** – removes and returns the first element from an array.
**pop** – removes and returns the last element from an array.
**unshift** – adds an element (or a list of elements) onto the start of an array.
**push** – adds an element (or a list of elements) onto the end of an array.

Here's a quick example that demonstrates the use of these subroutines:

```
#! /usr/bin/perl -w

# The 'pushpop' program - pushing, popping, shifting
# and unshifting.

@sequences = ( 'TTATTATGTT', 'GCTCAGTTCT', 'GACCTCTTAA',
```

```
                        'CTATGCGGTA', 'ATCTGACCTC' );

    print "@sequences\n";
    $last = pop @sequences;
    print "@sequences\n";
    $first = shift @sequences;
    print "@sequences\n";
    unshift @sequences, $last;
    print "@sequences\n";
    push @sequences, ( $first, $last );
    print "@sequences\n";
```

Which results in the following output:

```
TTATTATGTT GCTCAGTTCT GACCTCTTAA CTATGCGGTA ATCTGACCTC
TTATTATGTT GCTCAGTTCT GACCTCTTAA CTATGCGGTA
GCTCAGTTCT GACCTCTTAA CTATGCGGTA
ATCTGACCTC GCTCAGTTCT GACCTCTTAA CTATGCGGTA
ATCTGACCTC GCTCAGTTCT GACCTCTTAA CTATGCGGTA TTATTATGTT ATCTGACCTC
```

After printing the original contents of the @sequences array, the pop subroutine removes the last element from the array and assigns it to the $last scalar. The @sequences array is printed again to show that it is now one element shorter. A call to the shift subroutine then removes the first element from the array and assigns it to the $first scalar. Again, the @sequences array is printed to confirm that this is indeed the case.

The unshift subroutine is then called, with the @sequences array and the $last scalar as its parameters. This results in the value of the scalar being added to the start of the array. The value that was originally in the last array element is now in the first. The @sequences array is again printed to confirm that this has happened. Finally, the push subroutine is called to add a two-element list onto the end of the @sequences array. Note that the two-element list contains the values of the $first and $last scalars. The parentheses around the two scalars are not strictly required here, but their use helps clarify what the programmer's intention is.

### 4.2.7 Processing every element in an array

The while statement from the previous chapter can be used to *iterate over* an entire array and process every element. Here's how to display each of the short DNA sequences from the @sequences array on a separate line:

```
#! /usr/bin/perl -w

# The 'iterateW' program - iterate over an entire array
# with 'while'.

@sequences = ( 'TTATTATGTT', 'GCTCAGTTCT', 'GACCTCTTAA',
```

```
                            'CTATGCGGTA', 'ATCTGACCTC' );

    $index = 0;
    $last_index = $#sequences;

    while ( $index <= $last_index )
    {
        print "$sequences[ $index ]\n";
        ++$index;
    }
```

When executed, the iterateW program produces the following results:

```
TTATTATGTT
GCTCAGTTCT
GACCTCTTAA
CTATGCGGTA
ATCTGACCTC
```

After the usual first line and a short comment, the @sequences array is populated with the list of short DNA sequences. A scalar called $index is initialised to zero, and the $last_index scalar is initialised to the largest array index value associated with the @sequences array, which, in this case, is 4.

The loop condition is true for as long as the value of $index is less than or equal to (<=) the value of $last_index. Obviously, at this stage, zero is less than or equal to 4, so the statements inside the loop execute. The value of the element at array index $index is printed and then the $index scalar is incremented using the ++ operator.

The $index scalar now has the value 1, and the loop condition is checked again to see if another iteration is to occur. In this way, each element in the array is processed until the value of $index exceeds the value of $last_index. When this occurs, the loops ends.

As processing arrays in this way is so common, Perl provides another looping mechanism in support of this activity: the foreach statement. Here's how to rewrite the iterateW program to use foreach instead of while:

```
#! /usr/bin/perl -w

# The 'iterateF' program - iterate over an entire array
# with 'foreach'.

@sequences = ( 'TTATTATGTT', 'GCTCAGTTCT', 'GACCTCTTAA',
                'CTATGCGGTA', 'ATCTGACCTC' );

foreach $value ( @sequences )
{
    print "$value\n";
}
```

## Arrays: Associating Data with Numbers

The results produced by `iterateF` are exactly the same as those produced by the `iterateW` program. Take a moment to compare this program with `iterateW`.

With each iteration, the `foreach` statement arranges to assign the value from each array element in the `@sequences` array to the `$value` scalar. Once assigned, the `$value` is used *as if referring to the actual array element*. The implication of this statement is that a change to the `$value` scalar is also reflected in the corresponding array element in the `@sequences` array.

Although the use of the `while` statement is perfectly acceptable, most perl programmers prefer `foreach`.

> **Maxim 4.6** *Use `foreach` to process every element in an array.*

### 4.2.8 Making lists easier to work with

Recall the *Lists in Perl are comma-separated collections of scalars* maxim. When the `@sequences` array was first defined, the advice given was to surround the initialising list values in single quotes ('), as follows:

```
@sequences = ( 'TTATTATGTT', 'GCTCAGTTCT', 'GACCTCTTAA',
               'CTATGCGGTA', 'ATCTGACCTC' );
```

The use of single quotes is not strictly necessary. If the list value is a string that contains no whitespace[2], the single quotes are optional. As the short DNA sequences contain no whitespace, the list can also be written as:

```
@sequences = ( TTATTATGTT, GCTCAGTTCT, GACCTCTTAA,
               CTATGCGGTA, ATCTGACCTC );
```

As a further relaxation of the rules, it is also acceptable to remove the commas (,) separating the list elements. Assuming, that is, that Perl's `qw` operator is used, where `qw` is shorthand for "quote words". The following is another (generally preferable) way to specify the list of items used to initialise the `@sequences` array:

```
@sequences = qw( TTATTATGTT GCTCAGTTCT GACCTCTTAA
                 CTATGCGGTA ATCTGACCTC );
```

which is generally preferable because it involves less typing[3], in addition to reducing the likelihood of typographical errors.

---

[2] Whitespace: a space, tab, line feed, carriage return or form feed character.

[3] Developing techniques that require less typing is a uniquely Perlish way of being lazy. This type of laziness is regarded as a *good thing* in the Perl programming community.

## 4.3 Hashes: Associating Data with Words

In addition to arrays, Perl provides another very useful variable container: the hash. Unlike arrays that associate scalars with numbers, hashes *associate scalars with words*.

> **Technical Commentary:** In Computer Science circles, variable containers that associate values with words are called *associative arrays*. In Perl circles, they are called *hashes*. Technically, a hash is a collection of *name and value pairings*[4]. Think of the *word* as the *name*, and the *data* as the *value*.

> **Maxim 4.7** *A hash is a collection of name/value pairings.*

Whereas scalars are prefixed with $ and arrays are prefixed with @, hashes are prefixed with %. Here are some good hash names:

```
%bases
%genomes
%nucleotide_bases
```

A hash is populated in a number of ways. The simplest method is to use a list, as follows:

```
%nucleotide_bases = ( A, Adenine, T, Thymine );
```

When assigned to a hash, the list is turned into a series of name/value pairings. The first element in the list is a name, the second is a value, the third is a name, the fourth is a value, and so on. Figure 4.3 shows pictorially the current state of the %nucleotide_bases hash, after the above line of code is executed by perl. The figure clearly shows the relationship between the words (names) and the data (values).

A *hash entry* has a name part and a value part. Referring to Figure 4.3, the name parts are "A" and "T". The value parts are "Adenine" and "Thymine".

Hashes have a restriction on how they are used: hash name parts must be unique.

> **Maxim 4.8** *Hash name parts must be unique.*

---

[4] Also referred to as "*key-value pairs*".

| A | Adenine |
| T | Thymine |

**Figure 4.3** The %nucleotide_bases hash.

### 4.3.1 Working with hash entries

Once a hash is populated, individual values in the hash can be accessed by referring to their names, as follows[5]:

```
print "The expanded name for 'A' is $nucleotide_bases{ 'A' }\n";
```

Given a name ("A" in this case), refer to the value associated with the name in the hash using the syntax shown. That is, start with $, then provide the individual hash name (nucleotide_bases), then provide the name inside curly braces ("{" and "}").

Hashes, just like arrays, store scalar values. So, when referring to an individual value associated with a name in a hash, prefix the hash name with $. Prefix the hash name with % when referring to the *entire* hash.

### 4.3.2 How big is the hash?

Perl has a built-in subroutine called keys that, when called in list context, returns a list of the names in the hash, as follows:

```
%nucleotide_bases = ( A, Adenine, T, Thymine );

@hash_names = keys %nucleotide_bases;

print "The names in the %nucleotide_bases hash are: @hash_names\n";
```

When executed, the preceding three lines of code produce the following result:

```
The names in the %nucleotide_bases hash are: A T
```

When the keys subroutine is called in scalar context (by assigning the result to $hash_size below), it returns the number of entries in the hash, that is, the hash size:

```
%nucleotide_bases = ( A, Adenine, T, Thymine );

$hash_size = keys %nucleotide_bases;

print "The size of the %nucleotide_bases hash is: $hash_size\n";
```

---

[5] Another popular name for the name part of a hash entry is "key". Your authors prefer "name".

| | |
|---|---|
| A | Adenine |
| T | Thymine |
| C | Cytosine |
| G | Guanine |

**Figure 4.4** The grown %nucleotide_bases hash.

Which, when executed by `perl`, produces:

```
The size of the %nucleotide_bases hash is: 2
```

### 4.3.3 Adding entries to a hash

Additional entries can be added to an existing hash one at a time. Here's how to add the other bases:

```
$nucleotide_bases{ 'G' } = 'Guanine';
$nucleotide_bases{ 'C' } = 'Cytosine';
```

Following the execution of these two statements, the %nucleotide_bases hash has grown to look like Figure 4.4. Notice anything strange about Figure 4.4? The bases were added to the hash in the following order: ATGC, whereas the figure shows the order as ATCG. What's going on?

It turns out that hashes in Perl are not maintained in *insertion order*, as *is* the case with arrays. This means it is not possible to rely on the hash being in any particular order when working with it. A strategy for dealing with this hash "shortcoming" is discussed later in this chapter.

As using a list is such a useful method for populating a hash, Perl offers a convenient alias for comma (","), which can be used to improve the human readability of hash assignments within a program. The "=>" combination can be used anywhere a comma is used, and is often used as follows:

```
%nucleotide_bases = ( A => Adenine, T => Thymine,
                      G => Guanine, C => Cytosine );
```

Compare this with the earlier use of a list to populate the hash. Notice how the use of "=>" accentuates which names associate with which values.

### 4.3.4 Removing entries from a hash

A hash entry can be removed from a hash using Perl's built-in `delete` subroutine, which removes both the name part and value part from the hash:

```
delete $nucleotide_bases{ 'G' };
```

The hash entry has now been removed, and the hash is one entry shorter. It is also possible to nullify the value part of an individual hash entry by setting the value part to an undefined value:

```
$nucleotide_bases{ 'C' } = undef;
```

Here, a special *undefined* value is assigned to the value part of the hash entry associated with "C".

Just what is undef? In actual fact, undef is a Perl subroutine that returns the *undefined value*, a special "nothing value" that can be assigned to any variable, be that variable a hash, array or scalar. When a variable has undef as its value, the variable exists but does not contain a value. Its value is undefined, or void.

### 4.3.5 Slicing hashes

As with arrays, it is also possible to slice a hash. When a hash is sliced, a list of hash value parts is returned, so prefix the hash name with @ when slicing, as opposed to $. Remember: what's returned from a slice is a list.

To slice from a hash, prefix the hash name with @, and provide a list of name parts between the curly braces. Here's some code that demonstrates hash slicing:

```
%gene_counts = ( Human                 => 31000,
                 'Thale cress'         => 26000,
                 'Nematode worm'       => 18000,
                 'Fruit fly'           => 13000,
                 Yeast                 => 6000,
                 'Tuberculosis microbe' => 4000 );

@counts = @gene_counts{ Human, 'Fruit fly', 'Tuberculosis microbe' };

print "@counts\n";
```

In addition to providing an example of hash slicing, these lines of code serve to highlight some other hash characteristics. Of note is the formatting that the programmer has chosen to use when populating the %gene_counts hash. The *comma alternative*, =>, helps identify the name and value pairings. Additionally, the alignment of the values also aids the reader's understanding. Take a closer look at the names. Some are enclosed in single quotes ('), while others are not. The rule is straightforward: if a hash name has no whitespace, the single quotes are optional, as is the case with "Human", otherwise they are required, as is the case with "Tuberculosis microbe". Of the three lines of code, the hash slice is of most interest:

```
@counts = @gene_counts{ Human, 'Fruit fly', 'Tuberculosis microbe' };
```

**64**    *Places to Put Things*

Note how the hash name is prefixed with @, not $ nor %. The names of the three values to be sliced are provided as a list within the curly braces and, once sliced, the values are assigned to the @counts array. This array is then printed, which results in the following:

    31000 13000 4000

### 4.3.6 Working with hash entries: a complete example

Here's a short program, called bases, which uses the %nucleotide_bases hash to expand a short DNA sequence into a list of base names:

```
#! /usr/bin/perl -w

# The 'bases' program - a hash of the nucleotide bases.

%nucleotide_bases = ( A => Adenine, T => Thymine,
                      G => Guanine, C => Cytosine );

$sequence = 'CTATGCGGTA';

print "\nThe sequence is $sequence, which expands to:\n\n";

while ( $sequence =~ /(.)/g )
{
    print "\t$nucleotide_bases{ $1 }\n";
}
```

When executed, the bases program produces the following results:

```
The sequence is CTATGCGGTA, which expands to:

        Cytosine
        Thymine
        Adenine
        Thymine
        Guanine
        Cytosine
        Guanine
        Guanine
        Thymine
        Adenine
```

Let's work through bases and see what's going on. After the usual first line and a short comment, the %nucleotide_bases hash is populated with names equal to the abbreviated bases and values equal to the associated baseword. A string, "CTATGCGGTA", is assigned to a scalar variable called $sequence and a message is displayed on screen:

```perl
%nucleotide_bases = ( A => Adenine, T => Thymine,
                      G => Guanine, C => Cytosine );

$sequence = 'CTATGCGGTA';

print "\nThe sequence is $sequence, which expands to:\n\n";
```

A loop then iterates over the string in the $sequence scalar:

```perl
while ( $sequence =~ /(.)/g )
```

The condition of the loop needs further explanation. The binding operator, =~, is used to check the value in $sequence against the pattern "/(.)/g". Perl's pattern-matching technology was introduced in the last chapter, and has an entire chapter devoted to it later. Here's what this pattern does:

1. The "." in the pattern tells Perl to find any character except newline (i.e. any character except the "\n" character).

2. The parentheses, the "(" and ")" characters, tell perl to remember the character found by "." and put it into a special variable called $1.

3. The "g" after the pattern[6] tells Perl to apply the pattern *globally*. The "g" is not technically part of the pattern, it's a qualifier that changes how the pattern works. In this case, the qualifier tells perl to apply the pattern globally, that is, to the entire string.

The significance of this last point is that when used within a loop condition, the pattern is applied at every possible location that it can be applied to within the string *on each iteration*. That is, each time through the loop, the "." pattern matches each of the characters in the string one at a time. The effect of this is that each time the loop iterates, the $1 scalar is assigned a character from the string contained in $sequence. This allows the program to process the string "CTATGCGGTA" one character at a time.

As the pattern matches each character, the $1 scalar contains the match. This is then used to refer to the value part associated with the name part in the %nucleotide_bases hash, which is then printed to the screen:

```perl
{
    print "\t$nucleotide_bases{ $1 }\n";
}
```

Once the loop has exhausted all the characters in the $sequence scalar, it ends, and the program terminates. Note the use of the *tab character*, "\t", to indent each line.

---

[6] Remember: patterns are delimited by the "/" character.

## 4.3.7 Processing every entry in a hash

Use either `while` or `foreach` to process every name/value pairing in a hash. The `genes` program, which processes a hash twice, demonstrates both looping mechanisms:

```
#! /usr/bin/perl -w

# The 'genes' program - a hash of gene counts.

use constant    LINE_LENGTH => 60;

%gene_counts = ( Human                  => 31000,
                 'Thale cress'          => 26000,
                 'Nematode worm'        => 18000,
                 'Fruit fly'            => 13000,
                 Yeast                  => 6000,
                 'Tuberculosis microbe' => 4000 );

print '-' x LINE_LENGTH, "\n";

while ( ( $genome, $count ) = each %gene_counts )
{
    print "'$genome' has a gene count of $count\n";
}

print '-' x LINE_LENGTH, "\n";

foreach $genome ( sort keys %gene_counts )
{
    print "'$genome' has a gene count of $gene_counts{ $genome }\n";
}

print '-' x LINE_LENGTH, "\n";
```

Before working through the `genes` program in detail, take a look at the results produced by this program:

```
------------------------------------------------------------
'Human' has a gene count of 31000
'Tuberculosis microbe' has a gene count of 4000
'Fruit fly' has a gene count of 13000
'Nematode worm' has a gene count of 18000
'Yeast' has a gene count of 6000
'Thale cress' has a gene count of 26000
------------------------------------------------------------
'Fruit fly' has a gene count of 13000
'Human' has a gene count of 31000
'Nematode worm' has a gene count of 18000
'Thale cress' has a gene count of 26000
```

```
'Tuberculosis microbe' has a gene count of 4000
'Yeast' has a gene count of 6000
----------------------------------------------------------
```

Take particular note of the order of the two sets of results: they are different.

The `genes` program begins with the usual first line, a comment, a constant definition and the population of a hash called `%gene_counts`. An unfamiliar-looking `print` statement comes next:

```
print '-' x LINE_LENGTH, "\n";
```

This `print` statement demonstrates another Perl operator: `x`, the *repetition operator*. Given something to do (in this case, `print '-'`) and a number of times to do it (in this case, `LINE_LENGTH`, which has a constant value of 60), this repetition operator arranges to display a dash 60 times on the screen. Once done, the `print` statement takes a newline with "\n".

The first loop processes the `%gene_counts` hash, one name/value pairing at a time. As with the `bases` program, understanding the loop condition is the key to understanding what is occurring here. Perl's `each` subroutine returns the name and value of the next entry in the hash and in this code, assigns the name part to the `$genome` scalar and the value part to the `$count` scalar. These scalars are then used in the `print` statement to display the gene count for each of the genomes in the hash:

```
while ( ( $genome, $count ) = each %gene_counts )
{
    print "'$genome' has a gene count of $count\n";
}
```

With each iteration, the `each` subroutine returns the next name/value pairing until there are no more name/value pairings left. In this way, every entry in the hash is processed by the loop.

The repetition operator is again used with a `print` statement to display 60 dashes and a newline before the second loop is executed. This loop is a `foreach` statement:

```
foreach $genome ( sort keys %gene_counts )
{
    print "'$genome' has a gene count of $gene_counts{ $genome }\n";
}
```

To understand what is going on here, concentrate on the loop condition. The `keys` subroutine is used to generate a list of names from the `%gene_counts` hash, then the list of names is sorted into alphabetical order by Perl's built-in `sort` subroutine. The list resulting from the call to `sort` is then assigned one element at a time to the `$genome` scalar, which is then used in the body of the `foreach` loop to display the gene count for each of the genomes in the hash.

This use of `sort` within the loop condition explains why the results from this program produces lists in two different orders. The `while` statement did not order the results, so the hash is displayed in the order that it is currently used by `perl`, whereas the `foreach` statement explicitly instructed `perl` to sort the hash names alphabetically prior to their use. This resulted in the internal hash order being overridden by the `foreach` statement.

The `genes` program concludes with another repeated `print` statement, displaying 60 dashes and a newline.

## Where to from Here

This chapter described Perl's arrays, lists and hashes. The population, removal and accessing of data in arrays and hashes was demonstrated with a small collection of programs. Together with scalar variable containers, arrays and hashes provide a useful collection of places to put things. Often, however, a more complex structure for data is required, and we return to this subject in Part II.

In the next chapter, subroutines are described. What are subroutines, and why are they useful? Read on to find out.

## The Maxims Repeated

Here's a list of the maxims introduced in this chapter.

- *Lists in Perl are comma-separated collections of scalars.*
- *Perl starts counting from zero, not one.*
- *There are three main contexts in Perl: numeric, list and scalar.*
- *To access a list of values from an array, use a slice.*
- *To remove a list of values from an array, use `splice`.*
- *Use `foreach` to process every element in an array.*
- *A hash is a collection of name/value pairings.*
- *Hash name parts must be unique.*

## Exercises

1. Define a hash called %genome_speak, which associates the following abbreviations with the phrase in parentheses: AA (amino acid), BAC (bacterial artificial chromosome), BLAST (basic local alignment search tool), cDNA (complementary DNA), DNA (deoxyribonucleic acid), EST (expressed sequence tag),

FISH (fluorescence in situ hybridization), mRNA (messenger RNA), rDNA (recombinant DNA), RNA (ribonucleic acid), STS (sequence tagged site), SNP (single nucleotide polymorphism) and YAC (yeast artificial chromosome)[7].

2. Create a small file, called **abbrevs**, with the following contents:

    DNA
    SNP
    rDNA
    AA
    BLAST
    RNA
    YAC
    mRNA

    Write a program to process **abbrevs** and display the correct phrase from the **%genome_speak** hash for each abbreviation.

3. On the basis of the results produced by your solution to the previous exercise, does it matter in which order the abbreviations are checked against the hash name parts?

---

[7] This list is taken from pages 135–136 of *The Human Genome*, Dennis, C. and Gallagher, R. (editors), published 2001 by *Nature Publishing Group*, ISBN: 0-333-97143-4.

# 5
# Getting Organised

*Subroutines, modules and the wonder of CPAN.*

## 5.1 Named Blocks

As programs get larger, they become harder to *maintain*. The process of maintaining an existing program can involve:

1. *fixing* existing problems
2. *adding* new functionality
3. *enhancing* existing functionality
4. *removing* obsolete functionality or
5. any combination of the above activities.

The trick – of course – is to maintain the program without adding any additional problems to it. Such problems are commonly referred to as *bugs*.

Recall the `genes` program from the last chapter (on page 66). This line of code occurs three times throughout the program:

```
print '-' x LINE_LENGTH, "\n";
```

Its purpose is pretty straightforward: it draws a line across the screen using the "-" character. The length of the line is determined by the LINE_LENGTH constant, which is set to 60 at the top of the program. The use of a constant in this way allows the length of the line to be changed *globally* for all lines in the program. For instance, changing the length of the line from 60 dashes to 40 is straightforward: just change the constant value.

Let's assume that a requirement exists to produce fancier lines. For example, in addition to the standard dashed line, thus:

```
------------------------------------------------------------
```

a collection of line styles need to be supported, such as:

```
============================================================
-oOo--oOo--oOo--oOo--oOo--oOo--oOo--oOo--oOo--oOo--oOo--oOo-
- - - - - - - - - - - - - - - - - - - - - - - - - - - - - -
>>==<<==>>==<<==>>==<<==>>==<<==>>==<<==>>==<<==>>==<<==>>==<<==
```

This functionality can be provided by changing the above line of Perl code. The single dash is changed to the character (or selection of characters), and the constant value can either be left as it is, or changed to an appropriate value. These next four lines of Perl code produce the lines shown above:

```
print "=" x LINE_LENGTH, "\n";
print "-oOo-" x 12, "\n";
print "- " x 30, "\n";
print ">>==<<==" x 8, "\n";
```

Look *closely* at the code. Although the four lines produced by these four program statements are different, the program statements are *similar*. Each statement takes a string of one or more characters and prints it a fixed number of times (followed by a newline). It would be great if these fours lines could be replaced with the following:

```
drawline "=", LINE_LENGTH;
drawline "-oOo-", 12;
drawline "- ", 30;
drawline ">>==<<==", 8;
```

That is, a new command (called `drawline`) allows one or more characters to be defined together with a count of the number of times to repeat the character(s). In fact, when shown in this context, "LINE_LENGTH" is a poor name for this particular constant. "REPEAT_COUNT" is a much better name.

A good question to ask at this point is: *Is it possible to create a new, custom command like drawline with Perl?* The answer is yes, new custom commands can be created using subroutines.

## 5.2 Introducing Subroutines

Think of a subroutine as a collection of statements that has been given a name. Because the collection is named, the subroutine can be called in much the same way as any in-built Perl command. In addition to the name, it is possible to send data to the subroutine as well as get results returned from the subroutine. Both these features are optional – you do not have to send data to the subroutine or accept any results.

> **Technical Commentary:** Other popular names for *subroutine* include *method*, *procedure* and *function*. Which is used often depends on the programming language in use at the time. In many programming languages, the word *function* is reserved for those subroutines that return a value, whereas the use of the word *procedure* or *subroutine* indicates that no value is returned. The word *method* is more closely associated with object-oriented programming technologies. But that's another story (which is very much beyond the scope of *Bioinformatics, Biocomputing and Perl*). Most Perl programmers freely mix the use of the words *subroutine* and *function*.

Here's a maxim to help you understand when to create subroutines.

> **Maxim 5.1** *Whenever you think you will reuse some code, create a subroutine.*

### 5.2.1 Calling subroutines

Let's assume that `drawline` already exists as a subroutine. To call (or *invoke*) `drawline`, use either of the following:

```
drawline "=", REPEAT_COUNT;
drawline( "=", REPEAT_COUNT );
```

Subroutines are invoked with or without parentheses[1]. It is also possible to invoke `drawline` like this:

```
drawline;
drawline();
```

Which may do nothing, print a blank line or produce some sort of default line (such as sixty dashes), depending on how the programmer has coded the subroutine. Note that it is also legal (more correctly, "official") Perl syntax to prefix the name of the subroutine with "&". However, as such usage is optional, the vast majority of Perl programmers do not bother. There are a number of places where the "&" is *required*, and `perl` is pretty good at providing a warning message when such requirements are violated.

---

[1] The reasons for the use of one style over the other are not something we need to go into right now. The important point is that both styles are OK, and both work. Some programmers prefer one style over the other. Our advice: pick one style and try to use it *consistently*. But be aware of the other style.

## 5.3 Creating Subroutines

Creating a subroutine is straightforward once a name has been decided upon. As with variable containers, the trick here is to use good, descriptive names for subroutines, such as:

```
drawline
find_a_sequence
convert_data
```

as opposed to:

```
my_subroutine
sub1
tempsub
```

which do not provide any clue as to the role of the subroutine, whereas the good names do. It is generally not a good idea to give a subroutine a name that is already used as a Perl command (such as "`print`"). Of course, Perl does not stop a programmer from doing this, but the subroutine cannot be expected to work the way it is supposed to.

With a good name decided upon, create a subroutine by prefixing the name with the word "`sub`", and postfixing it with the block of statements to execute, recalling that blocks are contained within curly braces. Here's an *empty* version of `drawline`:

```
sub drawline {

}
```

This subroutine is "empty" because the block contains nothing.

Note the style of indentation used here. The *opening* curly brace appears immediately after the subroutine name (on the same line), while the *closing* curly brace appears below, and aligned to, the "s" in "sub". This is the indentation style preferred by the majority of Perl programmers. However, as with most Perl things, style is personal, so alternative indentation techniques are common, for example:

```
sub drawline { }

sub drawline
{

}
```

```
sub drawline
    {

    }
```

Pick an indentation style and try to use it consistently. To have the drawline subroutine do something, add statements to the block, as follows:

```
sub drawline {
    print "-" x REPEAT_COUNT, "\n";
}
```

To use drawline within a program, simply include the subroutine within the program's code, then invoke it as needed. When invoked, the program goes off and executes the statement(s) in the subroutine, then returns to the statement immediately after the invocation. Here's a program called first_drawline that does just that[2]:

```
1.   #! /usr/bin/perl -w

2.   # first_drawline - the first demonstration program for "drawline".

3.   use constant REPEAT_COUNT => 60;

4.   sub drawline {
5.       print "-" x REPEAT_COUNT, "\n";
6.   }

7.   print "This is the first_drawline program.\n";
8.   drawline;
9.   print "Its purpose is to demonstrate the first version of drawline.\n";
10.  drawline;
11.  print "Sorry, but it is not very exciting.\n";
```

When executed by perl, this program prints the following:

```
This is the first_drawline program.
------------------------------------------------------------
Its purpose is to demonstrate the first version of drawline.
------------------------------------------------------------
Sorry, but it is not very exciting.
```

The placement of the drawline subroutine within the program is worth explaining. By and large, perl does not care where in your program a subroutine is included. Some programmers like to include all of their subroutines near the top of their program (as is the case here), while others prefer to place subroutines

---

[2] Note that line numbers are included here for illustrative purposes only; they are not part of the program code.

at the bottom. Yet others place them in any arbitrary location (which is syntactically legal, but hard to justify). Although `perl` typically does not care where the subroutine is placed, programmers should care. As already advised, consistency is important here, so pick either near the top of your program or at the bottom[3].

No matter where a subroutine is placed, its code is not executed until it is invoked by some calling code. So, even though `drawline` is *defined* on line 4 of `first_drawline`, it is not executed until it is invoked on line 8.

As is stands, `drawline` works, but is not very flexible. For instance, if `drawline` is invoked like this:

```
drawline "=== ", 10;
```

it still prints 60 dashes instead of ten copies of the "=== " pattern. The reason for this is that `drawline` has not been told what to do with any data that is sent to it. When working with subroutines, this data is referred to as *parameters* or *arguments*.

### 5.3.1 Processing parameters

When parameters are sent to a subroutine, `perl` puts the individual data items into a special array, the *default* array, called `@_`[4]. Once there, the parameters can be accessed using standard array indexing syntax, as follows:

```
print "$_[0]";    # The first parameter.
print "$_[1]";    # The second parameter.
print "$_[2]";    # The third parameter, and so on.
```

That is, prefix the name of the array (which is "_") with a dollar sign, then indicate the array index that is to be accessed within the square brackets. This may look strange, but is perfectly fine Perl syntax. Armed with this information, let's rewrite the `drawline` subroutine to process some parameters:

```
sub drawline {
    print $_[0] x $_[1], "\n";
}
```

This version of `drawline` has the advantage of supporting any character pattern and any value for the repeat count, but the disadvantage of no longer supporting the invocation of `drawline` without parameters[5]. It is no longer enough to call `drawline` like this:

```
drawline;
```

---

[3] And it's your choice. Just because we place our subroutines near the top of our programs does not mean that we are right and everyone else is wrong. It just means that this is our preference.

[4] We already know that the default scalar is $_ and now we know that the default array is @_. So, does this mean that the default hash is %_? No, it does not. There is no default hash in Perl.

[5] And if you try to, `perl` complains quite loudly.

It has to be invoked like this:

```
drawline "-", REPEAT_COUNT;
```

or like this:

```
drawline( "-", REPEAT_COUNT );
```

Which is a bit of a drag, until you realize that it can now be called like any of these:

```
drawline "=", REPEAT_COUNT;
drawline( "-o0o-", 12 );
drawline "- ", 30;
drawline( ">>==<<==", 8 );
```

This new version of drawline is used to replace the one from first_drawline, producing a new program, called second_drawline. The four example invocations of the drawline subroutine above are also added to the program:

```
#! /usr/bin/perl -w

# second_drawline - the second demonstration program for "drawline".

use constant REPEAT_COUNT => 60;

sub drawline {
    print $_[0] x $_[1], "\n";
}

print "This is the second_drawline program.\n";
drawline "-", REPEAT_COUNT;
print "Its purpose is to demonstrate the second version of drawline.\n";
drawline "-", REPEAT_COUNT;
print "Sorry, but it is still not exciting.  However, it is more useful.\n";

drawline "=", REPEAT_COUNT;
drawline "-o0o-", 12;
drawline "- ", 30;
drawline ">>==<<==", 8;
```

Which, when executed, produces the following output:

```
This is the second_drawline program.
------------------------------------------------------------
Its purpose is to demonstrate the second version of drawline.
------------------------------------------------------------
Sorry, but it is still not exciting.  However, it is more useful.
============================================================
-o0o--o0o--o0o--o0o--o0o--o0o--o0o--o0o--o0o--o0o--o0o--o0o-
- - - - - - - - - - - - - - - - - - - - - - - - - - - - - -
>>==<<==>>==<<==>>==<<==>>==<<==>>==<<==>>==<<==>>==<<==
```

By accessing the individual elements of the default array (@_), the `drawline` subroutine is now able to support repeatedly printing any sequence of characters, which is quite useful. In fact, "accessing the individual elements of the default array" is so common that Perl provides an alternative technique for achieving the same thing without having to use the strange looking $_[0] syntax. Recall the `shift` function from the *Places To Put Things* chapter that when invoked, *removes* and *returns* the first element from a named array. When used on its own (that is, *without* referring to a named array), as follows:

```
shift;    # Or like this: shift();
```

the function returns and removes the first element from the default array[6]. It is now possible to rewrite `drawline` for the third time to take advantage of the `shift` function:

```
sub drawline {
    print shift() x shift(), "\n";
}
```

Note the use of the parentheses after `shift`, which are required in this instance by `perl`.

It is left as an exercise for the reader to replace the `drawline` from the second_drawline program with this new subroutine, creating `third_drawline`. When executed, `third_drawline` produces the same output as that provided by second_drawline.

### 5.3.2 Better processing of parameters

The third version of `drawline` is good, but it can be made better. Specifically, let's arrange to support default behaviour that ensures something sensible happens when no (or incomplete) parameters are provided to `drawline`. This sensible behaviour involves printing 60 dashes. Take a look at this, the fourth version of `drawline`:

```
sub drawline {
    $chars = shift || "-";
    $count = shift || REPEAT_COUNT;

    print $chars x $count, "\n";
}
```

---

[6] If you were able to guess this, congratulations: you are beginning to think like a Perl programmer. If you did not guess this but understand what's going on, congratulations: you are beginning to think like a Perl programmer. If you are somewhat lost, congratulations: go back and reread this section.

There are a few (new) things going on here. Two scalar variables are used within the subroutine, and they take their value from the two parameters supplied to drawline. The `$chars` scalar is assigned the first parameter (thanks to `shift`), and the `$count` scalar is assigned the second parameter (again, thanks to `shift`). These scalars are then used with the `print` command to draw the appropriate linestyle.

The `||` symbol is another Perl operator, and it means "or". Here, it is used to provide default values to the two scalars. The assignment to the `$chars` scalar takes its value either from the first parameter or, if no parameter is provided, is set to a single dash. The assignment to the `$count` scalar takes its value either from the second parameter or, if no parameter is provided, is set to the value of REPEAT_COUNT. So, it is now possible to invoke drawline like this:

```perl
drawline "=== ", 10;
```

to print ten copies of the "=== " pattern or, like this:

```perl
drawline;
```

to print the default line of 60 dashes. It is also possible to print 60 of any pattern by not providing a value for the second parameter, as follows:

```perl
drawline "=";      # Prints sixty equal signs.
drawline "*";      # Prints sixty stars.
drawline "$";      # Prints sixty dollars.
```

Which is neat. However, the ordering of the parameters is important, as this version of drawline expects the character(s) first, and the repeat count second. So, invoking drawline like any of these causes a problem:

```perl
drawline 40;       # Does NOT print forty dashes!
drawline 20, "-";  # Does NOT print twenty dashes!
```

Consider that this most recent version of drawline agrees a *contract* with programmers that use it. If the parameters are in the correct order, the contract holds. If they are not, the contract is broken (and who knows what will happen?). Here's another program, called fourth_drawline, that uses the fourth version of the drawline subroutine (and honours the contract):

```perl
#! /usr/bin/perl -w

# fourth_drawline - the fourth demonstration program for "drawline".

use constant REPEAT_COUNT => 60;

sub drawline {
    $chars = shift || "-";
```

80    *Getting Organised*

```
        $count = shift || REPEAT_COUNT;

        print $chars x $count, "\n";
}
print "This is the fourth_drawline program.\n";
drawline;
print "Its purpose is to demonstrate the fourth version of drawline.\n";
drawline;
print "Sorry, but it is still not exciting.  However, it is more useful.\n";

drawline "=", REPEAT_COUNT;
drawline "-oOo-", 12;
drawline "- ", 30;
drawline ">>==<<==", 8;
```

When executed, the output from `fourth_drawline` is similar to that of both `second_drawline` and `third_drawline`. This version is better, but still not the best. It would be helpful if `drawline` allowed the parameters to be supplied in any order. With Perl, this too is possible.

### 5.3.3  Even better processing of parameters

To provide a means whereby parameters are provided in any order, a parameter naming mechanism is required. Look at these invocations of `drawline`:

```
drawline;
drawline( Pattern => "*" );
drawline( Count => 20 );
drawline( Count => 5, Pattern => " -oOo- " );
drawline( Pattern => "===", Count => 10 );
```

The first invocation prints the default 60 dashes. The second prints 60 stars. The third prints 20 dashes. The fourth invocation prints five copies of the " -oOo- " pattern. And the sixth invocation prints ten copies of the "===" pattern.

Note that with these invocations of the next version of `drawline`, the subroutine now supports zero, one or two parameters. In addition, the parameters are named and, as such, can appear in any order. When parameters are missing, the subroutine does the most sensible thing by substituting reasonable default values. It's up to the programmer calling the subroutine to use whichever parameter ordering makes most sense. How's that for a flexible contract?

When data is passed to any subroutine, `perl` takes the data and populates the default array. So, if the invocation of `drawline` looks like this:

```
drawline( Count => 5, Pattern => " -oOo- " );
```

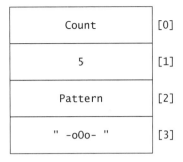

**Figure 5.1** The default array, @_, with assigned values.

the default array is assigned a list of items that looks like this[7]:

```
"Count", 5, "Pattern", " -oOo- "
```

That is, four values are assigned to the default array, which looks like Figure 5.1. The default array is now available within the subroutine. What happens next is the key to the entire parameter naming mechanism. Recall from the *Places To Put Things* chapter that an array can be used to initialise a hash. If the first statement in a subroutine is this:

```
%arguments = @_;
```

the %arguments hash is assigned the values in the default array, namely, the string "Count" as a name, with 5 as its value, and the string "Pattern" as a name, with " -oOo- " as its value. After the assignment, the hash looks like Figure 5.2. The %arguments hash now associates values with parameter names. When combined with the || operator, the %arguments hash can be used to assign values to scalar variables:

```
$chars = $arguments{ Pattern } || "-";
$count = $arguments{ Count }   || REPEAT_COUNT;
```

| Count   | 5        |
|---------|----------|
| Pattern | " -oOo- " |

**Figure 5.2** The %arguments hash, with assigned values.

---

[7] If this looks strange, recall that Perl's => symbol is another representation for comma. Additionally, note that perl is correctly (and sensibly) surrounding the words "Count" and "Pattern" with double quotes, since they are strings.

The first statement sets the $chars scalar to equal the value associated with "Pattern" or, if no value is associated, sets it to a single dash. The second statement sets the $count scalar equal to the value associated with "Count" or, if no value is associated, sets it to the value of REPEAT_COUNT. Here's the fifth version of drawline:

```perl
sub drawline {
    $chars = $arguments{ Pattern } || "-";
    $count = $arguments{ Count }   || REPEAT_COUNT;

    print $chars x $count, "\n";
}
```

The fifth_drawline program shows the latest version of drawline in action:

```perl
#! /usr/bin/perl -w

# fifth_drawline - the fifth demonstration program for "drawline".

use constant REPEAT_COUNT => 60;

sub drawline {
    %arguments = @_;

    $chars = $arguments{ Pattern } || "-";
    $count = $arguments{ Count }   || REPEAT_COUNT;

    print $chars x $count, "\n";
}

print "This is the fifth_drawline program.\n";
drawline;
print "Its purpose is to demonstrate the fifth version of drawline.\n";
drawline;
print "Things are getting a little more interesting.\n";

drawline( Pattern => "*" );
drawline( Count => 20 );
drawline( Count => 5, Pattern => " -oOo- " );
drawline( Pattern => "===", Count => 10 );
drawline;
```

which, when executed, produces the following output:

```
This is the fifth_drawline program.
------------------------------------------------------------
Its purpose is to demonstrate the fifth version of drawline.
------------------------------------------------------------
Things are getting a little more interesting.
************************************************************
--------------------
 -oOo-  -oOo-  -oOo-  -oOo-  -oOo- 
==============================
------------------------------------------------------------
```

which confirms that drawline can be invoked with zero, one or two named parameters, supplied in any order. When parameters are not supplied, this version of drawline does the sensible thing.

## 5.3.4 A more flexible drawline subroutine

As a final twist to this subroutine, consider the inclusion of the newline at the end of the print command on the subroutine's last line. This ensures that any line drawn includes a newline, which is a reasonable assumption to make until an attempt is made to draw something like this:

In an attempt to produce the first line, it is not possible to do something like this:

```
print "+";
drawline( Count => 15 );
print "+";
```

as these three statements produce the following:

```
+---------------
+
```

which is not what is required because of the inclusion of the newline at the end of the print command within drawline. Removing the newline results in a version of drawline that is more flexible (in that it fixes this particular problem) at the expense of requiring the programmer to worry about newlines. As long as this forms part of the contract, this is OK. Assuming a non-newline version of drawline, here's the code to draw the first line of the box:

```
print "+";
drawline( Count => 15 );
print "+\n";
```

A common temptation is to rewrite these three lines as one single print command:

```
print "+", drawline( Count => 15 ), "+\n";
```

which seems like a reasonable thing to do, until the statement is executed and it produces this:

```
---------------+1+
```

**84**   *Getting Organised*

Whoops! What's going on here? Well, let's take a look at what the statement is doing: it's a single `print` command that includes a literal "+", an invocation of `drawline`, another literal "+" and a newline.

When `perl` processes this statement, it looks at the parameters to `print`[8], and determines that `drawline` is a subroutine call, so `perl` invokes `drawline` *before* continuing with the invocation of `print`. This results in 15 dashes appearing on screen. Having finished with `drawline`, `perl` then returns to the `print` command and starts printing. The first thing printed is the literal "+". Next up is the *result* of the invocation of `drawline`. As `drawline` succeeded in printing its 15 dashes, the result of the invocation was true, which in Perl is the value 1, so "1" is printed. The `print` command concludes by printing the second literal "+" and a newline. This explains the unexpected output. Of course, Perl has a solution to this problem, as discussed below.

### 5.3.5   Returning results

Rather than have `drawline` actually draw (that is, print) the line, let's have `drawline` return the correctly formatted line to the caller. The calling code can then do what it likes with the line, which may or may not include printing it. The change is trivial: change the call to `print` within `drawline` to a call to `return`.

The `return` command, when invoked, causes the current subroutine to terminate immediately. When provided with a value, `return` sends the value to the caller. Here's another version of `drawline` that implements these changes:

```perl
sub drawline {
    %arguments = @_;

    $chars = $arguments{ Pattern } || "-";
    $count = $arguments{ Count }   || REPEAT_COUNT;

    return $chars x $count;
}
```

A program, called `boxes`, uses this version of `drawline` to draw the box from the last section:

```perl
#! /usr/bin/perl -w

# boxes - the box drawing demonstration program for "drawline".

use constant REPEAT_COUNT => 15;

sub drawline {
    %arguments = @_;
```

---
[8] Note: `print` is a subroutine, too.

```
        $chars = $arguments{ Pattern }  ||  "-";
        $count = $arguments{ Count   }  ||  REPEAT_COUNT;

        return $chars x $count;
}
print "+", drawline, "+\n";
print "|", drawline( Pattern => " " ), "|\n";
print "|", drawline( Pattern => " " ), "|\n";
print "|", drawline( Pattern => " " ), "|\n";
print "|", drawline( Pattern => " " ), "|\n";
print "|", drawline( Pattern => " " ), "|\n";
print "+", drawline, "+\n";
```

Note the clever adjustment to the value of the REPEAT_COUNT constant.

## 5.4 Visibility and Scope

No discussion of subroutines would be complete without describing *scope*. Scoping relates to the *visibility* of a variable throughout the lifetime of a program (that is, as it *runs*). It is best described by way of example. However, first a maxim.

> **Maxim 5.2** *When determining the scope of a variable, think about its visibility.*

Consider this small program, called global_scope:

```
#! /usr/bin/perl -w

# global_scope - the effect of "global" variables.

sub adjust_up {
    $other_count = 1;
    print "count at start of adjust_up: $count\n";
    $count++;
    print "count at end of adjust_up: $count\n";
}

$count = 10;
print "count in main: $count\n";
adjust_up;
print "count in main: $count\n";
print "other_count in main: $other_count\n";
```

When executed, the global_scope program prints these messages on screen:

```
count in main: 10
count at start of adjust_up: 10
```

```
count at end of adjust_up: 11
count in main: 11
other_count in main: 1
```

The $count scalar is accessible within the entire program, including any subroutines (such as adjust_up). Additionally, the $other_count scalar, which is assigned a value of 1 within the adjust_up subroutine is *also* accessible within the entire program. Both of these statements are confirmed by the messages produced by global_scope. The reason for this behaviour is that by default, and if not told otherwise, perl treats all variables as being global in scope, where "global" is defined as *accessible from anywhere within the disk-file that contains the program code.*

This behaviour can be very convenient, but also very dangerous. For instance, it might well be the case that the programmer who wrote the adjust_up subroutine did not intend the $other_count scalar to be visible outside adjust_up. After all, it is created within adjust_up, so perhaps the intention was to have the value visible only within that subroutine.

Also, consider trying to track accesses and adjustments to a global. Within a small program, like global_scope, keeping track of each of the variables is not a difficult task. However, consider a program that is 10,000 lines long. It is much more difficult to "keep track of things" when programs grow large. Is it possible to be sure that *all* of the program's variables (which are all global by default) are being accessed appropriately? Probably not.

Many computer scientists are aghast at Perl's default attitude regarding variable scope. This is due to the fact that the vast majority of programmers are taught from a very young age that "globals are bad" and are told "don't use globals". As a result, programmers grow up avoiding the use of globals at all costs, which is a shame. The truth is that globals are bad if used incorrectly but when used correctly, can greatly simplify some programs. In an effort to be as flexible as possible, the creators of Perl went against the groove and made all variables global by default. Despite this, it is possible to limit the likelihood of error in this area, as described below.

### 5.4.1 Using private variables

To help keep things organised, Perl provides a number of mechanisms that limit the visibility of a variable. The most common is the use of the my command, which limits the scope of a variable to within its enclosing block (curly braces), subroutine or file. When you use my to declare a variable, it tells perl to treat the variable as private to its enclosing scope.

> **Technical Commentary:** Within the Perl world, such variables are often referred to as *lexicals* or *mys*, as in "use a lexical" or "set the my variable to 10".

## Visibility and Scope 87

As will be shown in the *Perl Grabbag* chapter, perl can force the programmer to always use my variables. For now, let's rewrite global_scope to use my and see what effect it has. This new program is called private_scope:

```perl
#! /usr/bin/perl

# private_scope - the effect of "my" variables.

sub adjust_up {
    my $other_count = 1;
    print "count at start of adjust_up: $count\n";
    $count++;
    print "other_count within adjust_up: $other_count\n";
    print "count at end of adjust_up: $count\n";
}

my $count = 10;
print "count in main: $count\n";
adjust_up;
print "count in main: $count\n";
print "other_count in main: $other_count\n";
```

The $other_count scalar within the adjust_up subroutine is now private to the subroutine, because of the use of my. The $count scalar within the main code is also declared as a my when it is set to 10. Now, take a look at the output produced by this program by perl:

```
count in main: 10
count at start of adjust_up:
other_count within adjust_up: 1
count at end of adjust_up: 1
count in main: 10
other_count in main:
```

The value of the private $count (which is set to 10) is printed. The adjust_up subroutine is then called. After creating a private scalar called $other_count and setting it to 1, the value of the global variable $count is printed. But look at what happened: no value was printed. Why did the value 10 not print?

Well, at this point in the program's execution cycle, the $count scalar does not yet exist. A private scalar (which also happens to be called $count) does exist, but its visibility has been limited to the main code by the use of my, which means it is *not accessible from within any subroutines*. It is a private variable, after all.

Recall that when a variable is first used, it is assigned a default value (usually nothing). When $count is first used within the adjust_up subroutine, it is created as a global variable (remember: the private $count cannot be accessed), and given a default value. When printed, nothing appears, as the default value is nothing. At this point, two $count scalars exist: one is private to the main code, and the

**88** *Getting Organised*

other is global. Within `adjust_up`, the global $count is increment with the ++ operator. This causes the value of nothing to have 1 added to it, so the global $count now has a value of 1. The private $other_count scalar is then printed, giving a value of 1, as is the global $count (which is also 1). The `adjust_up` subroutine then ends.

Returning to the main code, the next statement prints the value of the private $count, which is 10 (not 1). Note that although a global $count variable now exists (thanks to `adjust_up`), the private variable of the same name takes precedence.

The final statement of the `private_scope` program attempts to print the value of $other_count. Again, nothing prints for the same reason as nothing printed for the access of $count at the start of the `adjust_up` subroutine. The $other_count scalar within the subroutine is private to the subroutine, which means its value cannot be accessed outside of the subroutine. So, in the main code, when $other_count is accessed, it is first created as a global, given a default value of nothing and then printed, which has the effect of printing nothing, as confirmed by the last line of output from the program.

Using my variables as a general rule is highly recommended. So much so, in fact, that it warrants another maxim.

> **Maxim 5.3** *Unless you have a really good reason not to, always declare your variables with my.*

### 5.4.2 Using global variables properly

There are times, of course, when a global is warranted. Rather than declare a variable without my and get a default global, a better practice is to specifically state that a global is required by use of the our command (which is only available in Perl version 5.6.0 and higher).

Let's look at a program that uses our, is based on `private_scope` and is called `hybrid_scope`:

```
#! /usr/bin/perl

# hybrid_scope - the effect of "our" variables.

sub adjust_up {
    my $other_count = 1;
    print "count at start of adjust_up: $count\n";
    $count++;
    print "other_count within adjust_up: $other_count\n";
    print "count at end of adjust_up: $count\n";
}

our $count = 10;
print "count in main: $count\n";
adjust_up;
```

```
print "count in main: $count\n";
print "other_count in main: $other_count\n";
```

The $count scalar within this program's main code is now declared with our. Let's see the effect this has on the output produced:

```
count in main: 10
count at start of adjust_up: 10
other_count within adjust_up: 1
count at end of adjust_up: 11
count in main: 11
other_count in main:
```

Unlike private_scope, the $count within hybrid_scope is deliberately global, so when accessed within the adjust_up subroutine, its value is accessible. The subroutine prints the value (10), increments $count and prints it again (11). Note that the value of $count within the main code is also 11, as it is *the same variable*. The $other_count variables behave exactly as they did in the private_scope program. Here's another maxim:

> **Maxim 5.4** *If you must use a global variable, declare it with our.*

It is also possible to use the local command to limit the visibility of variables, but it has been superseded by my and common wisdom appears to suggest that the use of local be avoided. So, the use of local is quietly avoided within *Bioinformatics, Biocomputing and Perl*.

From this point on, all of the presented code will use my and our as appropriate.

### 5.4.3 The final version of drawline

Now that scoping is understood, let's return to the drawline subroutine and produce a final version that uses my[9].

Referring back to page 84, there really is no reason why the %arguments, $chars and $count variables within drawline need to have global scope, so let's make them private:

```
sub drawline {
    my %arguments = @_;

    my $chars = $arguments{ Pattern } || "-";
    my $count = $arguments{ Count }   || REPEAT_COUNT;

    return $chars x $count;
}
```

---

[9] We promise that this is in fact the *final* version of drawline. However, we reserve the right to change our minds.

This tidies up the coding considerably. As the three variables within `drawline` are private to the subroutine, the use of `my` instructs `perl` to ensure that they are.

Thinking about the visibility of variables and the scope within which they operate is important and when used properly, leads to good software design.

## 5.5 In-built Subroutines

In addition to letting programmers create their own subroutines, Perl also has a large collection of in-built subroutines. The entire collection is documented in the `perlfunc` on-line documentation, which comes with Perl. Use this command on Linux to view the document:

    man perlfunc

Alternatively, use the `perldoc` program (which also comes with Perl) to search the `perlfunc` document for the documentation specific to a subroutine. For instance, to view the documentation for the in-built `sleep` subroutine, use this command:

    perldoc -f sleep

The in-built subroutines take a varying number of parameters, so always check the documentation for specifics. Be aware that some in-built subroutines can perform differently on the basis of how they are invoked and used. This book has already used some of the more popular in-built subroutines (which have been referred to as "Perl commands"). Here is an abbreviated list[10]:

`alarm` – Signals an alarm to occur a number of seconds in the future.

`chomp` – Deletes the trailing newline character from a scalar.

`chop` – Deletes the last character from a scalar.

`close` – Closes a previously opened filehandle.

`defined` – Returns "true" if a variable has a value associated with it.

`delete` – Deletes elements/entries from an array/hash.

`die` – Exits the current program after displaying a user-specified message.

`do` – Executes a block of statements as one, or reads in a collection of statements from another disk-file and executes them.

---

[10] Which is based on a similar list from Paul's first book, *Programming the Network with Perl*, Wiley, 2002.

`each` – Used to iterate over a hash.

`eof` – Tests for the end-of-file condition when working with disk-files.

`eval` – Evaluates a block of code and provides for exception handling.

`exists` – Returns "true" if a specific array element or hash entry exists.

`exit` – Exits the current program.

`gmtime` – Returns the date and time relative to GMT.

`join` – Joins a list of strings together.

`keys` – Returns a list of keys for a specified hash.

`last` – Exits from the current loop.

`length` – Returns the length of a scalar variable.

`localtime` – Returns the date and time relative to the local time zone.

`my` – Marks a variable as being lexically scoped.

`next` – Starts the next iteration of the current loop.

`open` – Opens a file, and associates a filehandle with it.

`our` – Declares a global variable.

`pack` – Converts a collection of variables into a string of bytes.

`package` – Declares a new namespace.

`pop` – Treats an array like a stack, and pops the last element off the end of the array.

`print` – Prints something.

`printf` – Prints to a particular format.

`push` – Treats an array like a stack, then pushes an element onto the end of the array.

`read` – Reads a specified number of bytes from a filehandle.

`redo` – Restarts the current loop iteration.

`ref` – Checks to see whether a scalar is a reference, and if it is, returns the type of reference as a string.

`return` – Returns a value from a subroutine.

`scalar` – Forces a list to be treated as if it were a scalar.

`shift` – Treats an array like a stack, and pops the first element off the start of the array.

sleep – Pauses execution for a specified number of seconds.

sort – Sorts a list using string comparison order (by default), or by using some user-specified ordering.

splice – Removes specified elements from an array.

split – Splits a delimited string into a list of individual elements.

sprintf – Like printf, above, except the result is assigned to a scalar.

sub – Declares a subroutine.

substr – Extracts a sub-string from a string.

system – Calls an operating system command, and returns its exit status to the calling program.

time – Returns the number of non-leap seconds since the operating systems "epoch"[11].

undef – Takes a previously defined variable, and undefines it.

unpack – The reverse of pack, above, which extracts a list of values from a string of bytes.

unshift – Treats an array like a stack, and pushes an element onto the start of the array.

wantarray – Returns "true" if a subroutine was called within a list context, "false" otherwise.

warn – Sends output to standard error (which may or may not be the screen).

write – Writes a specified number of bytes to a filehandle.

To reiterate, this list is not complete. See perlfunc for the complete list.

## 5.6 Grouping and Reusing Subroutines

The drawline subroutine is quite useful, and if a requirement exists to draw lines in a lot of different programs, it may well be a subroutine in each. Although this is a strategy that works, that is, cutting 'n' pasting the subroutine into every program that needs it, it introduces a problem. What happens if, at some stage in the future after the subroutine has been used in 443 programs, a decision is made to change how drawline works[12]. Furthermore, imagine a decision is

---

[11] What the operating system thinks is the start of time. It varies from system to system. This means that the start of time is different on Linux, Windows and Mac OS. Yes, they could not even agree on that!

[12] Don't worry, we aren't going to. A promise is a promise, after all.

made to ensure that the change is a global one, in that every program that uses `drawline` needs to be changed. That's 443 changes, or 443 cuts and 443 pastes, or whatever. Let's further assume that each change takes (on average) one minute – that's 443 minutes! And, that's before each of those 443 programs are re-tested now that they have changed.

Let's face it. Given such a situation, a reason will be found *not* to make the change. But it need not be like this. Most modern programming languages, including Perl, provide a mechanism to reuse a subroutine in multiple programs while maintaining a master copy of the code. If a change is required, the master copy of the subroutine is changed, tested and then released. The other programs are not even touched.

Say "hello" to the Perl module.

## 5.6.1 Modules

At its most basic (and useful), a *module* in Perl is a place to put subroutines that can then be used by many programs.

> **Maxim 5.5** *When you think you will reuse a subroutine, create a custom module.*

Let's assume that the `drawline` subroutine is part of a module called `UsefulUtils`. A program can access `drawline` by first using the module, then calling the subroutine, with code something like this:

```
#! /usr/bin/perl -w

use lib "$ENV{'HOME'}/bbp/";
use UsefulUtils;

drawline;
```

As `drawline` is part of the `UsefulUtils` module, it does not appear in the source code of the program that wishes to use it. The `use` statement "pulls in" the `drawline` subroutine as required.

The `use lib` statement tells `perl` where to find custom modules. More on this later.

Creating a custom module in Perl has been standardised. Every module starts with the following "blank" template:

```
package;

require Exporter;

our @ISA           = qw( Exporter );
```

```
our @EXPORT        = qw();
our @EXPORT_OK     = qw();
our %EXPORT_TAGS   = ();

our $VERSION       = 0.01;

1;
```

The module template starts with a `package` statement. This introduces and names the module, and creates a new *namespace*. To start creating `UsefulUtils`, start with a `package` statement like this:

```
package UsefulUtils;
```

The rest of the module template needs only minor changes. The `require` statement, together with the `our @ISA` statement, tells `perl` that the custom module draws on the facilities offered by a standard Perl module, called `Exporter`. It is not important that module writers understand what these two lines do. However, it is really important that they are included within each custom module, so leave these lines as they are.

Three "export" variables are then declared. The first, an array called @EXPORT, is set to the empty list. By adding to this list, it is possible to have a subroutine automatically imported into the program that uses the custom module.

The second variable, an array called @EXPORT_OK, is again set to the empty list. By adding to this list, it is possible to allow the program that uses the custom module to specify the subroutine(s) to import.

The final "export" variable, a hash called %EXPORT_TAGS, can be used to group related subroutines within the module into tagged categories. These can then be used to import the groups into the program that uses the custom module.

The `UsefulUtils` module is designed to house a collection of loosely related utilities. As such, there's no requirement to automatically export any subroutines, so the @EXPORT array is left empty. Users of the module are required to specifically identify the utility subroutine they wish to import into a program, so let's add the name of the only subroutine that we have to the @EXPORT_OK array:

```
our @EXPORT_OK     = qw( drawline );
```

The $VERSION scalar is used to track the maturity of the module. Typically, when the module is under development, the version number is less than 1. As the module matures, the version number is increased. For now, the version number is set to 0.01, as the module is under development and very new.

The "1;" at the end of the module ensures that the module returns true when it is used. Again, this is a detail that must be included, so make sure the last statement of every custom module is "1;". The subroutines to be included in the module are placed between the $VERSION scalar and the "1;".

Here's the first version of the UsefulUtils module, which includes the drawline subroutine:

```perl
package UsefulUtils;

# UsefulUtils.pm - the useful utilities module from "Bioinformatics,
#                  Biocomputing and Perl".

require Exporter;

our @ISA          = qw( Exporter );

our @EXPORT       = qw();
our @EXPORT_OK    = qw( drawline );
our %EXPORT_TAGS  = ();

our $VERSION      = 0.01;

use constant REPEAT_COUNT => 60;

sub drawline {
    # Given:  a character string and a repeat count.
    # Return: a string that contains the character string
    #         "repeat count" number of times.
    #
    # Notes:  For maximum flexibility, this routine does NOT include
    #         a newline ("\n") at the end of the line.

    my %arguments = @_;

    my $chars = $arguments{ Pattern } || "-";
    my $count = $arguments{ Count }   || REPEAT_COUNT;

    return $chars x $count;
}

1;
```

Note the inclusion of a number of comments, which is always a good idea. Note also the requirement to give the module a name that ends in ".pm". The REPEAT_COUNT constant is also included in the module.

Before continuing, let's put any custom modules that are created into a standard location. Create a directory under your home directory called "bbp", and copy the module there:

```
mkdir  ~/bbp/
cp  UsefulUtils.pm  ~/bbp/
```

The UsefulUtils module can now be used by any program. Here's another version of the boxes program, called boxes2, that uses the module:

**96**  *Getting Organised*

```perl
#! /usr/bin/perl -w

# boxes2 - the box drawing demonstration program for "drawline".

use lib "$ENV{'HOME'}/bbp/";
use UsefulUtils qw( drawline );

print "+", drawline( Count => 15 ), "+\n";
print "|", drawline( Pattern => " ", Count => 15 ), "|\n";
print "|", drawline( Pattern => " ", Count => 15 ), "|\n";
print "|", drawline( Pattern => " ", Count => 15 ), "|\n";
print "|", drawline( Pattern => " ", Count => 15 ), "|\n";
print "|", drawline( Pattern => " ", Count => 15 ), "|\n";
print "+", drawline( Count => 15 ), "+\n";
```

The `use lib` statement tells `perl` where to find any custom modules. When `UsefulUtils` is used, a list of subroutines to import is provided. Note that unlike `boxes`, this program has to explicitly provide a value for the "Count" parameter, as the default repeat count is 60, not 15.

In later chapters, the `UsefulUtils` module is extended with more subroutines.

## 5.7 The Standard Modules

The Perl programming environment comes with a large collection of standard modules. Take a look at the `perlmodlib` document for a complete list using this command:

```
man perlmodlib
```

Each standard module described in the `perlmodlib` document comes with its own documentation. It is highly recommended that every Perl programmer take as much time as is needed to become familiar with the standard modules. The reason for this advice is simple: if a standard module already implements a particular piece of functionality, it is always better to use the standard module than to attempt to write a subroutine or module that provides the same functionality.

> **Maxim 5.6** *Don't reinvent the wheel;*
> *use or extend a standard module whenever possible.*

So, don't waste time – take advantage of the excellent functionality that is included with Perl by way of the standard modules.

## 5.8 CPAN: The Module Repository

In addition to the standard modules, the Perl programming community, via a large web-site, provides a facility for programmers to share their work with others.

What sets the Perl community apart from others is the extent to which this sharing is coordinated and practised. The *Comprehensive Perl Archive Network*, more commonly referred to as "CPAN"[13], contains a vast collection of modules pertaining to every conceivable task that Perl can be put to. So, whether the requirement is to manipulate graphics images or interact with a web server, a CPAN module more than likely exists to assist in the task. To start exploring what CPAN has to offer, visit this web-site and start reading:

    http://www.cpan.org

All the modules on CPAN are donations made by a growing community of Perl programmers from around the globe, and any Perl programmer is free to download a module of interest and use it. As with the standard modules, the advice is straightforward: use a CPAN module rather than creating a custom module to do the same thing. Even if a CPAN module does not do exactly what is needed, it may be worthwhile downloading and tweaking it to meet a specific requirement. After all, CPAN modules are distributed in source code form, so it is simply a matter of editing the module and including the changes that are required.

> **Maxim 5.7** *Don't reinvent the wheel;*
> *use or extend a CPAN module whenever possible.*

If the changes made to a CPAN module are considered of use to others, programmers are encouraged to submit the changed module to CPAN so that the entire community can benefit from the enhancements. In this way, the truly useful modules available on CPAN get better with time. Programmers who submit their work to CPAN are referred to as *CPAN authors*.

> **Maxim 5.8** *When you think others might benefit from a custom*
> *module you have written, upload it to CPAN.*

## 5.8.1 Searching CPAN

In addition to the main CPAN web-site, another interface to the repository can be found at

    http://search.cpan.org

This web-site provides a mechanism to search CPAN by module, keyword, author and so on. It also provides a browseable categorisation of all of the modules that are available. Take some time to experiment and explore this web-site.

---

[13] Pronounced "see-pan".

## 5.8.2 Installing a CPAN module manually

To use a CPAN module, it needs to be installed into your Perl installation. This process has been standardised. Let's assume that a requirement exists to install a module called `ExampleModule` into a Perl installation[14]. To begin, decompress and unpack the downloaded file:

```
tar zxvf ExampleModule-0.03.tar.gz
```

CPAN modules are typically distributed in a packed, compressed format. The above command decompresses the disk-file and then unpacks its contents. A directory called `ExampleModule-0.03` is created, and all of the disk-files needed to install `ExampleModule` are put into this directory. To prepare for the install, change into this directory and use Perl to create the required `makefile`:

```
cd ExampleModule-0.03
perl Makefile.PL
```

It should now be possible to build and test `ExampleModule` using the standard Linux `make` command:

```
make
make test
```

It is usual for a collection of messages to appear on screen as a result of issuing these commands. If things go well, the module is now ready to install. For the next command to succeed, *superuser* privilege is required. If not already logged in as `root`, temporarily become the superuser as follows:

```
su
```

The `root` password will be required. As superuser, finish the install by issuing this command:

```
make install
<Ctrl-D>
```

Note the use of the `<Ctrl-D>` key-combination after the `make install` command. This logs out the superuser. As a general rule, work in superuser mode (i.e., as `root`) only for as long as is needed. It is generally a bad idea to do regular work logged in as `root`. Trust us when we tell you that if you spend a lot of time logged in as `root`, sooner or later, bad things will happen.

The installation of the module can be tested using either of these two commands:

---

[14] As with the other example commands in this book, the assumption is that you are running Linux. These commands should also work with Mac OS X and UNIX. If you are running Windows, use the Perl Package Manager (ppm) to install CPAN modules.

```
man ExampleModule
perl -e 'use ExampleModule'
```

The first command should display the documentation for `ExampleModule`. The second command should display *nothing* – the Linux command-prompt should reappear after a short delay. If the second command displays a message something along the lines of the following:

```
# Can't locate ExampleModule.pm in @INC.
# BEGIN failed--compilation aborted at -e line 1.
```

this means that the module has been installed incorrectly. If the on-line documentation is missing, this too means that the module has been installed incorrectly. Check that the above instructions have been followed correctly. If they have, check any `README` and `INSTALL` disk-files that came with the module for additional installation instructions.

### 5.8.3 Installing a CPAN module automatically

In addition to manually installing CPAN modules, it is also possible to have `perl` do the work for you. The fictitious module from the previous section can be installed into Perl with this *single* command:

```
perl -MCPAN -e "install 'ExampleModule'"
```

Use of this command assumes the following: the computer upon which this command is executed has to have an active Internet connection, and the command is issued by the superuser. This command downloads the `ExampleModule` distribution disk-file, decompresses and unpacks it, then installs it into Perl. And it does it automatically, allowing you to sit back and relax[15].

Which begs the question: why spend time describing the manual installation technique when CPAN modules can be installed automatically with commands such as this one? The answer to this reasonable question is that not all modules successfully install automatically. So, it is important to understand the manual installation process should anything go wrong during an automatic install.

### 5.8.4 A final word on CPAN modules

Not all CPAN modules are created equal. Some are well supported, have an active user community and are of high quality. Some are test modules, proof of concepts and are (sometimes) of dubious quality. It is important to test any

---

[15] Well, almost. If this is the first time you have executed a command like this, you will be prompted to answer a series of questions. Read the prompts and pick the appropriate answer from those provided. Also, keep an eye on the messages produced by the automatic installation, just in case an error occurs.

module downloaded from CPAN to ensure it works the way it is expected to. Do not blindly trust that a module works in a certain way simply because it says so in its documentation. Test, test and test again.

> **Maxim 5.9** *Always take the time to test downloaded CPAN modules for compliance with specific requirements.*

Having said that, CPAN really is a wonderful resource. It is the collected wisdom and work of the Perl community as a whole. Many Perl programmers, when asked why they continue to use and favour Perl, respond with a single word: CPAN.

## Where to from Here

In this chapter, the idea of code reuse was explored, first with subroutines and then with modules. Additionally, the standard modules and CPAN were described. This was a large chapter, and it covered a lot of important material. In the next chapter, another important topic is introduced: *input/output*.

## The Maxims Repeated

Here's a list of the maxims introduced in this chapter.

- *Whenever you think you will reuse some code, create a subroutine.*
- *When determining the scope of a variable, think about its visibility.*
- *Unless you have a really good reason not to, always declare your variables with my.*
- *If you must use a global variable, declare it with our.*
- *When you think you will reuse a subroutine, create a custom module.*
- *Don't reinvent the wheel; use or extend a standard module whenever possible.*
- *Don't reinvent the wheel; use or extend a CPAN module whenever possible.*
- *When you think others might benefit from a custom module you have written, upload it to CPAN.*
- *Always take the time to test downloaded CPAN modules for compliance with specific requirements.*

# Exercises

1. Write a subroutine, called drawbox, that draws boxes. The subroutine should accept two named parameters, Height and Width, which are used to specify the dimensions of the box. In the absence of either or both of the named parameters, the drawbox subroutine should substitute appropriate default values.

2. Add the drawbox subroutine to the UsefulUtils module.

3. The Plain Old Documentation (POD) technology, included with Perl, supports the addition of documentation to a program or module. Use the "man perlpod" command to access the POD documentation to read and learn about POD. Use POD to add appropriate documentation to the UsefulUtils module.

4. Explore the CPAN repository. Find a module that interests you, then download and install it into your Perl installation. Examine the disk-files that are included with the module, paying particular attention to the CPAN author's use of POD. Be sure to test the CPAN module to ensure it works the way you expect it to.

5. Use the perldoc command to search the perlfunc document for information on a subroutine called wantarray. Experiment with wantarray, then write a small subroutine that accepts a list of words as its parameters. Have the subroutine return the list of words when invoked in list context. When called in scalar context, the subroutine returns a count of the number of words provided as parameters to it. Can you think of a good name for such a subroutine?

# 6
# About Files

*Input, output and other things.*

## 6.1 I/O: Input and Output

Data entering a program is referred to as its *input*, while data produced by a program is its *output*. Rather than refer to (and write) "input/output", most programmers simply say "IO", which is written as I/O.

The majority of the programs presented in the previous chapters have been self-contained, in that they do not rely on data from any external source in order to work. On top of this, these programs have been happy to send their results to the screen. This is fine when all that is important is the demonstration of programming concepts. However, *real* programs work on *real* data, and data is typically stored in disk-file. In this chapter, the I/O facilities provided by Perl are described.

I/O facilities are often referred to as *streams*. It is possible to have many streams associated with a program, with some of them classed as *input streams* and others classed as *output streams*. As a minimum, every Perl program has three standard streams available to it.

### 6.1.1 The standard streams: STDIN, STDOUT and STDERR

Before creating streams of our own, let's review the standard streams, which were touched on briefly in Chapter 3.

The standard input stream (STDIN) is the default place from which data enters a program. Typically, STDIN is the keyboard, but it can also be a disk-file. To read data from STDIN, use the input operator:

    my $data = <STDIN>;

As STDIN is the default input stream, the following is identical to the above:

    my $data = <>;

perl is smart enough to know that an "empty" input operator actually refers to STDIN by default.

The standard output stream (STDOUT) is the default place to which data is sent by a program. Typically, STDOUT is the screen, but it can also be a disk-file. To write data to STDOUT, use print:

    print STDOUT $data;

As STDOUT is the default output stream, the following is identical to the above:

    print $data;

Again, perl is smart enough to know that print sends data to STDOUT by default.

The standard error stream (STDERR) is the default place to send error messages to. As with STDOUT, STDERR is typically the screen, but error messages can be sent to any other output stream (with a disk-file the most common case). To write data to STDERR, use print:

    print STDERR "Something terrible has happened ... aborting.\n";

As this is such a common requirement, Perl provides a special subroutine (called warn) that makes this more convenient:

    warn "Something terrible has happened ... aborting.\n";

Why provide two output streams, namely, STDOUT and STDERR? Surely, both streams do essentially the same thing, that is, output data? Well, "essentially the same" does not mean "exactly the same". The former output stream is designed to be used for data, the latter is reserved for error messages, and data and error messages are *not the same thing*. If a program is happily outputting data, then spots an error and generates a message and then continues to happily output data again, it is reasonable to expect that the error message will not corrupt the output data, especially if the output data is being written to a disk-file. By separating the error message from standard output, the error message can display on screen while the (uncorrupted) data can safely end up in a disk-file.

## 6.2 Reading Files

The `getlines` program from *The Basics* chapter (on page 41) demonstrates a standard technique for taking data one line at a time from a disk-file and feeding it *as standard input* to a program one line at a time. By specifying one or more filenames on the command-line, `perl` is able to process disk-files sequentially (that is, one disk-file at a time), taking the data one line at a time from a disk-file whenever the input operator is used.

Now, imagine a requirement exists to *merge* two disk-files. For the purposes of this discussion, imagine that merging the two disk-files is defined as reading a line from the first disk-file and sending it to STDOUT, then reading a line from the second disk-file and sending it to STDOUT, then reading the next line from the first disk-file and sending it to STDOUT, then reading the next line from the second disk-file and sending it to STDOUT and so on, until there are no more lines to read from either of the disk-files. If the first disk-file contains these lines:

```
This is the first disk-file, line 1.
This is the first disk-file, line 2.
This is the first disk-file, line 3.
This is the first disk-file, line 4.
This is the first disk-file, line 5.
```

and the second disk-file contains these lines:

```
This is the second disk-file, line 1.
This is the second disk-file, line 2.
This is the second disk-file, line 3.
```

then the merged output should look like this:

```
This is the first disk-file, line 1.
This is the second disk-file, line 1.
This is the first disk-file, line 2.
This is the second disk-file, line 2.
This is the first disk-file, line 3.
This is the second disk-file, line 3.
This is the first disk-file, line 4.
This is the first disk-file, line 5.
```

Given this requirement, the sequential behaviour of `getlines` will not work when merging two disk-files, as `getlines` is programmed to deal with a disk-file *in its entirety* before processing another. Some other strategy is required when two disk-files are to be merged. Let's start by trying a strategy based on these steps:

- Determine the names of the two disk-files to be merged.
- Open the two disk-files to enable data to be read from them.

**106** *About Files*

- Read a line from the first disk-file, then write it to STDOUT.
- Read a line from the second disk-file, then write it to STDOUT.
- Repeat the last two steps until there are no more lines to read.

Each of these steps is discussed in the subsections that follow. The merge program is called `merge2files`.

### 6.2.1 Determining the disk-file names

One technique for determining the names of the disk-files is to avoid determining them at all, and write the program in such a way that the two names are always the same. The user of the program must make sure that the two disk-files are named in a way that the program expects. Such a practice is referred to as *hard coding*, and is best avoided as it tends to lead to inflexible solutions. The `merge2files` program needs to work with *any* two disk-files. The two disk-files to be merged are named on the command-line as parameters to the `merge2files` program.

As with subroutines, parameters passed to a program are made available in a special array. Unlike subroutines, where the `@_` default array is used, the array of command-line parameters is called `@ARGV`, and is automatically populated by `perl`. Here's a small program, called `determine_args`, that determines the values of one or two command-line parameters:

```
#! /usr/bin/perl -w

# determine_args - print out the names of the disk-files named on
#                  the command-line.

if ( $#ARGV != 1 )
{
    warn "Please supply the names of two disk-files on the command-line.\n";
    exit;
}

my ( $first_filename, $second_filename ) = @ARGV;

print "first disk-file name is:  $first_filename\n";
print "second disk-file name is: $second_filename\n";
```

As `@ARGV` is an array like any other, the `$#` prefix is used to provide the value of the last array index. Recalling that Perl starts counting from zero, a value of 1 for `$#ARGV` means that the array contains two values. A check for this is performed at the start of the `determine_args` program, and if anything other than two command-line parameters are provided, the program displays an error message (thanks to `warn`) and then terminates by invoking the `exit` subroutine.

As the `warn` and `exit` combination is so common, Perl provides the `die` subroutine, which does the same thing. The two lines from `determine_args` can be replaced with this single line:

```
die "Please supply the names of two disk-files on the command-line.\n";
```

After determining that exactly two command-line parameters are provided to the program, the elements of the @ARGV array are assigned to two lexical variables with this line:

```
my ( $first_filename, $second_filename ) = @ARGV;
```

By surrounding the two lexical variables with parentheses, a *temporary* list is created on the left-hand side of the assignment operator. As an array is already on the right-hand side of the assignment operator, `perl` is smart enough to take the individual elements of the @ARGV array and assign each of them (in turn) to the named lexicals in the list. This is a very convenient and compact shorthand technique, which is equivalent to this:

```
my $first_filename  = $ARGV[0];
my $second_filename = $ARGV[1];
```

and to this:

```
my $first_filename  = shift;
my $second_filename = shift;
```

as `shift`, when used within the main code of a program, works with @ARGV by default. The `determine_args` program concludes by displaying the two disk-file names on STDOUT. When invoked as follows:

```
perl determine_args first_file.txt second_file.txt
```

the program produces this output:

```
first disk-file name is:  first_file.txt
second disk-file name is: second_file.txt
```

When invoked with anything other than two command-line parameters, the program complains:

```
Please supply the names of two disk-files on the command-line.
```

## 6.2.2 Opening the named disk-files

The `determine_args` program, despite the messages that it produces, does not actually check that the command-line parameters supplied to it refer to existing disk-files. It is certainly the intention that they do, but "intentions" are difficult, if not impossible, to program. Ideally, the parameters need to be *checked* to ensure that they refer to a disk-file and that if they do, the disk-file can be opened and read from. It is not sensible to try to open a disk-file that does not exist, or that cannot be read from.

Perl provides a number of *file test operators* that can help here[1]. Here is an expanded version of `determine_args`, renamed `check_args`, that uses three file test operators to:

1. Check that a disk-file associated with the name *exists*, and
2. Check that the disk-file is a *plain* disk-file, and
3. Check that the disk-file can in fact be *read from*.

Here is the entire source code to the `check_args` program:

```
#! /usr/bin/perl -w

# check_args - check that the disk-files named on the command-line exist.

if ( $#ARGV != 1 )
{
    die "Please supply the names of two disk-files on the command-line.\n";
}

my ( $first_filename, $second_filename ) = @ARGV;

unless ( -e $first_filename && -f $first_filename && -r $first_filename )
{
    die "$first_filename cannot be accessed.  Does it exist?\n";
}

unless ( -e $second_filename && -f $second_filename && -r $second_filename )
{
    die "$second_filename cannot be accessed.  Does it exist?\n";
}
```

The key statement is this one:

```
unless ( -e $first_filename && -f $first_filename && -r $first_filename )
```

which checks if the disk-file named in the `$first_filename` scalar *exists* using the `-e` file operator. This statement also checks to see if the disk-file is a *plain* disk-file using the `-f` file operator in combination with the `&&` "and" operator. Finally, the statement checks to see if the disk-file can be *read from*, using the `-r` file operator together with `&&`. If all three of these conditions hold, then the disk-file can be opened. If any one of these three conditions is false, the disk-file

---

[1] These are sometimes referred to as the *dash operators*.

cannot be opened, which explains the use of `unless` instead of the more usual `if`, that is, unless all three conditions hold, the program dies.

With the status of the disk-files determined, it is now possible to open each of them and assign them to *filehandles*:

```
open FIRSTFILE, "$first_filename";
open SECONDFILE, "$second_filename";
```

The filehandles in these examples are `FIRSTFILE` and `SECONDFILE`. These are a convenient way of referring to a disk-file, and can be thought of as variable names for open disk-files. Any word can be used as a filehandle, and it is a convention to use all UPPERCASE when naming filehandles.

> **Technical Commentary:** The use of the word "handle" is sometimes confusing. Think of a handle as a simple way of referring to something that is often more complex. CB radio hams know all about handles: they think up simple names to refer to themselves, as opposed to using a more complex identification mechanism.

The `open` subroutine is part of Perl, and it expects two parameters: the name of a filehandle to be created and the name of a disk-file to be opened. The invocations of `open` can also be written like this:

```
open FIRSTFILE, "<$first_filename";
open SECONDFILE, "<$second_filename";
```

The "less-than" symbol tells `open` to open the disk-file for reading. When no symbol is provided, it is assumed to be a less-than symbol. That is, `open` defaults to reading disk-files.

Even though care was taken to ensure that the disk-file could be opened, something can still go wrong[2]. It is prudent to check that the `open` on the disk-file succeeded, but appending a check to the `open` invocation:

```
open FIRSTFILE, "$first_filename"
    or die "Could not open $first_filename.  Aborting.\n";
open SECONDFILE, "$second_filename"
    or die "Could not open $second_filename.  Aborting.\n";
```

Now, if either disk-file cannot be opened, the program aborts with an appropriate error message.

In addition to opening disk-files prior to using them, it is prudent to close them as soon as they are no longer needed. Closing disk-files is straightforward: invoke Perl's `close` subroutine and pass the filehandle as its sole parameter:

```
close FIRSTFILE;
close SECONDFILE;
```

---

[2] Perhaps the disk-file does not belong to you and, as a result, the operating system is happy to let you check that it exists, is a plain disk-file and can be read from. However, the operating system may not be happy to let you open the disk-file since it is not yours.

**110**  *About Files*

Surprisingly, it is not absolutely necessary to specifically close disk-files when finished with them, as `perl` automatically closes any open filehandles when a program ends, thereby closing the associated disk-file. Despite this convenience, many programmers advise that a disk-file should be kept open only for as long as it is needed. This is sensible advice, so here's a maxim to highlight its importance:

> **Maxim 6.1** *Open a disk-file for as long as needed, but no longer.*

And here's another, just to be sure:

> **Maxim 6.2** *If you open a disk-file, be sure to close it later.*

### 6.2.3 Reading a line from each of the disk-files

To read from STDIN, place STDIN inside the input operator. To read from any filehandle, place the filehandle name inside the input operator, as follows:

```
my ( $linefromfirst, $linefromsecond );

$linefromfirst  = <FIRSTFILE>;
$linefromsecond = <SECONDFILE>;
```

which could not be any more straightforward, could it? Two lexical variables are declared using my, then each is assigned a line from each of the filehandles.

### 6.2.4 Putting it all together

With the disk-file names determined, the filehandles opened and a mechanism in place to read from them, it's now time to attempt to bring everything together to create `merge2files`. The only missing piece is a loop to read from the filehandles until there are no more lines to read:

```
#! /usr/bin/perl -w

# merge2files - merge the two disk-files named on the command-line.

if ( $#ARGV != 1 )
{
    die "Please supply the names of two disk-files on the command-line.\n";
}

my ( $first_filename, $second_filename ) = @ARGV;

unless ( -e $first_filename && -f $first_filename && -r $first_filename )
{
    die "$first_filename cannot be accessed.  Does it exist?\n";
}

unless ( -e $second_filename && -f $second_filename && -r $second_filename )
```

```
    {
        die "$second_filename cannot be accessed.  Does it exist?\n";
    }

    open FIRSTFILE,  "$first_filename"
        or die "Could not open $first_filename.  Aborting.\n";

    open SECONDFILE, "$second_filename"
        or die "Could not open $second_filename.  Aborting.\n";

    my ( $linefromfirst, $linefromsecond );

    while ( $linefromfirst  = <FIRSTFILE> )
    {
        $linefromsecond = <SECONDFILE>;

        print $linefromfirst;
        print $linefromsecond;
    }

    close FIRSTFILE;
    close SECONDFILE;
```

The while loop keeps going while there are lines to read from the FIRSTFILE filehandle. Every time through the loop, the $linefromfirst scalar is set to a line from the first disk-file, and the $linefromsecond scalar is set to a line from the second disk-file. Let's make the merge2files program executable and run it against the two disk-files to see what happens:

```
chmod   +x   merge2files
./merge2files   first_file.txt   second_file.txt
```

The following messages appear:

```
This is the first disk-file, line 1.
This is the second disk-file, line 1.
This is the first disk-file, line 2.
This is the second disk-file, line 2.
This is the first disk-file, line 3.
This is the second disk-file, line 3.
This is the first disk-file, line 4.
Use of uninitialized value in print at merge2files line 35, <SECONDFILE> line 3.
Use of uninitialized value in print at merge2files line 35, <SECONDFILE> line 3.
This is the first disk-file, line 5.
Use of uninitialized value in print at merge2files line 35, <SECONDFILE> line 3.
Use of uninitialized value in print at merge2files line 35, <SECONDFILE> line 3.
```

Oh dear, some of these messages were not expected.

The "Use of uninitialized value in print ... " messages are generated by perl, not by merge2files. The line numbers are significant. The 35 refers to the line 35 in the merge2files program, specifically, this one:

```
print $linefromsecond;
```

and the 3 refers to the number of lines that have been read from the SECONDFILE filehandle. The problem is that the two disk-files being merged are of differing lengths: there are five lines in the first and three in the second. By continuing to loop while there are lines in the first disk-file, the program gets into difficulty when it runs out of lines in the second disk-file, which results in the messages from perl. This is bad enough, but look what happens when the order of the disk-files is reversed on the command-line:

```
./merge2files  second_file.txt  first_file.txt
```

This execution of merge2files produces the following output:

```
This is the second disk-file, line 1.
This is the first disk-file, line 1.
This is the second disk-file, line 2.
This is the first disk-file, line 2.
This is the second disk-file, line 3.
This is the first disk-file, line 3.
```

Which looks OK, as the error messages are gone. However, this is not correct either: lines four and five from the first disk-file are missing in the output! The reason for this is that the loop terminates when it runs out of lines from second_file.txt, so lines four and five from the first_file.txt are never read. This is actually worse than the previous execution of the merge2files program, as this output looks OK, when, in fact, it is not.

There are a number of strategies that can be used to solve this problem[3]. One is to stop the offending line from printing if the second file has already reached the end of the file. Perl's eof subroutine returns true if a named filehandle has run out of lines. The next version of the merge program, called merge2files_v2, replaces the offending statement (line 35) with these:

```
if ( !eof( SECONDFILE ) )
{
    $linefromsecond = <SECONDFILE>;
    print $linefromsecond;
}
```

This code checks to see that the SECONDFILE has reached the end of the file or not. If it has not, another line is read and assigned to the $linefromsecond scalar and then printed. If the filehandle has reached the end of the file, nothing happens, as there are no more lines to read. When executed with this command-line:

```
./merge2files_v2  first_file.txt  second_file.txt
```

---

[3] The implementation of one of them is included as one of this chapter's exercises.

this version of the merge program produces this output:

```
This is the first disk-file, line 1.
This is the second disk-file, line 1.
This is the first disk-file, line 2.
This is the second disk-file, line 2.
This is the first disk-file, line 3.
This is the second disk-file, line 3.
This is the first disk-file, line 4.
This is the first disk-file, line 5.
```

which is correct. When the order of the disk-files is reversed on the command-line, the following output is produced:

```
This is the second disk-file, line 1.
This is the first disk-file, line 1.
This is the second disk-file, line 2.
This is the first disk-file, line 2.
This is the second disk-file, line 3.
This is the first disk-file, line 3.
```

which is *still* incorrect. In fact, this version of the merge program produces correct output only when the first disk-file has the same or more lines than the second disk-file, which, clearly, will not do.

The problem is that when the second disk-file has more lines than the first, the merge program finishes too early. By introducing another loop, immediately after the existing loop, it is possible to ensure that any "forgotten" lines are processed. Here's the second loop, which continues to process the second disk-file after the lines from the first are exhausted. The loop ends when the second disk-file reaches the end of the file:

```perl
while ( !eof( SECONDFILE ) )
{
    $linefromsecond = <SECONDFILE>;
    print $linefromsecond;
}
```

This code is added to merge2files_v2, creating merge2files_v3, as follows:

```perl
#! /usr/bin/perl -w

# merge2files_v3 - third version of merge2files: merge the disk-files named
#                  on the command-line (with some help from eof).
#                  Make sure all lines are read from both disk-files.

if ( $#ARGV != 1 )
{
    die "Please supply the names of two disk-files on the command-line.\n";
}

my ( $first_filename, $second_filename ) = @ARGV;
```

**114**   *About Files*

```perl
unless ( -e $first_filename && -f $first_filename && -r $first_filename )
{
    die "$first_filename cannot be accessed.  Does it exist?\n";
}

unless ( -e $second_filename && -f $second_filename && -r $second_filename )
{
    die "$second_filename cannot be accessed.  Does it exist?\n";
}

open FIRSTFILE,  "$first_filename"
    or die "Could not open $first_filename.  Aborting.\n";

open SECONDFILE, "$second_filename"
    or die "Could not open $second_filename.  Aborting.\n";

my ( $linefromfirst, $linefromsecond );

while ( $linefromfirst = <FIRSTFILE> )
{
    print $linefromfirst;
    if ( !eof( SECONDFILE ) )
    {
        $linefromsecond = <SECONDFILE>;
        print $linefromsecond;
    }
}

while ( !eof( SECONDFILE ) )
{
    $linefromsecond = <SECONDFILE>;
    print $linefromsecond;
}

close FIRSTFILE;
close SECONDFILE;
```

The merge2files_v3 program now satisfies the merging requirement as specified at the start of this chapter, and the ordering of disk-file names on the command-line no longer matters. Any two disk-files (of varying lengths) can be merged with this program. Take a few moments to run this program against a number of different disk-files to check this claim.

### 6.2.5 Slurping

The ability to read a line of data from a disk-file is very useful. However, there are occasions when it is convenient to read an entire disk-file in one go. This is referred to as "slurping".

To slurp a disk-file, use the input operator and, instead of assigning what's read to a scalar, assign what's read to an array:

```perl
@entire_file = <>;
```

As the input operator is invoked in list context, perl reads the entire disk-file and assigns it, one line at a time, to an array called entire_file. Here is a program, called slurper, which uses this technique. Note the inclusion of drawline from the UsefulUtils module:

```
#! /usr/bin/perl -w

# slurper - a program which demonstrates disk-file "slurping".

use lib "$ENV{'HOME'}/bbp/";
use UsefulUtils qw( drawline );

open FIRSTSLURPFILE, "first_file.txt"
    or die "Could not open first slurp disk-file.  Aborting.\n";

open SECONDSLURPFILE, "second_file.txt"
    or die "Could not open second slurp disk-file.  Aborting.\n";

my @linesfromfirst  = <FIRSTSLURPFILE>;
my @linesfromsecond = <SECONDSLURPFILE>;

print drawline( Count => 40 ), "\n";
print @linesfromfirst;
print drawline( Count => 40 ), "\n";
print @linesfromsecond;
print drawline( Count => 40 ), "\n";

close FIRSTSLURPFILE;
close SECONDSLURPFILE;
```

Unlike the merge2files programs, this program ignores the maxims from earlier and *hard codes* the names of the disk-files as part of the open statements, a practice that is best avoided[4]. When executed, that slurper program produces the following:

```
----------------------------------------
This is the first disk-file, line 1.
This is the first disk-file, line 2.
This is the first disk-file, line 3.
This is the first disk-file, line 4.
This is the first disk-file, line 5.
----------------------------------------
This is the second disk-file, line 1.
This is the second disk-file, line 2.
This is the second disk-file, line 3.
----------------------------------------
```

---

[4] Rest assured, your authors have slapped themselves on the wrist for this.

In addition to the single statement that reads the entire disk-file into its associated array, look at how the single `print` statement is used to display the entire array on screen (STDOUT). Recall that `print` takes a "list of things to display" as its parameters. The *single* array is a *list of lines*.

Be careful when slurping disk-files, especially if the disk-file contains a lot of material. When data is read into a variable, it is allocated a small piece of the computer's memory. This memory is a *finite space* and, while a program is running, will start to fill-up with the data that the program is using. A small slurped disk-file is (usually) easily accommodated. A large slurped disk-file may cause a "memory overflow", resulting in a program terminating because of this. Consequently, it is often better to read one line at a time from a large disk-file. You have been warned, so be careful.

## 6.3 Writing Files

To write to a disk-file, use the "greater-than" symbol when invoking `open`, as follows:

```
my $file_to_open = "errors.log";

   ...

open( LOGFILE, ">$file_to_open" )
    or die "Could not write to/create errors log disk-file.\n";
```

This `open` statement creates the `errors.log` disk-file and opens it for writing. If the disk-file already exists, it is discarded. This is affectionately known as *clobbering*[5]. If the disk-file does not already exist, the call to `open` creates it.

It is often the case that a disk-file needs to retain the content that it already has, something that is particularly true of logs. When `open` is called like this:

```
open( LOGFILE, ">>$file_to_open" )
    or die "Could not append to/create errors log disk-file.\n";
```

the disk-file is opened in *append* mode (note the double greater-than symbols, known as *chevrons*). Anything written to the disk-file is appended to the content that already exists within it. In this way, the disk-file grows (and is not clobbered). To write to a disk-file, use `print` to send data to it:

```
print LOGFILE "Error: something terrible has happened.\n";
```

The filehandle to write to is the first parameter to `print`. By including the filehandle, `perl` sends data to a disk-file as opposed to STDOUT.

---

[5] "Slurping" ... "clobbering" ... who says Perl's not *fun*?

## 6.3.1 Redirecting output

It is also possible to create a disk-file as a result of executing any program. For example, by executing the `merge2files` program with the following command-line, anything written to STDOUT is sent to a disk-file called `merge.out`, clobbering it if it already exists. Unlike the example code just described, be aware that the operating system, not `perl`, is creating, and then writing to, the disk-file:

```
./merge2files  first_file.txt  second_file.txt  >  merge.out
```

To append to the disk-file, use chevrons on the command-line, as opposed to the greater-than symbol:

```
./merge2files  first_file.txt  second_file.txt  >>  merge.out
```

The above commands take anything written to STDOUT and *redirect* it to the named disk-file. Anything written to STDERR still appears on screen. To redirects anything written to STDERR, use a command-line like this:

```
./merge2files  first_file.txt  second_file.txt  >  merge.out  2>  merge.err
```

which redirects standard output to `merge.out` and any error messages (standard error) to `merge.err`. Chevrons can be used to append to the disk-files as opposed to clobbering them:

```
./merge2files  first_file.txt  second_file.txt  >>  merge.out  2>>  merge.err
```

## 6.3.2 Variable interpolation

This is an appropriate place to discuss the process of *variable interpolation*. When a variable is written to a filehandle, what prints depends on whether or not the variable is enclosed within single or double quotes. Consider these program statements:

```
my $sequence = "TTATTATGTT GCTCAGTTCT GACCTCTTAA CTATGCGGTA";

print "The sequence is: $sequence\n";
print 'The sequence is: $sequence\n';
```

When executed by `perl`, the following output displays:

```
The sequence is: TTATTATGTT GCTCAGTTCT GACCTCTTAA CTATGCGGTA
The sequence is: $sequence\n
```

By using double quotes to surround that which is to be printed, `perl` knows to replace the `$sequence` scalar with its contents, as well as turn the "\n" into a newline. This process is known as *interpolation*. When single quotes surround that which is to be printed, no interpolation occurs, and the string prints as-is (that is, it prints literally).

## 6.4 Chopping and Chomping

When working with input data, two useful in-built subroutines are chop and chomp[6].

The chop subroutine, when provided with a scalar variable, removes the last character from the scalar and returns it. Consider this code:

```
my $dna = "ATGTGCGGTATTGCTGACCTCTTA\n";

my $last = chop $dna;

# $dna is now "ATGTGCGGTATTGCTGACCTCTTA";

my $next = chop $dna;

# $dna is now "ATGTGCGGTATTGCTGACCTCTT";
```

With each invocation of chop, the $dna scalar is shortened by one character, and the character is assigned to the named scalar. The $last scalar is assigned "\n" and $next is assigned "A".

The chomp subroutine, when provided with a scalar variable, removes the last character from the scalar if, and only if, that character is the newline, "\n". Consider this code:

```
my $dna = "ATGTGCGGTATTGCTGACCTCTTA\n";

my $last = chomp $dna;

# $dna is now "ATGTGCGGTATTGCTGACCTCTTA";

my $next = chomp $dna;

# $dna is still "ATGTGCGGTATTGCTGACCTCTTA";
```

There is no point in the chomp subroutine returning the character removed, as it can only ever remove the newline. Instead, chomp returns the number of characters removed as a result of its invocation. After executing the above statements, the $last scalar has a value of 1, whereas the $next scalar has the value 0. The $dna scalar has been *chomped*.

---

[6] As if the ability to *slurp* and *clobber* were not enough, with Perl we can *chop* and *chomp*, too. Who do we thank for such linguistic frivolity? Larry Wall, the creator of Perl. Larry's use of such terminology is enough to make him an honorary Yorkshireman, as such words were in common use where Michael grew up.

## Where to from Here

Perl's I/O mechanisms are very convenient and, once mastered, not difficult to use. Many programmers have extended the in-built mechanisms with custom modules, which are available on CPAN. A number of standard I/O modules are included with the Perl environment, and these include `IO::File` and `IO::Handle`. Take some time to read the documentation describing these modules.

In addition to the wonder that is CPAN and convenient, easy to use I/O, there is one other Perl feature that endears the language to more programmers than any other language: Perl's support for *regular expressions*. Perl's regular expression, pattern-matching technology forms the basis of the next chapter.

## The Maxims Repeated

Here's a list of the maxims introduced in this chapter.

- *Open a disk-file for as long as needed, but no longer.*
- *If you open a disk-file, be sure to close it later.*

## Exercises

1. Amend the `merge2files` program to slurp the two disk-files prior to performing the merge. Does this make the merge any easier?

2. Amend the `merge2files` program to enable it to merge three disk-files. Call your new program `merge3files`.

3. Create a new program, called `merge2alpha`, that can merge two disk-files alphabetically. Given a disk-file like this:

   ```
   a start
   is generally not the end
   of a disk-file
   it's the start
   ```

   and another one like this:

   ```
   the end of a disk-file
   usually comes after
   the start of a disk-file
   but, not always
   ```

your program should produce the following output:

```
a start
but, not always
is generally not the end
it's the start
of a disk-file
the end of a disk-file
the start of a disk-file
usually comes after
```

4. Create a program called `reverse_it` that takes a named disk-file (the input) and reverses the order of the lines contained therein. The output from this program is written to a new disk-file (the output), which takes its name from the inverse of the name of the input disk-file. That is, if the input disk-file is called "`input.data`", the output disk-file should be called "`atad.tupni`".

# 7

# Patterns, Patterns and More Patterns

*Exploiting Perl's built-in regular expression technology.*

## 7.1 Pattern Basics

Earlier in this book, a simple regular expression searched for and found a collection of characters within an input stream (recall the `patterns` program on page 44). At the time, the regular expression details were glossed over, and a discussion of them was deferred to this chapter.

Before exploring regular expressions, be advised that many people find the technology strange at first. Work slowly through this chapter, taking time to understand and learn the technology as it is presented. Regular expressions are very powerful, in that an awful lot of work results from very little effort on the part of the programmer. Regular expressions often seem quite cryptic at first, but persevere: the reward is worth the extra effort required to understand them. Of course, once understood, you'll wonder how you ever managed without them.

> **Technical Commentary:** Perl is one of a small, select group of programming languages that directly embed pattern-matching, regular expression technology into their core. Some programming languages provide similar technologies as bolted-on, optional add-ons. However, with Perl, regular expressions are an integral component. Programmers often refer to Perl's regular expression *engine*, which is the piece of technology that provides all this, but the engine is not really a separate component – it's built-in. It is this feature that *defines* Perl as much as any other.

What are regular expressions and what's so special about them?

### 7.1.1 What is a regular expression?

From the perspective of a Perl programmer, a regular expression is, first and foremost, a *pattern*. The pattern tells `perl` to look for something, and this "something" can be any sequence of characters. The `patterns` program looked for the word "even" within an input stream using this regular expression:

```
/even/
```

Typically, the pattern that makes up a regular expression is enclosed within two forward-leaning slash characters, as is the case above. It is important to realise that the pattern is just a sequence of characters to `perl`. Even though "even" is a word (for us), it is four individual characters (for `perl`). Specifically, the pattern `/even/` looks for the character "e", followed by the character "v", followed by the character "e", followed by the character "n". When the pattern is compared against something (such as an input stream or a string), it is said to *match* if this sequence of four characters appears. Here are some successful matches to the `/even/` pattern:

```
eleven              # matches at end of word
eventually          # matches at start of word
even Stevens        # matches twice: an entire word and within a word
```

And here are some unsuccessful matches (or non-matches):

```
heaven              # 'a' breaks the pattern
Even                # uppercase 'E' breaks the pattern
EVEN                # all uppercase breaks the pattern
eveN                # uppercase 'N' breaks the pattern
leave               # not even close!
Steve not here      # space between 'Steve' and 'not' breaks the pattern
```

Most regular expression technologies (and Perl's is no exception) are extended by a collection of *special* characters, referred to as *metacharacters*. Metacharacters influence how the pattern is matched, and are described later.

> **Technical Commentary:** The term "regular expression" is often shortened to "regex", and is pronounced "reg", as in "beg", and "ex" as in ... well, "x".

### 7.1.2 What makes regular expressions so special?

Let's answer this question with a demonstration. Imagine the requirement to write a subroutine to find the first occurrence of a pattern, such as "even", within a given string. Given the Perl that has been covered thus far, a reasonable strategy is to approach the problem in the following way (note that *before* any processing occurs, `perl` starts from the beginning of the string to search and has yet to read any characters from it):

1. Examine the next character of the string.
2. If the character under consideration is *not* "e", return to step 1.
3. If the character under consideration is "e", consider the next character of the string.
4. If the character under consideration is *not* "v", go back one character (that is, back to the found "e"), and return to step 1.
5. If the character under consideration is "v", consider the next character of the string.
6. If the character under consideration is *not* "e", go back two characters (that is, back to the first found "e"), and return to step 1.
7. If the character under consideration is "e", consider the next character of the string.
8. If the character under consideration is *not* "n", go back three characters (that is, back to the first found "e"), and return to step 1.
9. If the character under consideration is "n" - rejoice! - a match has been found.

Using a pencil and some paper, use this strategy to search for the pattern "even" in the strings "Steven", "heaven" and "eleven", convincing yourself that it does indeed work[1].

Now, imagine further that a subroutine called find_it is based on the above strategy and searches a given string for a given pattern, returning "true" upon success. The subroutine could be invoked like this:

```
my $pattern = "even";
my $string = "do the words heaven and eleven match?";

if ( find_it( $pattern, $string ) )
{
    print "A match was found.\n";
}
else
{
    print "No match was found.\n";
}
```

Assuming, of course that the subroutine did indeed exist, which it does not. The reason it does not exist is that no Perl programmer, even the most masochistic, would ever dream of creating a subroutine such as find_it. Writing such a subroutine is tedious, tricky and totally unnecessary. The Perl programmer uses a regular expression, and writes the above code like this:

---

[1] Even though it is not the most efficient strategy. Can you think of an improvement?

```perl
my $string = "do the words heaven and eleven match?";

if ( $string =~ /even/ )
{
    print "A match was found.\n";
}
else
{
    print "No match was found.\n";
}
```

And then the Perl programmer promptly gets on with whatever else needs doing. The requirement to write a subroutine to perform the searching is nullified.

The key point is that by using a regular expression, Perl programmers are able to specify what it is they are interested in finding, *without* having to spell out how it should be found. The "how" is left to `perl`, which performs the search on the basis of the specified regular expression.

> **Maxim 7.1** *Use a regular expression to specify what you want to find, not how to find it.*

At first glance, many think that this is not such an important thing. However, finding things in other things is such a common occurrence that any programming technology that makes it quick and easy is to be welcomed. And the significance of this for Bioinformaticians should be clear: *finding patterns in sequences is a very big deal indeed!*

Simple patterns, such as "even", are known as *concatenations*. To concatenate is to *link together* or *form a sequence of*. So, any sequence of characters is a pattern, specifically a concatenation pattern. There are other types of patterns. Unlike concatenations, the other types of patterns are associated with a particular pattern metacharacter.

## 7.2 Introducing the Pattern Metacharacters

In addition to concatenations, patterns can represent *repetitions* and *alternations*. It is also possible to state that a pattern may or may not be there, in that it is *optional*.

### 7.2.1 The + repetition metacharacter

The + metacharacter is read as *one or more of*. The following regular expression matches one or more occurrence of the letter "T":

/T+/

Which matches any of the following:

    T
    TTTTTT
    TT

But does not match any of these:

    t
    this and that
    hello
    tttttttttt

Repetitions can be combined with concatenations. This next pattern matches the letter "e", followed by the letter "l", followed by *one or more* occurrences of the letter "a":

    /ela+/

In the above example, the repetition is said to *bind more closely* than the concatenation, in that only the letter immediately preceding the + symbol is repeated. So, these strings successfully match the pattern:

    elation
    elaaaaaaaa

If a requirement exists to bind the repetition to more than one character (i.e., to a concatenation), use parentheses to indicate how many characters to repeat. Consider this regular expression:

    /(ela)+/

Now, if the combination of the letter "e", followed by the letter "l", followed by the letter "a" occurs one or more times, there's a match, as with these strings:

    elaelaelaela
    ela

This means that the "(" and ")" characters are also metacharacters, which is fine until a requirement exists to match either of these characters (or any other metacharacter, for that matter). When such a requirement exists, a metacharacter can have its special meaning *switched off* by the use of the \ character (which is known as *escaping*). Consider this regular expression:

    /\(ela\)+/

which now matches an opening parenthesis, "(", followed by the letter "e", followed by the letter "l", followed by the letter "a", followed by one or more occurrences of the closing parenthesis, ")". So, this string matches:

    (ela))))))

and this does not:

    (ela(ela(ela

## 7.2.2 The | alternation metacharacter

Another important metacharacter is the vertical bar, |, which indicates *alternation*. Alternation offers choice. Here's an example that matches any one of the digit characters:

    /0|1|2|3|4|5|6|7|8|9/

That is, the digit 0 or, alternatively, the digit 1 or, alternatively, the digit 2 or, alternatively, the digit 3, and so on, up to and including the digit 9 can match. So, if a single digit occurs *anywhere in the string*, there's a match. All of these strings match (as they all contain at least one digit):

    0123456789
    there's a 0 in here somewhere
    My telephone number is: 212-555-1029

As can be imagined, looking for any digit is a common requirement, as is trying to match any single lowercase or uppercase letter. It is possible to match any lowercase letter with this regular expression:

    /a|b|c|d|e|f|g|h|i|j|k|l|m|n|o|p|q|r|s|t|u|v|w|x|y|z/

Just as it is possible to use this regular expression to match any single uppercase letter:

    /A|B|C|D|E|F|G|H|I|J|K|L|M|N|O|P|Q|R|S|T|U|V|W|X|Y|Z/

Both seem like an awful lot of work just to match a single character. Perl's regular expression shorthand to the rescue!

### 7.2.3 Metacharacter shorthand and character classes

In order to reduce the amount of work required, Perl provides the *character class*, which is a shorthand notation for a long list of alternatives. Rather than using this regular expression to match any digit:

/0|1|2|3|4|5|6|7|8|9/

it is possible to define a character class, which means the same thing, as follows:

/[0123456789]/

That is, place the digits (or letters or whatever) between the "[" and "]" characters, to indicate a series of alternations. This regular expression:

/[aeiou]/

is exactly the same as this one:

/a|e|i|o|u/

Most Perl programmers prefer the character class version of the regular expression. When the first character of a character class is the ^ symbol (known as *hat*), the character class is *inverted*. This regular expression:

/[^aeiou]/

matches any single character that is *not* one of the five vowels. The ^ character can be included within a character class as a *literal character* by positioning it anywhere but the first position. Ranges can also be specified within character classes using the - symbol. This character class:

/[0123456789]/

can also be written as:

/[0-9]/

which is shorter, more convenient and less prone to a typing error[2]. As the letters are also ranges, the long "any letter" regular expressions from earlier in this section can be rewritten as

/[a-z]/

---

[2] Although we never make any of thsoe ... em, eh, sorry ... *those*.

which matches any single lowercase letter, and like this:

    /[A-Z]/

to match any single uppercase letter. If a requirement exists to match a literal "-" character, position the dash at the start of the character class:

    /[-A-Z]/

The above regular expression now matches for any single uppercase letter or the dash. Combining character classes defines very specific concatenations. Consider this regular expression:

    /[BCFHST][aeiou][mty]/

which matches any three-letter word that starts with an uppercase letter from the first character class, has a vowel in the middle (the second character class) and ends in either the letter "m", "t" or "y" (the third character class). Each of the following words matches this regular expression:

    Bat
    Hit
    Tot
    Cut
    Say

while these words do not:

    Hog
    Can
    May
    bat

Note the last word, "bat", which *almost* matches but does not as regular expressions are *case-sensitive*. To match words that start with either an uppercase or a lowercase word, rewrite the regular expression like this:

    /[BbCcFfHhSsTt][aeiou][mty]/

which now allows for both "bat" and "Bat" to match.

## 7.2.4 More metacharacter shorthand

The character classes that match any single digit and any single letter (either lowercase or uppercase) are so common that Perl provides further convenient shorthand related to them. Rather than using this character class to match any single digit:

    /[0-9]/

Perl provides the *slash-d* shorthand:

/\d/

So, \d means the same as [0-9], and it is easy to remember, as "d" is short for *digit*.

When it comes to lowercase and uppercase letters, Perl groups these together with the digits and the underscore character to form the word character class. Instead of having to specify this character class:

/[a-zA-Z0-9_]/

all that's required is Perl's *slash-w* shorthand:

/\w/

as both regular expressions mean the same thing. Again, this is easy to remember, as "w" is short for *word*.

Another special character class is the *slash-s* shorthand, where "s" is short for *space*. This regular expression:

/\s/

is short for this regular expression:

/[^ \t\n\r\f]/

These characters are generally referred to as the space (or whitespace) characters.

Each of these special character classes[3] has an inverted form. To match any single character that is *not* a digit, use this regular expression:

/\D/

That is, the "\D" regular expression is "\d" inverted. Likewise, "\W" is "\w" inverted and "\S" is "\s" inverted.

The beauty of these special shorthands becomes clear when they are seen in action. Consider a regular expression that must match a digit, followed by any whitespace character, followed by two word characters and then any other character that is not a digit. Without the *specials*, the following regular expression does the trick:

/[0-9][^ \t\n\r\f][a-zA-Z0-9_][a-zA-Z0-9_][^0-9]/

---

[3] Sometimes referred to as the *classic* character classes.

**130**  *Patterns, Patterns and More Patterns*

Here's the above regular expression rewritten to use the specials:

/\d\s\w\w\D/

Note: less typing, less chance of error and more convenience. As this is such an important point, here's a new maxim to drive the message home.

> **Maxim 7.2** *Use regular expression shorthand to reduce the risk of error.*

### 7.2.5 More repetition

Character classes can be combined with the repetition metacharacter to great effect. This regular expression matches a word of any length:

/\w+/

and is read as *one or more word characters*. Knowing this, the regular expression from the last section *could* be rewritten as:

/\d\s\w+\D/

However, this matches any number of word characters, not *exactly* two as was the requirement. Perl provides a facility to match a specific number of occurrences of something. The { and } metacharacters are used to specify the number of occurrences to match. Here's the above regular expression rewritten to match exactly two word characters, as required:

/\d\s\w{2}\D/

If a requirement exists to match two but not more than four word characters, use this regular expression:

/\d\s\w{2,4}\D/

And finally, if the requirement is to match at least two metacharacters (or characters) with no upper limit on the number of word characters to match, use this:

/\d\s\w{2,}\D/

### 7.2.6 The ? and * optional metacharacters

The *optional metacharacters* are used to specify that some part of a regular expression may or may not be there. Consider this example:

/[Bb]art?/

which matches any of the following words:

    bar
    Bar
    bart
    Bart

That is, the letter "t" is optional. More correctly, Perl programmers read the ? metacharacter as: *match zero or one time.* In other words, it is either there or it is not; it's *optional.*

The * metacharacter matches *zero or more times.* Rewriting the above regular expression as follows has the effect of matching any number of occurrences of the letter "t", including *not matching it at all*:

    /[Bb]art*/

Any of the following now match this regular expression:

    bar
    Bart
    barttt
    Bartttttttttttttttttttttt!!!

Note that even though the last example appends three exclamation marks, there's still a match, as regular expressions match *anywhere in a string.* More on this behaviour later.

Care is needed when using the * metacharacter. Consider this regular expression, which *always* matches successfully:

    /p*/

When applied against any string, the p* regular expression always matches, as the pattern is looking for zero or more occurrences of the letter "p". If the string matched against contains a "p", there's a match. Equally, if the string does not contain a "p", there is also a match! Remember: the * matches zero or more times, and something – whether it is the letter "p" or anything else, for that matter – is always *not there.*

## 7.2.7 The any character metacharacter

There is often a requirement to match any character, regardless of whether it is a word, digit or whitespace character. The . metacharacter does just that:

    /[Bb]ar./

The use of the *any character metacharacter* allows the above pattern to successfully match any of these strings[4]:

```
barb
bark
barking
embarking
barn
Bart
Barry
```

Appending the ? optional metacharacter to the pattern, thus:

```
/[Bb]ar.?/
```

allows words such as "bar" and "Bar" also to match.

## 7.3 Anchors

The last example from the last section highlights, once again, the fact that the match is successful if the pattern is found *anywhere* in the string under consideration. Note that "bark", "barking" and "embarking" are all successful matches. This can often result in patterns matching when they were not expected to, which can sometimes be a surprise. But what if a requirement exists to match an entire word, such as "bark", but not match "barking" and "embarking" (as the word "bark" is embedded in them)?

The *word boundary* metacharacters allow a regular expression to be *anchored* at a word boundary – that is the space between a word and something else, which is defined in Perl as the position between "\w" and "\W".

### 7.3.1 The \b word boundary metacharacter

To match an entire word, surround the word to be matched with the \b word boundary metacharacter, as follows:

```
/\bbark\b/
```

This string now successfully matches:

```
That dog sure has a loud bark, doesn't it?
```

as the word "bark" is surrounded by word boundaries, whereas it does not match:

```
That dog's barking is driving me crazy!
```

---

[4] The temptation to use the letter "f" in this example was great, but you'll be glad to know we resisted.

The \b metacharacter has an inverse in \B, which matches at any position that is *not* a word boundary. Note that this regular expression:

    /\Bbark\B/

matches "embarking" but not "bark" or "barking".

## 7.3.2 The ˆ start-of-line metacharacter

To anchor the regular expression to the start of a string (or line), use the ˆ metacharacter:

    /^Bioinformatics/

which states that a successful match to a string must begin with the word "Bioinformatics", as follows:

    Bioinformatics, Biocomputing and Perl is a great book.

The next string does not match, as the match cannot be made at the start of the string:

    For a great introduction to Bioinformatics, see Moorhouse, Barry (2004).

## 7.3.3 The $ end-of-line metacharacter

To anchor the regular expression to the end of a string (or line), use the $ metacharacter:

    /Perl$/

which matches successfully with this string:

    My favourite programming language is Perl

but not this one:

    Is Perl your favourite programming language?

A common regular expression to match against a blank line is:

    /^$/

That is, the line has a start, an end and nothing between the two: it's blank.

## 7.4 The Binding Operators

Consider this simple program, called `simplepat`:

```
#! /usr/bin/perl -w

# The 'simplepat' program - simple regular expression example.

while ( <> )
{
    print "Got a blank line.\n"          if /^$/;
    print "Line has a curly brace.\n"    if /[}{]/;
    print "Line contains 'program'.\n"   if /\bprogram\b/;
}
```

The `simplepat` program keeps reading lines of input from STDIN until there are no more lines to read. As perl has not been told otherwise, the line is assigned to the default scalar, $_. Three print statements form the body of the loop, with each statement qualified with an if conditional statement. Each of the if statements tries to match to a specific regular expression[5].

Let's execute the `simplepat` program, specifying the program's disk-file as the input to the program, with this command-line:

```
perl simplepat simplepat
```

Here's the output produced by the above command-line:

```
Got a blank line.
Line contains 'program'.
Got a blank line.
Line has a curly brace.
Line has a curly brace.
Line contains 'program'.
Line has a curly brace.
```

This program demonstrates that in the absence of any named scalar, perl uses $_ as the *thing* to match against. If a match is successful, perl returns *true* to the program. The `simplepat` program exploits this behaviour in each of its print statements.

It is often the case that the *thing* to match against is the value of some scalar variable, not $_. The *binding operator*, written as =~, is used to tell perl that a regular expression is to be applied (or *bound*) to a named scalar. For example, this statement:

```
if ( $line =~ /^$/ )
```

---

[5] The meaning of each should be clear. If they are not, you are advised to go back to the beginning of this chapter and start again.

checks to see if the $line scalar contains a blank line. In addition to =~, there's also a *not binding operator*, !~, which is the logical negation of =~. This statement:

    if ( $line !~ /^$/ )

checks to see if the $line scalar contains anything other than a blank line. The binding operators are very useful, but really come into their own when combined with *grouping parentheses*.

## 7.5 Remembering What Was Matched

The grouping parentheses were introduced earlier, when they were used to group a number of letters together so that they could be repeated:

    /(ela)+/

It wasn't mentioned then, but when the parentheses are used to group in this way, perl remembers the value that matched that part of the regular expression, often referred to as a subpattern. For each set of parentheses, perl creates a special scalar variable to hold what matched. These special variables, often referred to as the *after-match variables*, are numbered upward from 1.

Here's a small program, called grouping that demonstrates how the after-match variables are used:

    #! /usr/bin/perl -w

    # The 'grouping' program - demonstrates the effect of parentheses.

    while ( my $line = <> )
    {
        $line =~ /\w+ (\w+) \w+ (\w+)/;

        print "Second word: '$1' on line $..\n" if defined $1;
        print "Fourth word: '$2' on line $..\n" if defined $2;
    }

Each line read into this program is assigned to the $line scalar, which is then bound against a regular expression. The pattern looks for a word, \w+, a space, another word that is to be remembered (note the use of parentheses), another space, another word, another space and another remembered word[6].

After a successful pattern match, the two remembered values are automatically assigned by perl to the special scalars $1 and $2. The print statement displays

---

[6] As you can see, it is often easier to write a regular expression using shorthand than it is to actually describe it *in words*.

what was found. Note the use of `if defined`, which ensures output is generated only as a result of a successful match. Note, too, the use of the "$." scalar, another internal Perl scalar, which contains the current line number of the input file being processed. Given the following input data (contained in the `test.group.txt` data-file):

```
This is a sample file for use with
the grouping program that is included
with the Patterns
Patterns and More Patterns chapter
from Bioinformatics, Biocomputing and Perl.
```

the following command-line:

```
perl grouping test.group.data
```

produces the following results:

```
Second word: 'is' on line 1.
Fourth word: 'sample' on line 1.
Second word: 'grouping' on line 2.
Fourth word: 'that' on line 2.
Second word: 'and' on line 4.
Fourth word: 'Patterns' on line 4.
```

There is no match on line 3 as there are only three words on that line, and the regular expression is trying to match four words (two of which are *remembered*). Line 5 does not match either, as the regular expression does not take into consideration the comma. Note the program is able to use the values that were remembered.

It is possible to *nest* parentheses. Consider this version of `grouping`, which has the rather imaginative name `grouping2`:

```
#! /usr/bin/perl -w

# The 'grouping2' program - demonstrates the effect of more parentheses.

while ( my $line = <> )
{
    $line =~ /\w+ ((\w+) \w+ (\w+))/;

    print "Three words: '$1' on line $..\n" if defined $1;
    print "Second word: '$2' on line $..\n" if defined $2;
    print "Fourth word: '$3' on line $..\n" if defined $3;
}
```

which when executed against the `test.group.txt` data-file produces the following output:

```
Three words: 'is a sample' on line 1.
Second word: 'is' on line 1.
Fourth word: 'sample' on line 1.
Three words: 'grouping program that' on line 2.
Second word: 'grouping' on line 2.
Fourth word: 'that' on line 2.
Three words: 'and More Patterns' on line 4.
Second word: 'and' on line 4.
Fourth word: 'Patterns' on line 4.
```

When working with nested parentheses, count the opening parentheses, starting with the leftmost, to determine which parts of the pattern are assigned to which after-match variables. That last sentence is worth another maxim:

> **Maxim 7.3** *When working with nested parentheses, count the opening parentheses, starting with the leftmost, to determine which parts of the pattern are assigned to which after-match variables.*

## 7.6 Greedy by Default

Consider this regular expression:

```
/(.+), Bart/
```

matched against this string:

```
Get over here, now, Bart!  Do you hear me, Bart?
```

The pattern matches one or more of any character, .+, a literal comma, a space character, then the word "Bart". The parentheses ensure that anything matched by .+ is remembered in the $1 after-match variable. After performing the match, $1 contains this string:

```
Get over here, now, Bart!  Do you hear me
```

This may come as a bit of a surprise, as it would be reasonable to think that the match succeeds when the first "Bart" is encountered, not the second. A reasonable assumption indeed, but incorrect. By default, `perl` performs *greedy matching*, in that an attempt is always made to match *as much of the string as possible*, that is, the longest possible match. To specify that non-greedy (or *lazy*) matching should be applied to part of the regular expression (or *subpattern*), qualify it with the ? character:

```
/(.+?), Bart/
```

Note that the ? character when used in this way does not mean *optional*. It means *non-greedy*. Rather than match as much as possible, this part of the regular expression now matches *as little as possible*. When matched against the string from earlier, this non-greedy regular expression remembers the following value in the $1 after-match variable:

    Get over here, now

In addition to the use of the ? non-greedy qualifier with the + metacharacter, it can also be used with the * and ? metacharacters. It can also be used with the {x}, {x,y} and {x,} repetition specifiers (where "x" and "y" specify the minimum and maximum number of matches, respectively). Being able to control when `perl` is and is not greedy is important.

## 7.7 Alternative Pattern Delimiters

The use of the / character as a regular expression delimiter suffices for most needs. However, consider writing a regular expression to match against a string like this:

    /usr/bin/perl

It is *not* possible to write the regular expression as follows:

    //\w+/\w+/\w+/

as `perl` will treat the second / character as the end of the pattern and ignore the \w+/\w+/\w+/ bit. Whoops! It is possible to *escape* the / characters that are part of the pattern:

    /\/\w+\/\w+\/\w+/

to ensure that the leftmost and rightmost / characters are treated as pattern delimiters. Unfortunately, the pattern is now harder to read and understand, and it gets worse when each of the matched words is remembered:

    /\/(\w+)\/(\w+)\/(\w+)/

In situations such as this, Perl allows alternative delimiters to be specified, where the delimiter character is drawn from the set that includes any non-alphabetic, non-whitespace character. To use an alternative delimiter, prefix the regular expression with the letter "m" to signify the start of the pattern. The above *escaped* example regular expression can be rewritten as:

    m#/\w+/\w+/\w+#

or if the matched words are to be remembered:

    m#/(\w+)/(\w+)/(\w+)#

There is now no confusion as to the inclusion of the / characters within the regular expression: they are to be treated literally, *not* as delimiters.

Other common delimiter characters include !, |, , and :. It is also possible to use any of the following bracket-pairings as delimiters:

```
m{ }
m< >
m[ ]
m( )
```

As the use of the "m" prefix signifies the start of a pattern, it is possible to use it with the standard delimiter characters:

```
/even/
```

is the same (and can be written) as:

```
m/even/
```

However, as the use of the "m" prefix is implied when used with /, the majority of Perl programmers omit it.

## 7.8 Another Useful Utility

Let's extend the UsefulUtils module from the *Getting Organised* chapter to include a subroutine that relies on a regular expression to get its work done.

Later in this book, a subroutine is required to convert from one date format to another. A date in DD-MMM-YYYY format needs to be converted to YYYY-MM-DD format. Here's a subroutine, called biodb2mysql, that performs the conversion:

```
sub biodb2mysql {
    #
    # Given:  a date in DD-MMM-YYYY format.
    # Return: a date in YYYY-MM-DD format.
    #
    my $original = shift;

    $original =~ /(\d\d)-(\w\w\w)-(\d\d\d\d)/;

    my ( $day, $month, $year ) = ( $1, $2, $3 );

    $month = '01' if $month eq 'JAN';
    $month = '02' if $month eq 'FEB';
    $month = '03' if $month eq 'MAR';
    $month = '04' if $month eq 'APR';
    $month = '05' if $month eq 'MAY';
```

```
        $month = '06' if $month eq 'JUN';
        $month = '07' if $month eq 'JUL';
        $month = '08' if $month eq 'AUG';
        $month = '09' if $month eq 'SEP';
        $month = '10' if $month eq 'OCT';
        $month = '11' if $month eq 'NOV';
        $month = '12' if $month eq 'DEC';

        return $year . '-' . $month . '-' . $day;
    }
```

As this subroutine only ever expects a single parameter, there's no need to go to the trouble of supporting named parameters. Any parameter supplied is assigned to the $original lexical variable by invoking shift. For the purposes of this discussion, the key statements are these:

```
$original =~ /(\d\d)-(\w\w\w)-(\d\d\d\d)/;

my ( $day, $month, $year ) = ( $1, $2, $3 );
```

The value of $original is matched against a regular expression that looks for two digits, followed by a dash, followed by three word characters, followed by another dash, followed by four digits. Three sets of parentheses ensure that the $1, $2 and $3 after-match variables remember the values matched by each subpattern. The three after-match variables are then assigned (as a list) to the $day, $month and $year lexicals. The rest of the subroutine performs the conversion and returns the converted date string. Note that the regular expression could just as easily be written as follows:

```
/(\d{2})-(\w{3})-(\d{4})/
```

As always with Perl, *there's more than one way to do it*. However, owing to greediness considerations, it would be considered dangerous to write the regular expression like this:

```
/(\d+)-(\w+)-(\d+)/
```

## 7.9 Substitutions: Search and Replace

In addition to the *match operator* (m//), Perl supports the *substitution operator* (s///) as part of its regular expression technology. Unlike the match operator, which surrounds the regular expression with a pair of delimiters, the substitution operator adds a third. The pattern to *search for* is delimited by the first two / characters, while the string to use as a *replacement* is delimited by the last two / characters. Here's an example that searches for a simple concatenation and replaces it with another:

```
s/these/those/
```

By default, the substitution stops replacing after the first successful match. So, if a string has this value:

    Give me some of these, these, these and these. Thanks.

the substitution as shown above transforms the string into this:

    Give me some of those, these, these and these. Thanks.

If a requirement exists to search for and replace *all* occurrences of a pattern, the behaviour of the substitution can be modified with a trailing "g". This is the *global* modifier[7], and it is used as follows:

    s/these/those/g

When applied to the string, a global search and replace is performed, resulting in this string:

    Give me some of those, those, those and those. Thanks.

Another important modifier is the *ignore case* modifier, which matches regardless of case. Modifiers can be combined, so if the substitution is:

    s/these/those/gi

any of the following words would match: "these", "These", "THESE", "ThEsE" and so on. That is, the capitalisation and case of the string are ignored.

## 7.9.1 Substituting for whitespace

A common use for the substitution operator is to remove unwanted whitespace from a scalar. This regular expression removes any leading whitespace:

    s/^\s+//

while this substitution removes any trailing whitespace:

    s/\s+$//

One final variation compresses (or collapses) any number of whitespace characters into a single space character:

    s/\s+/ /g

Note the use of "g", the global modifier.

---

[7] This can be applied to any regular expression, not just substitutions. This is true of most modifiers.

## 7.10 Finding a Sequence

Let's conclude this chapter with a complete example that demonstrates a real-world usage of regular expressions.

The data for this example is taken from the *EMBL Nucleotide Sequence Database*[8], is identified as ID AF213017 and is described as: *Acinetobacter calcoaceticus KHP18 partial pKLH2 plasmid including aberrant mercury resistance transposon TnPKLH2, truncated insertion sequence IS1011.D1 and determinants for CinH resolution system.* The data contained in the EMBL database associated with this entry looks like this:

```
gccacagatt acaggaagtc atatttttag acctaaatca ctatcctcta tctttcagca        60
agaaaagaac atctacttgg tttcgttccc tatccaagat tcagatggtg aaacgagtga       120
tcatgcacct gatgaacgtg caaaaccaca gtcaagccat gacaaccccg atctacagtt       180
...
gcatctgtct gtatccgcaa cctaaaatca gtgctttaga agccgtggac attgatttag      6660
gtacgtgtag agcaagactt aaatttgtac gtgaaactaa aagccagttg tatgcattag      6720
cttttcaat ttgtataacg tataacgtat ataatgttaa ttttagattt tcttacaact        6780
tgatttaaaa gtttaagatt catgtatttа tattttatgg ggggacatga atagatct        6838
```

There is a readily identifiable pattern here. Each line starts with some whitespace. The sequence data is then presented in groups of up to ten letters (with up to six groups on a line) and the line ends with a count of the size of the sequence up to that point.

The real-world requirement is to take any arbitrary sequence and determine if it appears within the sequence as extracted from the EMBL database. This is complicated by the fact that the sequence is presented as a group of ten-letter strings over multiple lines, as opposed to one single stringed sequence on one line. If the sequence data was on one line, and contained in a scalar called $sequence, a line of code similar to this:

```
if ( $sequence =~ /acttaaatttgtacgtg/ )
```

would do the trick. Unfortunately, the sequence data is presented as groups of ten-letter strings over multiple lines, as opposed to one single stringed sequence on one single line. All of those spaces complicate things somewhat and need to be removed if there is to be any chance of matching a sequence that is greater than ten letters in length. Even if the matching sequence is less than ten letters long, it may fail if the spaces remain, as part of the matching sequence may match the end of one group and continue into the next (or worse still, straddle two separate lines). Look closely at the sample data, above, for examples of these problems.

Rather than try to solve all of the challenges posed by this example in one go, let's deal with each one at a time. The only assumption made is that the sequence

---
[8] We cover this in detail later.

data has been extracted from the EMBL database entry and stored in a data-file called embl.data[9].

Removing the number at the end of each line is accomplished with the following substitution:

    s/\s*\d+$//

The regular expression matches zero or more whitespace characters (\s*), followed by one or more digits (\d+) positioned at the end of the line ($). When it is found, it is replaced with nothing, that is, removed.

Removing the unwanted spaces within a line involves another substitution, which looks for any amount of whitespace (\s*) and replaces it with nothing:

    s/\s*//g

Note the use of the global modifier, which ensures all space characters are removed. Let's use these two substitutions within a program, called prepare_embl:

```
#! /usr/bin/perl -w

# The 'prepare_embl' program - getting embl.data ready for use.

while ( <> )
{
    s/\s*\d+$//;
    s/\s*//g;
    print;
}
```

The two substitutions are performed on each line that is read in. In the absence of a named variable, the line is assigned to the $_ default variable. After the two substitutions have been performed, the adjusted line is printed to STDOUT. Use this command-line to process the embl.data disk-file, and redirect STDOUT to another data-file, producing embl.data.out:

    perl prepare_embl embl.data > embl.data.out

Use the Linux wc utility to request some statistics on the newly created disk-file:

    wc embl.data.out

which produces the following results:

        0       1    6838 embl.data.out

---

[9] Available for download from the *Bioinformatics, Biocomputing and Perl* web-site.

The wc utility reports that the newly created disk-file contains 6838 characters. This matches the total at the end of the extracted EMBL database entry, so it appears as if the entire sequence has been processed. Interestingly, wc also reports that the disk-file contains a single line. This is an interesting, and somewhat pleasing, side effect of executing the prepare_embl program.

With the sequence now stored as a single line of data in the embl.data.out data-file, it is a relatively straightforward exercise to produce a small program to check arbitrary sequences against the EMBL database entry. This program is called match_embl:

```perl
#! /usr/bin/perl -w

# The 'match_embl' program - check a sequence against the EMBL
#                            database entry stored in the
#                            embl.data.out data-file.

use constant TRUE => 1;

open EMBLENTRY, "embl.data.out"
    or die "No data-file: have you executed prepare_embl?\n";

my $sequence = <EMBLENTRY>;

close EMBLENTRY;

print "Length of sequence is: ", length $sequence, " characters.\n";

while ( TRUE )
{
    print "\nPlease enter a sequence to check.\nType 'quit' to end: ";

    my $to_check = <>;

    chomp( $to_check );
    $to_check = lc $to_check;

    if ( $to_check =~ /^quit$/ )
    {
        last;
    }

    if ( $sequence =~ /$to_check/ )
    {
        print "The EMBL data extract contains: $to_check.\n";
    }
    else
    {
        print "No match found for: $to_check.\n";
    }
}
```

Let's review a sample usage session with this program. Execute the match_embl program with this command-line:

```
perl match_embl
```

Here's a captured usage session, showing the messages produced and the input provided by the user (which is shown in *italics*):

```
Length of sequence is: 6838 characters.

Please enter a sequence to check.
Type 'quit' to end: aaatttgggccc
No match found for: aaatttgggccc.

Please enter a sequence to check.
Type 'quit' to end: acttaaatttgtacgtg
The EMBL data extract contains: acttaaatttgtacgtg.

Please enter a sequence to check.
Type 'quit' to end: TATCATGAT
No match found for: tatcatgat.

Please enter a sequence to check.
Type 'quit' to end: accttaaatttgtacgtg
No match found for: accttaaatttgtacgtg.

Please enter a sequence to check.
Type 'quit' to end: cagcaagaaaa
The EMBL data extract contains: cagcaagaaaa.

Please enter a sequence to check.
Type 'quit' to end: caGGGGGgg
No match found for: caggggggg.

Please enter a sequence to check.
Type 'quit' to end: tcatgcacctgatgaacgtgcaaaaccacagtcaagccatga
The EMBL data extract contains: tcatgcacctgatgaacgtgcaaaaccacagtcaagccatga.

Please enter a sequence to check.
Type 'quit' to end: quit
```

Even on a relatively slow computer[10], the matching occurs in the blink of an eye. Imagine how long this matching would take to do by hand!

Rather than describe the workings of the match_embl program in detail, it is left as an exercise for the reader to work out what is going on. At this stage in the book, most of what you need to know has already been covered. Attention is drawn to the use of the in-built lc subroutine, which takes a scalar variable and converts it to lowercase, returning the lowercase version to the caller. Note also the use of the $to_check scalar as the regular expression against which to match. The value of the scalar is used as the regular expression against which to match and, in this case, it is a concatenation.

---

[10] By the standard of the day (Summer 2003). The match_embl program was tested on a "slow" Pentium III.

## Where to from Here

Perl's regular expression technology is often referred to as *a programming language within a programming language*. This is a cute sound bite. However, it masks the fact that the integration of a full-featured regular expression engine into the core of Perl is what makes it the programming language it is. There is a lot more to regular expressions than described in this chapter, which presented the core technology to whet your appetite. The *Suggestions for Further Reading* appendix contains pointers to more thorough treatments. As can be expected, example uses of Perl's regular expression technology appear throughout the remainder of *Bioinformatics, Biocomputing and Perl*.

## The Maxims Repeated

Here's a list of the maxims introduced in this chapter.

- *Use a regular expression to specify what you want to find, not how to find it.*
- *Use regular expression shorthand to reduce the risk of error.*
- *When working with nested parentheses, count the opening parentheses, starting with the leftmost, to determine which parts of the pattern are assigned to which after-match variables.*

## Exercises

1. Work through the `perlretut` documentation included with Perl, trying out each of the example regular expressions as they are presented. This is a large document that will take some time to read. Pay particular attention to the regular expression example that matches DNA stop codons.

2. In addition to Perl, many other programs and tools utilise regular expressions in interesting ways. One such tool (included with Linux) is `grep`, the "generalised regular expression parser", which can be used to search for a pattern within any selection of disk-files. Read the manual page for `grep` to learn how to use it, then use `grep` to search for the existence of arbitrary sequences in the `embl.data.out` disk-file. [Be advised that upon success and by default, `grep` prints the matching line to `STDOUT`. Check the options, documented in the manual page, to learn how to change this default behaviour.]

# 8

# Perl Grabbag

*Some useful bits 'n' pieces that every Perl programmer should know.*

## 8.1 Introduction

Rather than discuss a specific topic or feature of Perl in detail, this chapter presents a collection of Perl topics. There's much more to Perl than what's been covered so far in this book and, even when this chapter is worked through, there's still more Perl to learn. However, the core of the language has been covered and, as demonstrated in the rest of *Bioinformatics, Biocomputing and Perl*, this core is more than enough to perform a varied number of programming tasks vital to the work of Bioinformaticians.

## 8.2 Strictness

Perl is often considered too loose a programming language to be taken seriously by "real" computing folk. True, in its own unique way, Perl happily breaks a number of the "golden rules" of the traditional programming language. For instance, Perl allows variables to be used before they are declared, magically assigning default values to variables and then making them immediately available to the program within which they appear. The same goes for subroutines: they can be invoked *before* they are defined. And then there are those *sticky* global variables: everything is a global by default[1].

---

[1] Many a computer scientist considers this to be an *unspeakable sin*.

*Bioinformatics, Biocomputing and Perl.* Michael Moorhouse and Paul Barry
© 2004 John Wiley & Sons, Ltd ISBN 0-470-85331-X

As described in Chapter 5, the use of my variables turns a global variable into a lexical. By default, the use of my variables is *optional*. However, it is possible to have `perl` insist on the use of my variables, making their use *mandatory*.

This insistence is referred to as *strictness*, and is switched on by adding this line to the top of a program:

```
use strict;
```

This is a *directive* that, among other things, tells `perl` to insist on all variables being declared before they are used[2], as well as requiring that all subroutines be declared (or defined) before they are invoked.

Why do such a thing? Why restrict the programmer, when Perl is all about freedom? The answer has to do with *scale*. As programs get bigger, they become harder to maintain. The use of `use strict` helps keep things organised and reduces the risk of errors being introduced into programs. And anything that helps reduce errors is a good thing, even if it is *sometimes* inflexible. Think of the `use strict` directive as a gentle reminder to take the time to limit the scope of any variables used in a program. Thinking about the scope of variables, and using my and our to control the visibility of variables, really becomes important as a program grows in size.

When strictness is enabled, `perl` takes the time to check the declaration of each of a program's variables *before* execution occurs. Consider this program, called `bestrict`:

```
#! /usr/bin/perl -w

# bestrict - demonstrating the effect of strictness.

use strict;

$message = "This is the message.\n";

print $message;
```

Note that the `$message` scalar is *not* declared as a lexical (my) or global (our) variable. When an attempt is made to execute the `bestrict` program, `perl` complains loudly that the strictness rules have been broken:

```
Global symbol "$message" requires explicit package name at bestrict line 7.
Global symbol "$message" requires explicit package name at bestrict line 9.
Execution of bestrict aborted due to compilation errors.
```

These "compilation errors" are *fixed* by simply declaring the `$message` scalar as a my variable, thus:

```
my $message = "This is the message.\n";
```

---

[2] Either as a my or our variable.

which really isn't that big a deal, is it? As a program grows in size (and, consequently, complexity), the benefits of switching on strictness far outweigh the disadvantage (and perceived inconvenience) of having to declare all variables with either my or our.

Defining all subroutines at the top of a program ensures that perl sees them before they are invoked. However, as discussed earlier in this book, the placement of subroutines is a personal preference: some programmers place them near the top, others near the bottom, while others place subroutines at any location within a program's disk-file. The use of use strict forces programmers to place subroutines near the top of their programs. Something that "forces" programmers to behave in a particular way is decidedly "unperlish" and inflexible.

To retain the flexibility of being able to place subroutines anywhere in a program's disk-file, while still taking advantage of the use strict directive, Perl provides the use subs directive that can be used in combination with use strict to *declare* a list of subroutines at the top of a program. Subroutine *definitions* can then appear anywhere in a program's disk-file. Here's an example:

```
use strict;
use subs qw( drawline biodb2mysql );
```

The use subs directive declares a list of subroutine names that are later defined *somewhere* in the program's disk-file.

Although the Perl documentation advises the use of use strict for everything but the most "casual" of programs, your authors' advice is, well, *more strict*: always use use strict.

> **Maxim 8.1** *Unless you have a really good reason not to, always switch on strictness at the top of your program.*

It is left as an exercise for the reader to think up a *really good reason* for not using use strict.

## 8.3 Perl One-liners

Most of the example programs seen thus far in *Bioinformatics, Biocomputing and Perl* start with the following line:

```
#! /usr/bin/perl -w
```

The -w *switch* is one of a large collection of directives that can be provided to perl on the command-line[3]. The "w" stands for "warnings", and instructs perl to warn the programmer when it notices any dubious programming practices

---
[3] Refer to the perlrun manual page, included with Perl, for all the gory details.

(such as defining a subroutine twice). It is always a good idea to switch on warnings, as it makes for better programs.

When discussing the installation of third-party CPAN modules during Chapter 5, the -e switch was used to check that the module had installed correctly, as follows:

```
perl -e 'use ExampleModule'
```

The "e" stands for "execute", and instructs `perl` to execute the program statements included within the single quotes. Here's another example command-line:

```
perl -e 'print "Hello from a Perl one-liner.\n";'
```

The ability to use the -e switch on the command-line in this way creates what's known in the Perl world as a *one-liner*. That is, a single line of Perl code is provided to `perl` to execute immediately from the command-line. Here's another one-liner that turns `perl` into a simple command-line calculator:

```
perl -e 'printf "%0.2f\n", 30000 * .12;'
```

which calculates 12% of 30,000 and displays the result (3600.00). The `printf` subroutine is a variant of the more common `print`, and prints to a specified *format*. Use these commands to learn more about `printf` and *formats*:

```
perldoc -f printf
perldoc -f sprintf
```

Another useful switch is -n, which, when used in combination with -e, treats the one-liner as if it is enclosed with a loop. Consider this one-liner:

```
perl -ne 'print if /ctgaatagcc/;' embl.data
```

which is equivalent to the following program statements:

```
while ( <> )
{
    print if /ctgaatagcc/;
}
```

That is, the code between the single quotes (the one-liner) is equivalent to the above loop. The `embl.data` part of the command-line is just that: part of the command-line, not part of the one-liner. When the one-liner is executed, the following output is generated:

```
attgtaatat ctgaatagcc actgattttg taggcaccctt tcagtccatc tagtgactaa      5880
```

as there's a match. Of course, for those readers who took the time to complete the exercises from the last chapter[4], they already know that there's an easier way to do this using the grep utility:

```
grep 'ctgaatagcc' embl.data
```

which produces output identical to that produced by the one-liner. Note: less typing, less risk of error.

When the -n switch is combined with -p, the loop has a print statement added to the end. Here's a one-liner that prints only those lines from the embl.data disk-file that *do not* end in four digits:

```
perl -npe 'last if /\d{4}$/;' embl.data
```

When executed, the following output is produced:

```
gccacagatt acaggaagtc atatttttag acctaaatca ctatcctcta tctttcagca         60
agaaaagaac atctacttgg tttcgttccc tatccaagat tcagatggtg aaacgagtga        120
tcatgcacct gatgaacgtg caaaaccaca gtcaagccat gacaaccccg atctacagtt        180
tgatgttgaa actgccgatt ggtacgccta cagtgaaaac tatggcacaa gtgaagaaaa        240
acgctttgtt aagtttgttg caactcaaat tgacgagctt aaatcacgct acaagggtgc        300
agagatttac ctgatacgga atgaactcga ttattggttg tttagcccta aagatggtcg        360
tagattcagc cctgactaca tgctgatcat taatgatgct gaaaatagtg aaatgtacta        420
tcaatgctta attgagccta aaggtggtca tttgcttgaa aaggatactt ggaaagagga        480
agtattgatt agtttggatg atgaaagcca aattgttttt gatgcagatc aagatgattc        540
acaaaactat gttgagttct taaatgaagt taaagagcat ggttataagg aagttaaatg        600
tttaggcttc aaattctaca ataccgaacc acgatctgaa tcagattttg ctattgattt        660
tcacaatagg atgccgagtt aatctaggtt tctcactgta acctgctgat tattatcttt        720
ttgtgaagtt gctacataat attgttttta agatcattga ataaaaaagc cagctctata        780
ctggcttttt tattgcttaa aattatattc cgatgcttgg tcaaactgc aagtatgcag         840
tcttgaccag gcatctaggg gtcgtctcag aattcggaaa ataaagcacg ctaaggcgta        900
gtcaccccgt gactcccccg cgccgatgca gcgagcttcg ttccgtcttg cagtgacgca        960
```

The above one-liner is equivalent to this program:

```
while ( <> )
{
    last if /\d{4}$/;
}
continue {
    print $_;
}
```

The one-liner is a little harder to do with grep. Your authors came up with this grep equivalent[5]:

```
grep -v '[0123456789][0123456789][0123456789][0123456789]$' embl.data
```

---

[4] You mean to say you *didn't* do the exercises?!? Quick: go back and do them now, before we ask again.

[5] This can probably be improved, depending on the version of grep available to you.

## 152  Perl Grabbag

Note: more typing for the grep equivalent this time. The Perl one-liner involves less typing and, consequently, less risk of error.

## 8.4 Running Other Programs from perl

Another feature that helps maintain Perl's prominent position as a "glue language" is its ability to execute other programs. There are two main ways to do this:

1. By invoking the program in such a way that after execution, the calling program can determine whether the called program successfully executed.

2. By invoking the program in such as way that after execution, any results from the called program are returned to the calling program.

Perl's in-built system subroutine behaves as described in point 1 above, while Perl's *backticks* and qx// operator behave as described in point 2.

Here's an example program, called pinvoke, that demonstrates each of these mechanisms by invoking the Linux utility program, ls, that lists disk-files in the current directory[6]:

```perl
#! /usr/bin/perl -w

# pinvoke - demonstrating the invocation of other programs
#           from Perl.

use strict;

my $result = system( "ls -l p*" );

print "The result of the system call was as follows:\n$result\n";

$result = `ls -l p*`;

print "The result of the backticks call was as follows:\n$result\n";

$result = qx/ls -l p*/;

print "The result of the qx// call was as follows:\n$result\n";
```

The invocation of system results in the ls program executing. Any output from ls is displayed on screen (STDOUT) as normal (as that's what ls does), then, as ls executed successfully, a value of zero is returned to pinvoke and assigned to

---

[6] Specifically, this invocation of the ls program lists, in long format, any disk-file whose name starts with the letter "p".

the $result scalar. The $result scalar is then printed to STDOUT as part of an appropriately worded message. If the ls program fails, the $result scalar is set to −1.

Perl's backticks (` and `) also execute external programs from within Perl. Unlike system, the results from the program are *captured* and returned to the program. In the pinvoke program, the results are assigned to the $result scalar, and then printed to STDOUT as part of an appropriately worded message.

The qx// operator is another way to invoke the backticks behaviour: it works exactly the same way as backticks, which is confirmed by the output produced by the pinvoke program (as shown and discussed below).

Note that the pinvoke program enforces strictness, requiring the programmer to declare any variables as lexicals. There is only one, $result, which is declared as a my variable. When executed, the pinvoke program produces the following output:

```
-rw-rw-r--   1 barryp      barryp          403 Aug 16 16:48 pinvoke
-rw-rw-r--   1 barryp      barryp          145 Aug  7 12:36 prepare_embl
-rw-rw-r--   1 barryp      barryp          422 Jul 22 15:10 private_scope
The result of the system call was as follows:
0
The result of the backticks call was as follows:
-rw-rw-r--   1 barryp      barryp          403 Aug 16 16:48 pinvoke
-rw-rw-r--   1 barryp      barryp          145 Aug  7 12:36 prepare_embl
-rw-rw-r--   1 barryp      barryp          422 Jul 22 15:10 private_scope

The result of the qx// call was as follows:
-rw-rw-r--   1 barryp      barryp          403 Aug 16 16:48 pinvoke
-rw-rw-r--   1 barryp      barryp          145 Aug  7 12:36 prepare_embl
-rw-rw-r--   1 barryp      barryp          422 Jul 22 15:10 private_scope
```

Note that the first invocation of ls results in the production of the first three lines on STDOUT, which are produced *before* the appropriately worded message (which is produced by pinvoke, not ls).

It is also possible to invoke a disk-file containing Perl code from within a Perl program. Use Perl's in-built do subroutine.

## 8.5 Recovering from Errors

It is not always appropriate to die whenever an error occurs. Sometimes it makes more sense to spot, and then recover from, an error. This is referred to as handling exceptional cases, or *exception handling*. Consider the following code:

```
my $first_filename = "itdoesnotexist.txt";

open FIRSTFILE, "$first_filename"
    or die "Could not open $first_filename.  Aborting.\n";
```

When executed, the above code produces the following message:

```
Could not open itdoesnotexist.txt.  Aborting.
```

This assumes that the itdoesnotexist.txt disk-file does not exist. The program terminates as a result of the invocation of die. It is possible to protect this code by enclosing it within an eval block.

The in-built eval subroutine takes a block of code and executes it (or *evaluates* it). This is exactly what perl does to code. When perl invokes eval, anything that happens within the eval block that would usually result in a program terminating[7] is *caught* by perl and does *not* terminate the program. Here's the above code surrounded by an eval block:

```
eval {
    my $first_filename = "itdoesnotexist.txt";

    open FIRSTFILE,  "$first_filename"
        or die "Could not open $first_filename.  Aborting.\n";
};
```

When executed, this code does not produce the error message from earlier, nor does it die. This has nothing to do with the fact that the itdoesnotexist.txt disk-file now exists. It has everything to do with the fact that the code is now protected by the eval block, which is a great facility, as potentially troublesome code can now be protected.

**Maxim 8.2** *Use eval to protect potentially erroneous code.*

What completes the eval facility is the addition of a mechanism to check if a fatal error did indeed occur during the eval block. If die is invoked within an eval block, the block immediately terminates and perl sets the internal $@ variable to the message generated by die. After the eval block, it is a simple matter to check the status of $@ and act appropriately. Adding this if statement after the above eval block:

```
if ( $@ )
{
    print "Calling eval produced this message: $@";
}
```

prints the following message to STDOUT when the itdoesnotexist.txt disk-file does not exist:

```
Calling eval produced this message: Could not open itdoesnotexist.txt.  Aborting.
```

---

[7] Perl programmers refer to the program *dieing*.

Typically, the code within the `if` block associated with the `eval` does more than print a message to STDOUT. That's where *recovery* comes in. It would be a good idea in this particular example to try to open another disk-file or, perhaps, create `itdoesnotexist.txt` as an empty disk-file, and attempt to open it again. The program can then continue as normal.

## 8.6 Sorting

Perl provides powerful in-built support for sorting[8]. Two subroutines, `sort` and `reverse`, can be used to sort lists of strings or numbers into ascending order, descending order or any other customized order. To demonstrate the power of Perl's sorting technology, let's step through a program, called `sortexamples`, that demonstrates what's possible.

The `sortexamples` program starts in the usual way: the *magic* first line is followed by a short comment, then strictness is switched on. A list of four short DNA sequences is assigned to an array called `@sequences`, which is then printed to STDOUT:

```
#! /usr/bin/perl -w

# sortexamples - how Perl's in-built sort subroutine works.

use strict;

my @sequences = qw( gctacataat attgttttta aattatattc cgatgcttgg );

print "Before sorting:\n\t-> @sequences\n";
```

This `print` statement produces the following output:

```
Before sorting:
        -> gctacataat attgttttta aattatattc cgatgcttgg
```

that is, the four short sequences are displayed in the order that they were assigned to the array. The next three lines of code produce three new arrays from the `@sequences` array:

```
my @sorted = sort @sequences;
my @reversed = sort { $b cmp $a } @sequences;
my @also_reversed = reverse sort @sequences;
```

The first array, `@sorted`, is created as a result of invoking the in-built `sort` subroutine, passing the `@sequences` array as its sole parameter. This sorts the

---

[8] Only readers who have had to implement the *Quicksort* algorithm in any other programming language can truly appreciate what a treat this actually is.

@sequences array in Perl's default order, which is to sort alphabetically in ascending order (from "a" through to "z").

The second array, @reversed, is also created as a result of invoking sort. However, in addition to providing an array to sort, this invocation also supplies a small block of code that is used to specify the *sort order* to be applied to the array. The small block of code on this line is:

```
{ $b cmp $a }
```

To understand this block of code, consider that the $a and $b scalars are special scalars, reserved for use with sort. On the basis of the *comparison operator* applied to the two scalars and the order in which it is applied, the block of code can customise the sort order. In this example, $b is being compared (cmp) to $a, which results in the sort being applied in descending order (from "z" through to "a").

The third array, @also_reversed, is created by first sorting the @sequences array (using the default sort order), then reversing the sorted list by invoking the in-built reverse subroutine. Note that the reverse subroutine reverses the order of elements in a list; it does not sort in reverse order. With the three sorted lists created and assigned to arrays, they are printed to STDOUT using these statements:

```
print "Sorted order (default):\n\t-> @sorted\n";
print "Reversed order (using sort { \$b cmp \$a }):\n\t-> @reversed\n";
print "Reversed order (using reverse sort):\n\t-> @also_reversed\n";
```

which results in the following output:

```
Sorted order (default):
        -> aattatattc attgttttta cgatgcttgg gctacataat
Reversed order (using sort { $b cmp $a }):
        -> gctacataat cgatgcttgg attgttttta aattatattc
Reversed order (using reverse sort):
        -> gctacataat cgatgcttgg attgttttta aattatattc
```

Note that the output shows the original unsorted list of sequences in the various sort orders. Both the second and the third array (@reversed and @also_reversed) contain the same list of sorted elements.

It is also possible to sort in numerical order using sort. To demonstrate the standard method of sorting in numerical order, the sortexamples program defines a list of chromosome pair numbers and assigns them to another array, called @chromosomes. The array is then printed to STDOUT:

```
my @chromosomes = qw( 17 5 13 21 1 2 22 15 );

print "Before sorting:\n\t-> @chromosomes\n";
```

This results in the following output:

```
Before sorting:
        -> 17 5 13 21 1 2 22 15
```

Two invocations of the sort subroutine sort the @chromosomes array into numerical order, with the first in ascending order (from 1 through to the largest number in the array) and the second in descending order (from the largest number in the array down to 1). These numerically sorted lists are assigned to the @sorted and @reversed arrays respectively:

```
@sorted = sort { $a <=> $b } @chromosomes;
@reversed = sort { $b <=> $a } @chromosomes;
```

Note the requirement to provide a block of code to each of the invocations of sort in order to define the correct sort order. Unlike earlier, the numerical comparison operator (<=>) is used, as opposed to cmp, as the requirement here is to sort numerically, not alphabetically. Note, too, the use of $a and $b when defining the sort order. Two print statements display the results of the numerical sorting to STDOUT, and conclude the sortexamples program:

```
print "Sorted order (using sort { \$a <=> \$b }):\n\t-> @sorted\n";
print "Reversed order (using sort { \$b <=> \$a }):\n\t-> @reversed\n";
```

The two print statements result in the following output:

```
Sorted order (using sort { $a <=> $b }):
        -> 1 2 5 13 15 17 21 22
Reversed order (using sort { $b <=> $a }):
        -> 22 21 17 15 13 5 2 1
```

Perl's ability to sort is powerful and highly customizable. Of course, there's much more to sort than is presented in this short example. To learn more, use the following command-line to read the on-line documentation for sort that comes with Perl:

```
perldoc -f sort
```

Here's a small program, called sortfile, that takes any disk-file and sorts the lines in the disk-file in ascending order[9]:

```
#! /usr/bin/perl -w

# sortfile - sort the lines in any file.

use strict;
```

---

[9] It should be easy for you to work out what's going on in this program. Everything you need to know has already been covered in *Bioinformatics, Biocomputing and Perl*.

```perl
    my @the_file;

    while ( <> )
    {
        chomp;
        push @the_file, $_;
    }

    my @sorted_file = sort @the_file;

    foreach my $line ( @sorted_file )
    {
        print "$line\n";
    }
```

Given a disk-file, called sort.data, with the following contents:

```
Zap! Zoom! Bang! Bam!
Batman, look out!
Robin, behind you!
Aaaaah, it's the Riddler!
```

The following command-line sorts the lines in sort.data into ascending order:

```
perl sortfile sort.data
```

and produces the following output:

```
Aaaaah, it's the Riddler!
Batman, look out!
Robin, behind you!
Zap! Zoom! Bang! Bam!
```

Of course, the savvy Linux user would use the sort utility to do the same thing, using this command-line:

```
sort sort.data
```

which illustrates, as with the grep examples from earlier, that some of the effort required in creating a custom program can be avoided when the operating system utilities are used instead.

> **Maxim 8.3** *Take the time to become familiar with the utilities included in the operating system.*

Refer to the *Suggestions for Further Reading* appendix (page 461) for some advice on learning more about the utilities included with Linux. A short list of Linux commands and utilities is provided in the appendix entitled *Essential Linux Commands*, beginning on page 467. Use this command-line to learn more about the sort utility:

```
man sort
```

## 8.7 HERE Documents

Consider the requirement to display the following text on screen in *exactly* the format shown from within a program:

```
Shotgun Sequencing

This is a relatively simple method of reading
a genome sequence.  It is ''simple'' because
it does away with the need to locate
individual DNA fragments on a map before
they are sequenced.

The Shotgun Sequencing method relies on
powerful computers to assemble the finished
sequence.
```

Utilising the Perl features already known, a sequence of `print` statements would do the trick, as follows:

```
print "Shotgun Sequencing\n\n";
print "This is a relatively simple method of reading\n";
print "a genome sequence.  It is ''simple'' because\n";
print "it does away with the need to locate\n";
print "individual DNA fragments on a map before\n";
print "they are sequenced.\n\n";
print "The Shotgun Sequencing method relies on\n";
print "powerful computers to assemble the finished\n";
print "sequence.\n";
```

By enclosing each line in double quotes and appending the appropriate number of newlines to the end of each line, the above sequence of `print` statements satisfies the requirement defined at the start of this section. Of course, there is a better way to do this using Perl's HERE document mechanism. Rather than try to describe what a HERE document is, let's look at an example:

```
my $shotgun_message = <<ENDSHOTMSG;
Shotgun Sequencing

This is a relatively simple method of reading
a genome sequence.  It is ''simple'' because
it does away with the need to locate
individual DNA fragments on a map before
they are sequenced.

The Shotgun Sequencing method relies on
powerful computers to assemble the finished
```

```
sequence.
ENDSHOTMSG

print $shotgun_message;
```

The above code assigns a HERE document to the `$shotgun_message` scalar. The HERE document starts with the `<<` chevrons, which has a programmer-chosen identifier (written in uppercase by convention) attached to it. Note that there should be no space character between the chevrons and the start of the identifier. Everything between the identifier and the repetition of the identifier is the HERE document. This means that the message describing Shotgun Sequencing is a HERE document assigned to the `$shotgun_message` scalar. It is then printed to STDOUT using a simple `print` statement.

Of note is the fact that the HERE document does not need to include all those newlines, as was the case above with the sequence of `print` statements. In addition, the double quotes surrounding each string are also missing from the HERE document. All that programmers using the HERE document have to worry about is formatting the text in the way that they wish it to display. It is possible to improve upon the HERE document example above by removing the need for the `$shotgun_message` scalar and printing the HERE document directly, as follows:

```
print <<ENDSHOTMSG;
Shotgun Sequencing

This is a relatively simple method of reading
a genome sequence.  It is ''simple'' because
it does away with the need to locate
individual DNA fragments on a map before
they are sequenced.

The Shotgun Sequencing method relies on
powerful computers to assemble the finished
sequence.
ENDSHOTMSG
```

HERE documents are surprisingly useful, especially when it comes to dynamically producing HTML documents. This use of HERE documents is discussed later in the *Working with the Web* part of *Bioinformatics, Biocomputing and Perl*.

## Where to from Here

This chapter ends Part I, *Working with Perl*. Readers who worked through this and the preceding five chapters now know enough Perl to confidently perform a variety of programming tasks. The remainder of this book builds upon this base and applies what has been learnt about Perl to a number of Bioinformatics tasks.

As the authors of *Programming Perl*, the classic Perl reference, advise at the end of the first chapter of their book: *Have the appropriate amount of fun.*

## The Maxims Repeated

Here's a list of the maxims introduced in this chapter.

- *Unless you have a really good reason not to, always switch on strictness at the top of your program.*
- *Use `eval` to protect potentially erroneous code.*
- *Take the time to become familiar with the utilities included in the operating system.*

## Exercises

1. Add the `use strict` directive to a selection of programs that you have written. What effect does the addition of the directive have?
2. Write a one-liner that scans a disk-file for any blank lines, printing the words "`Got one!`" as soon as a blank line is found.
3. Write a program to do the same thing as the one-liner from the last question.
4. Can `grep` be used to perform the same task as the one-liner? Why or why not?
5. Write a program that invokes the `ls` utility in long format, captures its output, then displays a total count for the number of bytes in all of the listed disk-files.
6. Write a program that writes another program, then uses `eval` to execute it.
7. Change the `sortexamples` program to sort the `@chromosomes` array alphanumerically, both in ascending and descending order. That is, given the following list of values: 17, 5, 13, 21, 1, 2, 22 and 15, your program should produce "1 13 15 17 2 21 22 5" and "5 22 21 2 17 15 13 1".
8. Consider the following HTML:

    ```
    <!DOCTYPE html PUBLIC "-//W3C//DTD HTML 4.01 Transitional//EN">
    <html>
    <head>
      <meta http-equiv="content-type"
     content="text/html; charset=ISO-8859-1">
      <title>Check out this great resource!</title>
    </head>
    ```

```
<body>
A great introduction to Bioinformatics Computing Skills and Practice is
to be had by reading <i>Bioinformatics, Biocomputing and Perl</i> by
Michael Moorhouse and Paul Barry, published by Wiley, 2004.
<p> Check out the book's web-site <a
 href="http://glasnost.itcarlow.ie/~biobook/index.html">here</a>. </p>
</body>
</html>
```

Write a program using `print` statements to produce the above HTML *exactly* as shown. Write a second program to do the same thing using a HERE document. Which technique do you prefer?

# Part II
# Working with Data

# 9

# Downloading Datasets

*Fetching datasets from the Internet.*

## 9.1 Let's Get Data

This chapter shows the reader how to download Bioinformatics datasets from the Internet[1]. A small selection of datasets is used in the chapters that follow this one, so it is best if the datasets are downloaded now, *before* they are required.

## 9.2 Downloading from the Web

While downloading individual data-files from a World Wide Web (WWW) site is often useful, there are times when downloading a large number of data-files makes the use of such a highly interactive mechanism cumbersome.[2]. Some technologies allow the easy integration of data sources across the Internet. Despite this, it is often convenient to download frequently used datasets and store them locally. The advantages of such a strategy are:

---

[1] Throughout this chapter, the terms "Internet" and "WWW" are used interchangeably to mean the same thing.

[2] Having said that, the *Web Automation* chapter, later in this book, shows how to automate interactive web browsing.

---

*Bioinformatics, Biocomputing and Perl.* Michael Moorhouse and Paul Barry
© 2004 John Wiley & Sons, Ltd ISBN 0-470-85331-X

**Ease of access** – It is easier to access data-files on a local hard disk than it is to write an interface routine to download them as needed from a – possibly congested – location on the Internet.

**Speed** – Local hard-disk access, even over a shared file system, is usually faster than operating through external networks to Internet locations. When the processing is performed locally, it may be possible to allocate extra computational resources to the analysis.

**Reliability** – Accessing local hard-disk copies of data-files is more reliable than network connections and WWW servers. This allows processing even in the event of network failures, as the network is not required to run the analysis.

**Stability** – If the data changes frequently, it is often helpful to "freeze" it by downloading a copy and using it locally until all analyses are completed.

**Flexibility** – Often the search facilities that exist on the WWW lack certain required functionality. With the datasets available locally, it is possible to develop bespoke search programs using, for example, Perl.

**Security** – Data or results are often sensitive, and sending them to a remote, third-party Internet site may be unacceptable.

There are also disadvantages to this strategy:

**Stale data** – The local copy is a one-time "snapshot" of the dataset at a particular point in time. At some stage, it will need to be updated or replaced by newer data.

**Storage** – The dataset has to be stored somewhere, and some datasets can be large. The Protein Databank (PDB), which is discussed in detail in the next chapter, is close to four gigabytes, and the PDB is one of the *smaller* databases! Consequently, storing multiple copies of the PDB is often impractical.

**Performance** – The centralised specialist services accessible from the WWW are often configured with dedicated parallelised systems, designed to service requests as quickly as possible. If the stored dataset is designed with such systems in mind, it is unlikely that a local system will be able to match this advanced processing capability. Consequently, some analyses may be slower locally when compared to those performed on the WWW.

Downloading datasets can be accomplished in a number of ways. Some of the more established sequence analysis programs, such as *EMBOSS*[3], which is available from:

    http://www.emboss.org

---

[3] The commercial equivalent of this program is called *GCG*.

## Downloading from the Web

have specific methods for performing downloads. Typically, datasets are accessed via a standard network connection to remote Internet sites. Frequently, downloads are automated to occur at regular intervals. The wget program, included with most Linux systems, can be used to do just this.

wget is an excellent example of GNU software as distributed by the *Free Software Foundation*. It is free, reliable and fully featured, yet simple to use. *The Administrators Guide*, written by *David Martin* for use with the *EMBOSS* program, uses wget within its automatic dataset update script. To learn about wget, issue this command at the Linux command-line:

```
man   wget
```

This displays the wget manual page. Use the arrow keys, PgUp and/or PgDn to scroll through the manual page. When done, press the "q" key to quit. As can be seen from reading the manual page, wget can accomplish a lot. Let's start with some simple examples.

### 9.2.1 Using wget to download PDB data-files

To download a single data-file via anonymous FTP, simply provide the URL[4] of the data-file required after the wget command. To download the two PDB structures used in the chapters that follow, use these commands:

```
mkdir   structures
cd   structures
wget   ftp://ftp.rcsb.org/pub/pdb/data/structures/all/pdb1m7t.ent.Z
wget   ftp://ftp.rcsb.org/pub/pdb/data/structures/all/pdb1lqt.ent.Z
```

Note that a directory called structures is first created (with the *make directory* command, mkdir) then entered (with the *change directory* command, cd) prior to invoking wget. An "ls -l" command confirms the download and creation of the two data-files in the structures directory[5]:

```
-rw-r-----   1 michael   users    574440 2003-11-04 16:05 pdb1lqt.ent.Z
-rw-r-----   1 michael   users    592220 2003-11-04 16:05 pdb1m7t.ent.Z
```

The ".Z" at the end of the downloaded data-files is significant. It indicates that the data-files have been compressed with the popular ZIP compression technology. The gzip program can *unzip* compressed data-files, as follows:

```
gzip   -d   pdb1m7t.ent.Z   pdb1lqt.ent.Z
```

---

[4] URL stands for "Uniform Resource Locator", the technical name for all those web addresses you type into the *Location Bar* of your favourite web browser.

[5] The disk-file sizes shown here may not match those you download, as there is every possibility that these entries will have changed by the time this book appears in print.

**168** *Downloading Datasets*

Another `ls -l` command confirms that the data-files have been decompressed:

```
-rw-r-----    1 michael  users     2470986 2003-11-04 16:05 pdb1lqt.ent
-rw-r-----    1 michael  users     2843181 2003-11-04 16:05 pdb1m7t.ent
```

## 9.2.2 Mirroring a dataset

The `wget` program can be used to mirror datasets. Here is all that is required to download the *entire* PDB, which is four gigabytes of data, stored in over 18,000 data-files:

```
wget --mirror ftp://ftp.rcsb.org/pub/pdb/data/structures/all/pdb
```

Obviously, such a command should be invoked only when there is a real need to mirror the PDB. Remember: a download of this size takes a considerable amount of time, not to mention disk space. If such a need exists, once complete, another invocation of the same command downloads only additions or updates to the PDB *since the last mirror*.

Before mirroring a dataset, check with other users on the network to see if a local mirror already exists. If a fellow researcher from "down the hall" has a PDB mirror, it is better to use that than download another copy. This important piece of advice warrants its very own maxim.

> **Maxim 9.1** *Download a dataset only when absolutely necessary. Consider the implications of doing so first.*

## 9.2.3 Smarter mirroring

While the `wget` command described in the previous subsection works, it results in a *deep* directory tree. The actual data-files are found in locations similar to this:

```
structures/ftp.rcsb.org/pub/pdb/data/structures/all/pdb
```

Such a deep directory structure can be very inconvenient and frustrating to navigate. Another `wget` invocation can help with this problem. Let's look at the command-line first, then describe what `wget` is being asked to do:

```
wget --output-file=log --mirror --http-user=anonymous          \
     --http-passwd=email@where.ever.net                        \
     --directory-prefix=structures/mmCIF                       \
     --no-host-directories                                     \
     --cut-dirs=6 ftp://ftp.rcsb.org/pub/pdb/data/structures/all/pdb
```

**Technical Commentary:** Note that the '\' characters at the end of each line are *continuation markers* used to indicate that the command continues on the next line. These are used here to allow your authors to fit this command onto this page.

When entering the command, remember to put it all on one line and *remove* the continuation markers. This technique of spreading a long line over multiple lines in order to fit the printed page is common practice within the computing world.

The above wget command sets a number of options:

- **--output-file** – a disk-file into which any message produced by wget is placed.
- **--mirror** – turns on mirroring.
- **--http-user** – sets the web username to use (if needed).
- **--http-passwd** – sets the web password to use (if needed).
- **--directory-prefix** – the place to put the downloaded data-files.
- **--no-host-directories** – the instruction *not* to use the hostname when creating a mirrored directory structure, which is the "ftp.rcsb.org" part.
- **--cut-dirs** – instructs wget to ignore the indicated number of directory levels. In the above example, six directory levels are to be ignored, that is, the "pub/pdb/data/structures/all/pdb" part.

## 9.2.4 Downloading a subset of a dataset

On many occasions, the entire contents of an FTP site might not be required, in which case wget can fetch a specific data-file, placing it in the current directory. Use a command similar to this:

```
wget ftp://beta.rcsb.org/pub/pdb/uniformity/data/mmCIF/all/1ger.cif.Z
```

While multiple URLs to data-files can be supplied on the command-line (separated by spaces), it is often more convenient to place the URLs in a data-file and use the "--input_file=" switch.

The pdbselect program takes the *PDB-Select* list produced in the *Non-Redundant Datasets* chapter (coming soon), builds a list of URLs, removes the duplicates (as more than one chain may be contained in the same PDB data-file) and then downloads them:

```perl
#! /usr/bin/perl

# pdbselect <list of PDB IDs> - a program that takes a list of PDB ID
#                               codes; build a list of URLs for them;
#                               and automates the downloading of them
#                               using 'wget'.

use strict;

my $Base_URL = "ftp://ftp.rcsb.org/pub/pdb/data/structures/all/pdb";

my $Output_Dir = "structures";

open URL_LIST, ">pdb_select_url.lst"
```

```perl
    or die "Cannot write to file: 'pdb_select_url.lst'\n";

while ( <> )
{
    if ( /Failed/ )
    {
        next;
    }

    s/ //g;

    my ( $Structure, $Length ) = split ( ":", $_ );
    my ( $ID, $Chain ) = split ( ",", $Structure );

    $ID =~ tr /[A-Z]/[a-z]/;

    print URL_LIST "$Base_URL/pdb$ID.ent.Z\n";
}

close URL_LIST;

if ( !-e $Output_Dir )
{
    system "mkdir $Output_Dir";
}

if ( !-w $Output_Dir or !-d $Output_Dir )
{
    die "ERROR: Cannot access directory: '$Output_Dir'. Exiting\n";
}

system "sort -u pdb_select_url.lst > unique_urls.lst";

system "rm $Output_Dir/* > /dev/null";

system "wget --output-file=log --http-user=anonymous        \
              --http-passwd=email\@some.where.net            \
              --directory-prefix=$Output_Dir -i unique_urls.lst";
```

This program takes a list of PDB ID codes from STDIN and downloads them from the URL specified in the scalar variable $Base_URL[6]. Those structures marked as "Failed" are skipped, otherwise a URL is built and written to the pdb_select_url.lst file. Duplicate structures are filtered out using the "sort -u" operating system utility, as it is pointless downloading the same structure more than once (even though it contains multiple chains that might be useful). It is easier to do it by using the system subroutine to invoke the sort utility, rather than to perform the same operation in Perl.

Error-checking is performed to see if the output directory exists (otherwise it is created) and that the directory can be accessed. All previous files in it are then

---

[6] In this case, this is the home RCSB FTP site. If you are going to do this for yourself on a regular basis, then use one of the geographically close mirror sites advertised on the RCSB homepage.

deleted using the rm system call[7]. Finally, wget is invoked with the list of URLs. It should then be a case of sitting back, relaxing and waiting for the download to complete. This may take some time: there are some hundreds of megabytes of disk-files to download!

This short demonstration of how to use a pre-packaged download tool should be useful not only for downloading PDB data-files but also in many other contexts. With wget, the difficult part is building the URLs, which can be automated by using Perl. The mirroring power of wget is best suited to bulk downloading of data-files.

## Where to from Here

This chapter introduced the powerful wget utility, which is used to download large collections of data-files from the Internet, specifically PDB data-files. Having secured the data, let's investigate what it comprises in the next chapter.

## The Maxims Repeated

Here's a list of the maxims introduced in this chapter.

- *Download a dataset only when absolutely necessary. Consider the implications of doing so first.*

## Exercises

1. Issue the command "man cron" to learn about the cron facility provided by most Linux distributions. Once you understand how cron works, write a small Perl script to automatically download a PDB structure of your choosing, then add a "crontab entry" to your system in order to download the structure once a week, every week.

2. Visit CPAN and download a copy of the libwww-perl library (also known as LWP). Install the library and its associated modules into Perl. Use the "man LWP" command to learn about the facilities provided by the library, then rewrite the script created in answering the previous exercise to use the facilities of LWP.

3. Rewrite the pdbselect program to use LWP instead of wget. Was it worth the effort?

---

[7] Even if there were none to start with: the redirection to /dev/null silences any complaining messages of protest from rm.

# 10

# *The Protein Databank*

*Working with protein data-files.*

## 10.1 Introduction

The similarity between the amino acid sequence of a "new" protein and one previously characterised can give an indication of the function of the new protein. Sequence search algorithms assume some groups of amino acids have similar functional roles and consequently, occur in both sequences. It is also assumed that these amino acids have similar local structures, where "structure" refers to the amino acids *arrangement in space*. It is these structures that determine the function of a protein. Although these assumptions are far from perfect and ignore many subtle details, they are useful as a working model.

Determining the detailed structure (or more technically, the *conformation*) of a protein is difficult for various technical reasons, especially compared to finding a DNA or amino acid sequence. Despite this, the wonderfully detailed knowledge that can result from determining the structure of the protein often justifies the hard work. The aim of some structural studies is often more than to know how the protein (or other biomolecule) "does what it does", it is also to *alter its function*. A classic application of this knowledge is to design a small molecule that *binds to the protein*, more commonly known as a "drug".

---

*Bioinformatics, Biocomputing and Perl.* Michael Moorhouse and Paul Barry
© 2004 John Wiley & Sons, Ltd   ISBN 0-470-85331-X

## 10.2 Determining Biomolecule Structures

There are many methods used for gaining information about the structure of a biomolecule[1], but the two major methods by which the location of atoms can be determined to a useful accuracy (against the overall shape) are *X-Ray Crystallography* and *Nuclear Magnetic Resonance* (NMR).

As the reader of *Bioinformatics, Biocomputing and Perl* may not be familiar with the underlying methods, the briefest of descriptions are given here. These descriptions are far from complete and a myriad of the finer details are intentionally omitted. The aim of these descriptions is to help the reader appreciate the strengths and weaknesses of the models found in *The Protein Databank* (PDB), so that they can be used intelligently.

Modern protein structures as found in the PDB are of very good quality. Those researchers working in this field go to great lengths to demonstrate that the data is a valid representation of the functional protein it describes.

### 10.2.1 X-Ray Crystallography

This is the most common method by which the 3D spatial locations of atoms within proteins are determined. As its name implies, both *crystals* and *x rays* are needed, although both of these cause problems.

The need for crystals is a major limiting factor for this method, as not all proteins crystallise easily. Some proteins, such as membranes, are nearly impossible to crystallise with current methods. A general rule is: *no crystal, no 3D structure*. Many of the world's: structural biologists, in their attempts to become *crystallographers*, get stressed as they "*Have not found any crystals yet*". Sometimes the process of finding the right set of conditions (initial concentration, salinity, co-factors, temperature and so on) under which a particular protein will crystallise well enough to be useful in a structural study can take years of effort. This seems to be a science-directed art and, in many cases, *luck* plays an important role. At times, suitable conditions for a useful protein crystal are not found in time, the money to do the trials runs out and there is frustration all around. That is life in the life sciences for you.

The use of x rays creates another problem in that they cannot be focused. The overall set-up of an X-Ray Crystallographic structural diffraction study is that x rays are directed at a protein crystal and some of them are said to be *reflected*[2]. *Bragg's Law* states:

$$2d \sin \theta = n\lambda$$

---

[1] In this chapter, the examples used are proteins. Many of the general points and principles apply equally to pure DNA or RNA structures, as well as DNA, RNA and protein complexes.

[2] The correct term is *diffracted*.

and when the conditions of the law are satisfied, *constructive interference* takes place. That is, for certain combinations of x-ray wavelength ($\lambda$), angles of deflection ($\theta$) and distance (d) between the planes of atoms in the protein structure, multiple *reflections* have taken place. These can be detected using, traditionally, electronic devices or photographic film. The pattern observed is specific to the particular set of experimental parameters, so rotating the crystal gives different *diffraction patterns* or sets of *reflections*. A protein crystal is required for two reasons:

1. The reflections from an individual protein are very weak. This means that lots of protein molecules arranged in the *same orientation*, as well as being *reasonably static*, are needed to ensure that the contribution from each atom, in every protein molecule, reinforces each other.

2. The reflection must be intense enough to be detectable.

A major problem with the use of x rays is that they cannot be *focused* using a lens, in a method similar to the way light can be in a microscope. Light cannot be used, as the protein molecules are too small to reflect them. This leads to the so-called *Phase Problem*. The reflections are unfocused images of the protein molecules in the crystal. Correcting this requires the use of complex mathematics to regenerate the actual image of an individual protein. In fact, what *is* observed is the x rays' interaction with the electrons that surround the atomic nuclei in the molecules, not the nuclei themselves[3]. The result of this "complex-maths-lens processing" is an electron density map inside which the protein structure can be fitted.

A series of iterations of *refinement* improves the electron density map, by fitting the atoms that are known to be present within the protein inside the exoskeleton mesh of the electron density map. The result is a series of *structural models* that (hopefully) fit progressively better inside progressively better electron density maps. As the electron density map improves, the reflections that contain more detailed information, that is, those that are observed further from the incident x-ray beam, can be included. At some stage, no more reflections can be observable because of the limitations of the crystal or apparatus used to collect the data. This limit is referred to as the *resolution* of the structure and it refers to the minimum observable distance between two objects in the structure. Any distances below this (such as 1.41 Angstrom bond lengths in a 2-Angstrom resolution structure) are educated guesses based on what has been observed in other molecular structures. For instance, it has been observed that bond lengths between carbon atoms do not change much.

Another important measure is the *R Factor*. This indicates how well the proposed structure matches the observed reflections. The recently devised *Free R*

---

[3] This relates to the "electronic environment" referred to in the NMR discussion that follows.

measure has advantages over *R Factor*, as the calculation of *Free R* uses reflections that are not used in producing the structural model, effectively avoiding the "over-refinement" of the model. A comparison of the *Free R* factor and the resolution is made in one of the PDB parsing examples discussed later in this chapter.

The result of an X-Ray Crystallographic study is a single protein structure in which the variations of individual atoms from their point locations are described by *temperature factors*. Some parts of the protein are distorted by crystallisation. This is especially true of *loops* that in solution "hang out" away from the core of the protein. When it is impossible to identify the actual locations of some amino acids, perhaps due to the fact that the electron density maps are so poor, these atoms are omitted. The electron density map is said to be *disordered* and such "omitted" parts are left out of the reported structure.

The size of the protein that can be studied is many hundreds of amino acid residues. The practical upper limit is set by the complexity of the data.

## 10.2.2 Nuclear magnetic resonance

NMR uses a very different approach to that of X-Ray Crystallography. No crystals are used in the process, and the protein remains in solution throughout the entire experiment. An intense and very linear magnetic field aligns the atomic nuclei of the protein into one of two *spin states*. A series of radio frequency pulses is used to perturb these by "flipping" some of the nuclei from one spin state to the other. As the total amount of energy absorbed is low, the protein remains undamaged and functions as normal. Eventually, the "flipped" spin state of the nuclei realigns to the normal state, emitting a radio frequency pulse as it does so. The timing of this re-emission of energy is determined by the *electronic environment* in which the nucleus is embedded. A feature of this environment is the electrostatic shielding effects of the surrounding nuclei. The nuclei, in addition to the bonds linking them, can be identified by their *spin decay properties*.

A series of preparation pulses can be used to probe different environments within the protein and eventually, a series of *constraints* is produced that describe *which* nuclei exist in *what* type of environments. A series of models is then proposed using the amino acid sequence of the protein, then the observed bond lengths and angles as compared to those observed in other proteins (or small molecules). The models are energy *minimised* using the identified constraints and act as a guide to which of the model structures are most consistent with the experimental data. Over time, some of the models converge upon a set of *similar structures*. These, if all is well, are the *series of configurations* of the protein as observed during the experiment. The models are, quite literally, the protein in motion and, often, the structures partition into two or more sets corresponding to different *functional states*.

As to which is "best", the answer is that they are all consistent with the experimental data and general assumptions about proteins (such as the bond lengths and angles)[4]. Also, as with X-Ray Crystallographic structures, some parts of the structure are more reliable than others. For instance, the number of restraints observed in the loop regions can be so low that the reported structure might be more a function of the minimisation process than of the experimental data. A problem with NMR methods is the size of the proteins that can be studied. Using current techniques, this equates to a maximum of 200 amino acids. This is low compared to the many hundreds of amino acids that can be studied using X-Ray Crystallography.

## 0.2.3 Summary of protein structure methods

The X-Ray Crystallography and NMR systems are complementary in many respects, as both determine, to a high accuracy, the coordinates of the atoms in protein structures. If protein structures determined by X-Ray Crystallography and NMR are compared, they are generally consistent with each other and moreover are biologically plausible. This should give the researcher confidence when using them.

## 10.3 The Protein Databank

The PDB contains a large collection of previously determined biological structures. For inclusion in the PDB, the spatial locations of the atoms have to be determined with sufficient accuracy to usefully describe protein structures. The PDB also includes experimental details of *how* the structure was determined, what publications and other databases to consult for more information on the structure, some "derived data" information (such as notable secondary structure features) and details of any *ill-defined regions*. While this information is meant to be included in the PDB, some of it may be missing, incomplete or - in extreme cases - incorrect for some database entries.

The PDB is one of the oldest bioscience data stores, dating back to 1971. It originally stored the 3D coordinates of protein structures as determined by the X-Ray Crystallography method. Prior to the PDB, structures were typically published in journals, and many researchers re-entered the information *manually* into their computers so as to facilitate further manipulation of them. As can be imagined, this was less than ideal!

The original PDB data-file format adopted was a "flat" textual disk-file that was 80 columns wide.

> **Technical Commentary:** The choice of 80 columns is no accident. Back in 1971, data was stored on paper cards that had to have the information *punched* onto them. Until the mid 1990s, most computer screens were still designed to display 80

---

[4] There might be a structure nominated as "the most representative" in some NMR PDB data-files.

columns, usually with 25 rows of information, so it made sense to restrict the width of PDB disk-files to 80 characters when the format was devised.

Today, the structures in the PDB are determined by either X-Ray Crystallography or NMR. Often, many years of effort go into determining an individual structure. This is reflected in the growth of the number of entries in the PDB over some 30 years. There are currently over 18,000 entries in the database, as shown in Figure 10.1 on page 178.

The structures of some macromolecules, such as the membrane-bound proteins that are thought to make up over 30% of the protein complement of cells, are particularly difficult to obtain using current methods. These macromolecules are poorly represented with less than five "good structures" in the entire databank[5]. Against this background, it is not surprising that the contents of the PDB are somewhat biased towards certain types of proteins. This point is returned to later.

The PDB has been through many changes since its inception. The two most notable are the inclusion of structural data from NMR studies (starting in the mid 1980s) and the transfer of the databank's administration from the original *Brookhaven National Laboratories* (BNL) to *Research Collaboratory Structural Biology* (RCSB) at the end of the 1990s. On the horizon is the adoption of the new `mmCIF` data format, which is designed to replace the legacy PDB flat data-file. It is

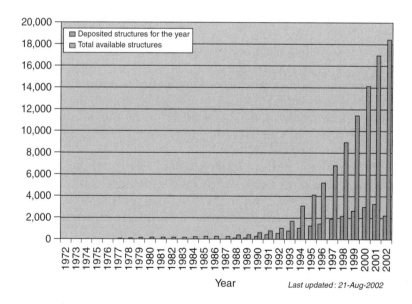

**Figure 10.1** The PDB growth chart.

---

[5] Though the non-membrane domains of some receptors have been determined by cutting off their membrane-bound parts and expressing them separately.

unclear how great or widespread the impact of the introduction of this new data format will be.

Some problems exist with the data in the PDB, mainly in terms of its overall quality and presentation. This is to be expected with a data-store as old as the PDB is. These problems are being addressed with great success by the valiant efforts of those involved in the *Data Uniformity Project*. More recently, there has been a tightening of the PDB procedures relating to the acceptance of structures, and new tools have been developed to help depositors ensure their data is consistent. Even so, be cautious when intending to make a large-scale survey of the biological structures available in the databank. The unwary researcher can fall into many traps, some of which are discussed later in this chapter. The biggest trap awaits those researchers who rely on the PDB *Header Section* information (akin to the *annotation* section in the sequence databases). As this is an important point, let's have a maxim to help keep everyone on the straight and narrow.

**Maxim 10.1** *Beware of anything in the PDB Header Section.*

## 10.4 The PDB Data-file Formats

Data from the PDB is available in one of two formats. To some degree, these formats are *inter-convertible*:

**PDB flat file** – The original, generic and highly unstructured PDB data-file format that is still widely used by researchers. When biologists talk of "PDB files" or "PDB format", they are referring to this data-file format. The current standard format is the 2.3 version.

**mmCIF** – The new PDB data-file format that is designed to offer a highly structured, modern replacement to the original PDB Flat File format. The mmCIF format is often informally referred to as the "new PDB format".

PDB data-files conforming to the mmCIF standard store their data as *key-value pairings*, with additional relationships between the data items defined in a separate data-file. This separation of document structure from document content allows researchers to verify that the data-file is complete and that no unauthorised additions have been made. A considerable amount of work has gone into defining the mmCIF standard, and it is returned to in detail later in this chapter.

For now, let's concentrate on the original PDB format. Newer structures, those added to the PDB after 1996, conform to the v2.3 standard. Despite this, most older structures do not. These "legacy" structures have been painstakingly converted[6] into modern data structures conforming to the mmCIF standard. To

---

[6] By those involved in the PDB Data Uniformity Project.

maintain compatibility with older software, researchers often convert the newer mmCIF data-files into v2.3 of the PDB format.

Within the PDB, a structure is identified by a unique code. The code has two parts: a single number, followed by three letters. Example unique codes are 1AFI and 1LQT. Structure 1AFI is used in the *Tools and Datasets* chapter, as is the NMR structure of "MerP", the mercury ion-binding protein. Structure 1LQT is a high quality X-ray Crystallographic structure entitled "*A covalent modification of NADP+ revealed by the atomic resolution structure of FPRA, a mycobacterium tuberculosis oxidoreductase*".

## 10.4.1 Example structures

For the purposes of explanation, a range of structures is used to illustrate the similarities and differences between PDB data-files. These differences are due mainly to the method by which the structure was determined, in addition to the details of the proteins they describe. The example PDB data-files are as follows:

**1LQT** – A modern, high-resolution "Oxidoreductase" enzyme structure produced using X-Ray Crystallographic techniques. This structure is shown in Figure 10.2 on page 180.

**1M7T** – A modern protein structure of "Thioredoxin" produced using NMR. This structure is shown in Figure 10.3 on page 181.

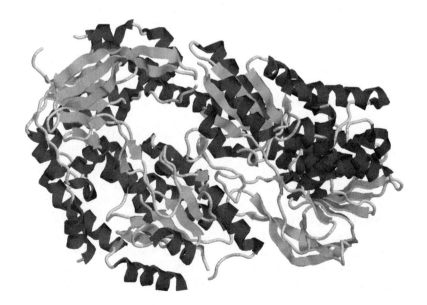

**Figure 10.2** Example PDB structure 1LQT.

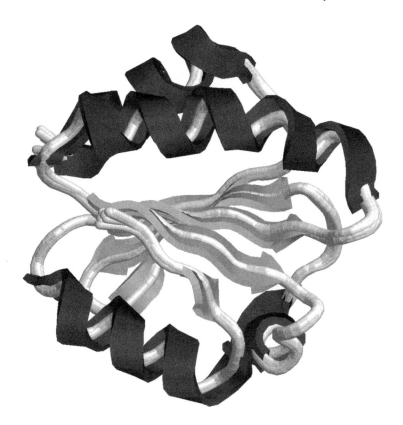

**Figure 10.3** Example PDB structure 1M7T.

Take the time to download the data-files containing 1LQT and 1M7T, as it is helpful to have them available while working through the remainder of this chapter. The overall arrangement of the structures is shown in "cartoon form" in the figures (thanks to the use of the *Open Rasmol* program). These figures highlight the helices (corkscrews) and the sheets (flat ribbons). Note that even these simple molecular graphics help give an appreciation of the structure.

### 10.4.2 Downloading PDB data-files

PDB structure data-files can be downloaded from many web-site locations on the Internet. As described in the previous chapter, the RCSB web-site is always a good place to start:

```
http://www.rcsb.org/pdb/
```

Alternatively, the EBI hosts a European mirror. Follow the links from:

```
http://www.ebi.ac.uk/services/
```

to access the PDB from the EBI.

> **Technical Commentary:** On the Internet, most busy web-sites are geographically replicated a number of times, primarily to lighten the load on the main, central web-site. Such replicated web-sites are known as *mirrors* within the Internet community. It is assumed that users of a busy web-site will contact the mirror geographically closest to their current location, as opposed to always sending requests to the central web-site. The RCSB has a number of such mirrors. Always try to use a mirror close to you: they are listed on the main RCSB web page (http://www.rcsb.org/pdb/).

There is quite a variation in the size of PDB data-files. Some contain no more than a few kilobytes, while others contain many megabytes of data. The NMR entries tend to be large, as instead of just one set of coordinates, they often contain 20 or more.

> **Technical Commentary:** A *kilobyte* is 1024 bytes, where a *byte* is commonly considered to be the amount of space required to hold a single character. A *megabyte* is 1,048,576 bytes (technically 1024 by 1024 bytes). It is common practice to refer to a kilobyte as 1000 bytes and a megabyte as 1,000,000 bytes. Although common, such practice is technically inaccurate and is best avoided.

If a small number of data-files is required, the extensive search facilities provided by the RCSB web-site are an excellent method for finding specific PDB entries. For a more extensive study, perhaps involving many different proteins, it is often more convenient to download a *PDB-Select* non-redundant data set (described in the next chapter).

## 10.5 Accessing Data in PDB Entries

There are some common sections to all PDB entries: those concerned with indexing, bibliographic data, notable features and 3D coordinates. Other sections are radically different from each other, as they depend on the experimental technique (X-Ray Crystallography or NMR) used to determine the structure. Rather than give a verbose description of each subsection, a summary of the most important sections is provided. Note that in a PDB data-file there is a *left-right split* (per line) and a *top-bottom split* (per data-file):

- **Left-right** – The left-most characters (a maximum of nine) on each line indicate what information is present on the right-hand side.

- **Top-bottom** – There is an upper HEADER section that contains the annotation about the structure (top) and a lower *coordinates* section that contains the 3D spatial locations of the atoms in the structure (bottom). The boundary between these is taken as the first "ATOM" or "HETATM" line found in the entry.

A short description of the most important fields in the PDB data-file is presented below. An important point is that the "REMARK" field contains most of the information about the structure in a series of subsections (or sub-remarks). The most important fields include:

**HEADER** – Contains a brief description of the structure, the date and the PDB ID code.

**TITLE** – The title of the structure.

**COMPND** – Brief details of the structure.

**SOURCE** – Identifies which organism the structure came from.

**KEYWDS** – Lists a set of useful words/phrases that describe the structure.

**AUTHOR** – The scientists depositing the structure.

**REVDAT** – The date of the last revision.

**JRNL** – One or more literature references that describe the structure.

**REMARK 1 through REMARK 999** – Details of the experimental methods used to determine the structure are contained in this subsection (see the example in the next section).

**DBREF** – Cross links to other databases.

**SEQRES** – The official amino acid sequence (protein, RNA or DNA) of the structure.

**HELIX/SHEET** – Details of the regions of secondary structure found in the protein.

**ATOM/HETATM** – The 3D spatial coordinates of particular atoms in the protein structure (the "ATOM" lines) or other molecules such as water or co-factors (the "HETATM" lines).

## 10.6 Accessing PDB Annotation Data

There are many examples of parsing data from the HEADER section of PDB data-files, all of which involve pattern matching. Perl is exceptionally good at this. Rather that repeating the same basic procedure over and over again, two representative examples are described in detail in this section. These examples explore:

1. The relationship between the *resolution* of a structure and its *Free R value*, both of which are measures of the quality of the X-Ray Crystallographic structures.

2. The database cross-referencing section used to link to other databases.

### 10.6.1 Free R and resolution

The REMARK tag, type 2 subsection stores *resolution*, whereas the *Free R value* is quoted in REMARK tag, type 3. Here's a small extract from the 1LQT entry:

```
REMARK   2
REMARK   2 RESOLUTION. 1.05 ANGSTROMS.
```

Note that in NMR structures, REMARK tag, type 2 and type 3 are present, but the data in them is "NOT APPLICABLE" for REMARK tag, type 2 and "NULL" or free text for REMARK tag, type 3. This is a historic quirk of the PDB. Originally, the requirement was for these fields to be filled in, which was the case when the PDB contained only crystallographic structures. When NMR structures started to be added, rather than leave the fields out (which in many cases would make more sense), the approach adopted specified that they be set to "NULL". By way of example, consider this "note" from the 1M7T structure's HEADER:

```
REMARK 215 NMR STUDY
REMARK 215 THE COORDINATES IN THIS ENTRY WERE GENERATED FROM SOLUTION
REMARK 215 NMR DATA.  PROTEIN DATA BANK CONVENTIONS REQUIRE THAT
REMARK 215 CRYST1 AND SCALE RECORDS BE INCLUDED, BUT THE VALUES ON
REMARK 215 THESE RECORDS ARE MEANINGLESS.
```

*Structural Refinement* is the process of iteratively fitting the model structure into the electron density map, and details of this refinement are stored in REMARK tag, type 3. Of these, the *Free R value* is very useful, as it measures the agreement between the model and the observed x-ray reflection data. The lower the *Free R Value*, the better the fit between the model and the observed data. Here's an extract:

```
         .
         .
         .
REMARK   3    FIT TO DATA USED IN REFINEMENT.
REMARK   3    CROSS-VALIDATION METHOD          : THROUGHOUT
REMARK   3    FREE R VALUE TEST SET SELECTION  : RANDOM
REMARK   3    R VALUE      (WORKING + TEST SET) : 0.134
REMARK   3    R VALUE            (WORKING SET) : 0.134
REMARK   3    FREE R VALUE                     : 0.153
REMARK   3    FREE R VALUE TEST SET SIZE   (%) : NULL
REMARK   3    FREE R VALUE TEST SET COUNT      : 2200
         .
         .
         .
```

Older structures may lack a *Free R Value*, as it was often not calculated.

A program, called free_res, extracts the *resolution* and *Free R Value* from any PDB data-files contained in a named directory. The entire source code to free_res is as follows:

```perl
#! /usr/bin/perl -w

# free_res - Designed to extract the 'Free R Value' and 'Resolution'
#            quantities from 'PDB data-files' containing structures
#            produced by 'Diffraction'.

use strict;

my $PDB_Path = shift;

opendir ( INPUT_DIR, "$PDB_Path" )
    or die "Error: Cannot read from mmCIF directory: '$PDB_Path'\n";

my @PDB_dir = readdir INPUT_DIR;

close INPUT_DIR;

my @PDB_Files = grep /\.pdb/, @PDB_dir;

foreach my $Current_PDB_File ( @PDB_Files )
{
    my $Free_R;
    my $Resolution;

    open ( PDB_FILE, "$PDB_Path/$Current_PDB_File" )
        or die "Cannot open PDB File: '$Current_PDB_File'\n";

    while ( <PDB_FILE> )
    {
        if ( /^EXPDTA    / and !/DIFFRACTION/ )
        {
            last;
        }
        if ( /^REMARK   2 RESOLUTION/ )
        {
            ( undef, undef, undef, $Resolution ) = split ( " ", $_ );
        }
        if ( /^REMARK   3   FREE R VALUE                   / )
        {
            $Free_R = substr ( $_, 47, 6 );
            $Free_R =~ s/ //g;

            if ( $Free_R =~ /NULL/ or $Resolution eq "" )
            {
                last;
            }
            else
            {
                printf ( "%7s %4.2f %7.3f \n", $Current_PDB_File,
                                 $Resolution, $Free_R );
                last;
            }
        }
    }
    close ( PDB_FILE );
}
```

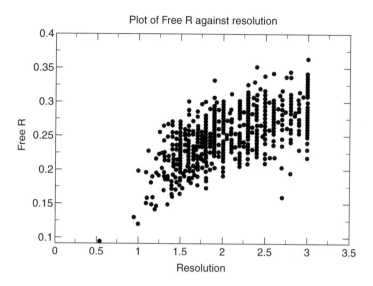

**Figure 10.4** Plotting free R values against resolution.

When executed against a directory containing PDB data-files, specified as a command-line parameter, the `free_res` program checks each data-file in turn as to whether the structure was determined by X-Ray Crystallography. It does this by looking for "DIFFRACTION" in the EXPDTA field. If there's no match, the program skips to the next disk-file. Otherwise it parses (and extracts) the resolution and *Free R values* from the current data-file. Before displaying the results in an "easy-to-parse" format, using Perl's `printf` subroutine, the program checks to see if both the $Free_R and $Resolution scalar variables actually contain data. The idea here is that the output from `free_res` be redirected to a disk-file.

When *Free R* and *Resolution* are plotted against each other, they show a good correlation of 0.666 (Pearson Correlation Coefficient). Figure 10.4 on page 186 presents the plot. This is an improvement on the poorer value of 0.36 between the standard *R value* and *Resolution* as found by others. The reason for the difference between the *R value* and *Free R* factors is multi-factorial, but is mainly due to the difficulty with which a low *Free R factor* can be obtained, relative to a standard *R factor*, from poorer x-ray resolution data.

## 10.6.2 Database cross references

The DBREF subsection gives a list of cross references to other Bioinformatics databases. This makes it easier for researchers to integrate biological datasets. The present deposition policy of the PDB requires that all proteins longer than ten residues *should* be cross referenced. This means that short peptides, which may be synthetic, are excluded.

The second value on the DBREF line is the PDB identifier. By examining this value, researchers and automatic parsing programs can tell to which structure the entry belongs. The DBREF lines from our example structures are shown here:

```
DBREF    1LQT A    1    456   GB     13882996  AAK47528      1     456
DBREF    1LQT B    1    456   GB     13882996  AAK47528      1     456

DBREF    1AFI      1     72   SWS    P04129    MERP_SHIFL   20      91

DBREF    1M7T A    1     66   SWS    P10599    THIO_HUMAN    0      65
DBREF    1M7T A   67    106   SWS    P00274    THIO_ECOLI   68     107
```

To what does SWS and GB from these extracts refer? The PDB publishes a table (reproduced below) of database names and their associated, abbreviated codes. It can be useful to have this table close at hand when working with cross references:

```
-----------------------------------------------------------
Database Name                              Database Code
-----------------------------------------------------------
BioMagResBank                              BMRB
BLOCKS                                     BLOCKS
European Molecular Biology Laboratory      EMBL
GenBank                                    GB
Genome Data Base                           GD
Nucleic Acid Database                      NDB
PROSITE                                    PROSIT
Protein Data Bank                          PDB
Protein Identification Resource            PIR
SWISS-PROT                                 SWS
TREMBL                                     TREMBL
-----------------------------------------------------------
```

The DBREF lines identify the following fields, working from left to right:

- PDB ID code.
- Chain identifier (if needed).
- The start of the sequence.
- Insertion code (absent in all our examples).
- End of the sequence.
- The external database to which the cross reference refers.
- The external database accession code.
- The database external accession name (the more human-memorable version of the accession code in many cases).

- The start, insertion (absent in all our examples) and end of the sequence in the external database.

The field boundary positions can be found in the PDB documentation. By way of illustration, use code similar to the following to extract the structure name, chain, accession code and the external database the accession code refers to from the DBREF line. This code assumes the DBREF is stored in Perl's default variable, $_:

```
my ( $Struct, $Chain, $Dbase, $AC_code ) = ( substr( $_,  7, 4 ),
                                             substr( $_, 12, 1 ),
                                             substr( $_, 26, 6 ),
                                             substr( $_, 33, 8 ) );
$Struct   =~ s/ //g;
$Chain    =~ s/ //g;
$Dbase    =~ s/ //g;
$AC_code  =~ s/ //g;
```

The four substitutions "clean up" the parsed data by removing any and all unwanted space. Note that for the 1M7T structure, the start and stop positions have *not* been extracted. This structure is a *chimera* between the two example entries, producing the following results if printed:

```
'1M7T', 'A', 'SWS', 'P10599'
'1M7T', 'A', 'SWS', 'P00274'
```

It is worth mentioning that the 1AFI structure contains a *Heavy Metal Associated sequence motif* that is indexed in the PROSITE database as PS01047. The depositors of the structure knew about this motif as it gives the protein its mercury ion scavenging ability. This also explains why much effort was expended on determining the protein's structure. A SITE entry is included later in the data-file, but a database reference to PROSITE is not. Be aware that just because a database cross-reference field is absent does not mean that a reference does not exist.

## 10.6.3 Coordinates section

The coordinate data for the locations of atoms in the macromolecular structure is straightforward, especially when compared to the annotation contained in the HEADER section of the PDB data-file. Recall that while the coordinates are presented as points in space, the atoms they represent are actually *in motion*. In crystallographic structures, *isotropic B-factors*, commonly referred to as "Temperature Factors", give us an idea of the vibration of the molecule. For very high-resolution structures, *Anisotropic Temperature Factors* may be included in the ANISOU lines. These provide an idea of the vibration of the molecule in the directions of the coordinate axes. In NMR structures, the variation in position of a particular atom between different models in the ensemble can be used as a similar measure of motion or as an indication of the error between the minimisation models. It is sometimes easy to tell the difference, while other times it is not.

Another major difference found in NMR structures is the inclusion of the ensemble of models delimited by the MDL and ENDMDL lines. There are some entries that contain a single NMR structure, but this is only the most representative model; others have a nominated "most representative structure". Here is an example from 1M7T:

```
REMARK 210
REMARK 210 BEST REPRESENTATIVE CONFORMER IN THIS ENSEMBLE : 21
REMARK 210
```

Referring to the 1LQT x-ray structure, an extract of lines from the coordinate section looks like this:

```
ATOM      1  N   ARG A   2      26.318  -8.010  39.090  1.00 20.71           N
ANISOU    1  N   ARG A   2      2040   3071   2755    114   -339   -393      N
ATOM      2  CA  ARG A   2      25.150  -8.702  38.505  1.00 18.85           C
ANISOU    2  CA  ARG A   2      2029   2677   2455     67   -321   -209      C
ATOM      3  C   ARG A   2      24.846  -8.176  37.123  1.00 17.23           C
ANISOU    3  C   ARG A   2      1689   2429   2429    143   -282   -258      C
ATOM      4  O   ARG A   2      25.151  -7.048  36.775  1.00 18.14           O
          .
          .
          .
TER    7215      GLY A 456
ATOM   7216  N   ARG B   2     -19.423  25.709   6.980  1.00 21.57           N
ANISOU 7216  N   ARG B   2      2476   3012   2707   -165   -370     95      N
ATOM   7217  CA  ARG B   2     -18.718  26.510   8.024  1.00 19.01           C
ANISOU 7217  CA  ARG B   2      2127   2672   2424    -63   -285     91      C
ATOM   7218  C   ARG B   2     -17.250  26.207   8.002  1.00 17.22           C
ANISOU 7218  C   ARG B   2      1955   2392   2196    -91   -299    121      C
ATOM   7219  O   ARG B   2     -16.851  25.158   7.535  1.00 18.15           O
          .
          .
          .
TER   14289      GLY B 456
HETATM14290  C   ACT  1866    -13.075   1.733  10.218  1.00 27.25           C
ANISOU14290  C   ACT  1866     3493   3560   3299    -39    -36    -44      C
          .
          .
          .
CONECT14290142911429214293
CONECT1429114290
CONECT1429214290
TER
          .
          .
          .
CONECT1469014663
MASTER      389    0   15   46   38    0    0  620280    2  401   72
END
```

Likewise for the 1M7T NMR structure, and extract of the coordinates looks like this:

```
MODEL        1
ATOM      1  N   MET A   1       3.110  -4.682  -3.025  1.00  0.00           N
ATOM      2  CA  MET A   1       2.546  -3.712  -2.053  1.00  0.00           C
ATOM      3  C   MET A   1       1.134  -3.295  -2.450  1.00  0.00           C
ATOM      4  O   MET A   1       0.882  -2.130  -2.758  1.00  0.00           O
ATOM      5  CB  MET A   1       3.466  -2.491  -2.002  1.00  0.00           C
ATOM      6  CG  MET A   1       3.781  -1.903  -3.370  1.00  0.00           C
ATOM      7  SD  MET A   1       4.256  -0.166  -3.285  1.00  0.00           S
ATOM      8  CE  MET A   1       6.004  -0.307  -2.920  1.00  0.00           C
ATOM      9 1H   MET A   1       2.906  -4.327  -3.980  1.00  0.00           H
ATOM     10 2H   MET A   1       2.650  -5.601  -2.859  1.00  0.00           H
ATOM     11 3H   MET A   1       4.134  -4.738  -2.858  1.00  0.00           H
ATOM     12  HA  MET A   1       2.517  -4.178  -1.079  1.00  0.00           H
ATOM     13 1HB  MET A   1       2.996  -1.724  -1.405  1.00  0.00           H
ATOM     14 2HB  MET A   1       4.397  -2.778  -1.536  1.00  0.00           H
ATOM     15 1HG  MET A   1       4.596  -2.461  -3.807  1.00  0.00           H
ATOM     16 2HG  MET A   1       2.907  -1.993  -3.998  1.00  0.00           H
ATOM     17 1HE  MET A   1       6.344  -1.302  -3.167  1.00  0.00           H
ATOM     18 2HE  MET A   1       6.169  -0.120  -1.869  1.00  0.00           H
ATOM     19 3HE  MET A   1       6.553   0.416  -3.505  1.00  0.00           H
ATOM     20  N   VAL A   2       0.215  -4.256  -2.446  1.00  0.00           N
  .
  .
  .
TER    1659      VAL A 107
ENDMDL
MODEL        2
ATOM      1  N   MET A   1       2.750  -6.779  -1.627  1.00  0.00           N
ATOM      2  CA  MET A   1       2.487  -5.475  -2.290  1.00  0.00           C
  .
  .
  .
TER    1660      VAL A 107
ENDMDL
```

In each ATOM line, the fields[7] are as follows:

```
COLUMNS        DATA TYPE       FIELD       DEFINITION
---------------------------------------------------------------------
 1 -  6        Record name     "ATOM(s)
 7 - 11        Integer         serial      Atom serial number.
13 - 16        Atom            name        Atom name.
               Character       altLoc      Alternate location indicator.
18 - 20        Residue name    resName     Residue name.
22             Character       chainID     Chain identifier.
23 - 26        Integer         resSeq      Residue sequence number.
27             AChar           iCode       Code for insertion of residues.
31 - 38        Real(8.3)       x           Orthogonal coordinates for X in
                                           Angstroms.
39 - 46        Real(8.3)       y           Orthogonal coordinates for Y in
```

---

[7] These are extracted from the PDB's on-line documentation.

```
                                          Angstroms.
47 - 54      Real(8.3)       z            Orthogonal coordinates for Z in
                                          Angstroms.
55 - 60      Real(6.2)       occupancy    Occupancy.
61 - 66      Real(6.2)       tempFactor   Temperature factor.
73 - 76      LString(4)      segID        Segment identifier, left-justified.
77 - 78      LString(2)      element      Element symbol, right-justified.
79 - 80      LString(2)      charge       Charge on the atom.
-----------------------------------------------------------------------------
```

Some fields are separated by whitespace, while others are not. Note that there is space between the second and third columns, while there is none between the last three *LString* fields. This can make the parsing of the data more difficult than it would normally be, although Perl's `substr` subroutine can work wonders here.

## 0.6.4 Extracting 3D coordinate data

Extracting coordinate data from PDB data-files, despite the lack of whitespace, is straightforward. The technique involves extracting the three substrings from each line that contains the X, Y and Z coordinates. Assuming the data is in $_, three invocations of Perl's `substr` subroutine do the trick:

```perl
my ( $X, $Y, $Z ) = ( substr( $_, 30, 8 ),
                      substr( $_, 38, 8 ),
                      substr( $_, 46, 8 ) );
```

The X, Y and Z coordinates are now held in appropriately named scalar variables for later use by a program. It is also good practice to remove any additional (and unwanted) whitespace from the three variables. The standard technique is demonstrated in the `simple_coord_extract` program, which follows:

```perl
#! /usr/bin/perl -w

# simple_coord_extract <PDB File> - Demonstrates the extraction of
#                                   C-Alpha co-ordinates from a PDB
#                                   data-file.

use strict;

while ( <> )
{
    if ( /^ATOM/ && substr( $_, 13, 4 ) eq "CA  " )
    {
        my ( $X, $Y, $Z ) = ( substr( $_, 30, 8 ),
                              substr( $_, 38, 8 ),
                              substr( $_, 46, 8 ) );

        $X =~ s/ //g;
        $Y =~ s/ //g;
        $Z =~ s/ //g;
```

```
        print "X, Y & Z: $X, $Y, $Z\n";
    }
}
```

This program binds against the "ATOM" at the start of the line, in addition to a value of "CA " at position 13 in the line. This latter test ensures that the atom is of type carbon-alpha. When both tests pass, that is, the pattern is found and the line represents a *carbon-alpha* atom, the X, Y and Z coordinates of the atom are extracted from the line as a result of the three invocations of `substr`. The resulting scalar variables ($X, $Y and $Z) have any spaces removed from them in the three substitution statements. Finally, the coordinates are displayed on STDOUT. When executed, the `simple_coord_extract` produces this output for the 1LQT structure:

```
X, Y & Z: 25.150, -8.702, 38.505
X, Y & Z: 23.675, -8.497, 35.069
X, Y & Z: 20.747, -6.252, 34.332
X, Y & Z: 17.545, -8.297, 34.292
X, Y & Z: 15.182, -7.484, 31.454
X, Y & Z: 11.736, -8.952, 30.942
X, Y & Z: 10.261, -9.014, 27.451
X, Y & Z:  6.507, -9.548, 27.173
```

Note that the program makes no attempt to test for the *protein chain marker*. Consequently, in the case of 1LQT, all of the coordinates for both the A and B chains are displayed.

## 10.7 Contact Maps

The `simple_coord_extract` program is amended in this section to create a *Contact Map*. In a contact map, the distances between *all* the amino acids are calculated (using the standard Pythagoras equation), then those within a certain distance of each other are marked with an "O" character. Those outside the distance are marked with a space character.

One aspect to consider is whether this is computationally possible: is the computer being asked to do too much? Calculating the distances between *all* possible amino acids seems to be complicated. How many calculations need to be performed? How much memory is needed? Will the program become much more complicated?

Although consideration of these questions is reasonable, there is no need to panic. Proteins at the level of abstraction of fixed 3D spatial coordinates are *computationally small*. As there are 450 carbon-alpha atom points in the test protein, there are 450 by 450 potential distance calculations, which gives a total of 202,500. Although large, performing this number of calculations is practical using modern PCs. It is also possible to omit half the calculations, as the distance from amino acid number 4 to amino acid number 5 is the same as that from number 5 to number 4. Also, it is possible to omit the diagonal, which calculates the distances between the same amino acid and itself which is, unsurprisingly, zero.

The strategy used is an extension of those from earlier programs: the "CA" (carbon-alpha) atoms are extracted from the PDB data-file. These are loaded into memory and a loop iterates over them, calculating the distance to a currently nominated *reference* atom. A second nested (or inner) loop changes this reference point as required. A test verifies whether the two amino acids are closer than a particular distance and, if they are, prints a "O" marker, otherwise a space is printed[8]. For the purposes of demonstration, a distance of 12 Angstroms is used. This value is used by many other contact maps and, while somewhat arbitrary, is close to the maximum distance that one part of a protein can directly affect another part. Additionally, it is close to the maximum distance across two closely packed secondary structures, such as $\alpha$ helices. Here is the entire source code to the Contact_map.pl program, which implements this strategy:

```perl
#! /usr/bin/perl -w

# Contact_map.pl - based on the CA_dist_calc.pl program.  Produces a
#                  triangular diagram of all the distances between c-alpha
#                  atoms under a certain threshold.
#
# Usage: Contact_map.pl <PDB FILE> [Chain]

use strict;

use constant  CONTACT_DEFINITION => 12;

my $Chain = "*";

my $Previous_Res = '';

if ( $#ARGV == -1 )
{
    die "Usage: CA_dist_calc.pl <PDB FILE> [Chain]\n";
}
elsif ( $#ARGV == 1 )
{
    $Chain = pop @ARGV;
}

my %Atoms;

my @Res_List;

while ( <> )
{
```

---

[8] This strategy, rather conveniently, preserves any existing spacing.

```perl
        if ( /^ENDMDL/ or /^TER/ )
        {
            last;
        }

        if ( !/^ATOM/ or substr( $_, 13, 3 ) ne "CA " )
        {
            next;
        }

        if ( ( substr( $_, 21, 1 ) ne $Chain ) and  ( $Chain ne "*" ) )
        {
            next;
        }

        my $Res_Number = substr( $_, 22, 4 );

        if ( $Res_Number eq $Previous_Res )
        {
            next;
        }
        else
        {
            $Previous_Res = $Res_Number;
        }

        $Res_Number =~ s/ //g;

        push @Res_List, $Res_Number;

        my ( $X, $Y, $Z ) = ( substr( $_, 30, 8 ),
                              substr( $_, 38, 8 ),
                              substr( $_, 46, 8 ) );

        $X =~ s/ //g;
        $Y =~ s/ //g;
        $Z =~ s/ //g;

        $Atoms{ $Res_Number }{ X } = $X;
        $Atoms{ $Res_Number }{ Y } = $Y;
        $Atoms{ $Res_Number }{ Z } = $Z;
    }

print "Number of Residues: ", $#Res_List+1, "\n";

foreach my $Current_Res_Column ( @Res_List )
{
    printf "%03d: ", $Current_Res_Column;

    foreach my $Current_Res_Row ( @Res_List )
    {
        my $Dist = sqrt( ( $Atoms{ $Current_Res_Column }{ X } -
                           $Atoms{ $Current_Res_Row }{ X } ) ** 2 +
                         ( $Atoms{ $Current_Res_Column }{ Y } -
                           $Atoms{ $Current_Res_Row }{ Y } ) ** 2 +
                         ( $Atoms{ $Current_Res_Column }{ Z } -
                           $Atoms{ $Current_Res_Row }{ Z } ) ** 2 );
```

```
            if ( $Dist < CONTACT_DEFINITION )
            {
                print "O";
            }
            else
            {
                print " ";
            }
        }
        print "\n";
    }
```

The Contact_map.pl program is executed against the 1LQT structure with the following command-line:

    perl  Contact_map.pl  pdb/1LQT.pdb

The first 25 lines of output are shown here:

```
Number of Residues: 452
002: 0000                             00000
003: 00000                  00   00   0000000
004: 000000                  00 00    0000000
005: 0000000           0   00    0    0000000
006:  0000000       0   00  000       0000000
007:    00000000       00 00   0      0000000
008:       0000000000000000 0         0000000
009:        00000000000 00             0000000    000
010:          0000000000000              000000000000
011:          0000000000000              0000000000000000         0   00 0
012:           000000000000                 00000 0000000000      0 00 000
013:           0000000000000                      000000 0        000000000 0
014:           00000000000000                 0       00000                0
015:         00000000000000000000          00000      00000                0  0
016:         00000000000000000000          00000     000 0            00000000
017:        0 000000000000000                 0         0             00 00000
018:         000000000000000000000            000                          0 0
019:         000000000000000000000000       000000                         00000
020:     0       000000000000000            000 0                         00000
021:    0         000000000000000            0 0                              0
022:     000000      000000000000000    000000                                0
023:     0000         000000000000      000000                               00
024:                    00000000000000 00000                                  0
025:       0            000000000000000000
```

Examine the grouping of the zeroes along the diagonal. These indicate that amino acids close together in sequence are close together in physical space, too. Of more interest are the off-diagonal contacts, which show how the protein has folded back on itself and which part associates with which. A far better representation than this textual printout is to plot an image using Perl's GD module and the gdlib library, as described later in the *Data Visualisation* chapter. By way of a taster of what is possible, a preview of the image is shown in Figure 10.5 on page 196.

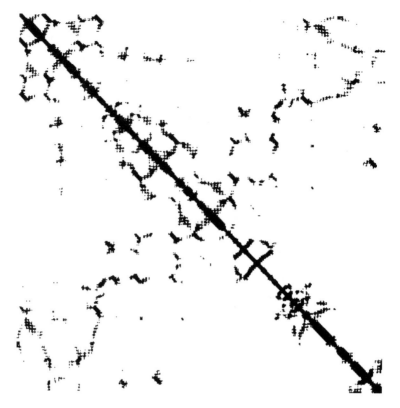

**Figure 10.5** The graphic image contact map.

## 10.8 STRIDE: Secondary Structure Assignment

In this section, the *STRIDE* program[9], maintained by *Dmitrij Frishman*, is used to find the *secondary structural elements* in the example proteins. The results should be similar to those in the HELIX, SHEET and TURN subsections of the HEADER annotation for the PDB data-file. Why go to the trouble of using *STRIDE* when the information is already available? There are a number of reasons:

1. The annotation may be missing or was, for some reason, never generated.

2. It is often easier to run *STRIDE* on the structure than reconstitute the assignments from the HEADER section.

3. The *STRIDE* output has a residue-by-residue assignment (as described below).

---

[9] Available from the http://mips.gsf.de/mips/staff/frishman/ web-site.

4. *STRIDE* can find "turns" that exist in a structure that are often not listed in the HEADER section.

5. *STRIDE* can produce extra derived information as part of its output. For example, the location of hydrogen bonds, the dihedral angles in backbone or the solvent accessibility. *STRIDE* can also report the amino acid sequence in protein.

> **Maxim 10.2** *It is often easier and desirable to regenerate database annotation than trawl through entries reconstituting the annotation using custom code.*

Do not assume that because it is not acknowledged in the database annotation the information is absent from the entire data set. Often, data can be found using better analysis tools.

## 0.8.1 Installation of STRIDE

The installation of *STRIDE* is straightforward. Either download one of the many pre-compiled binaries or compile the program from source. As the compilation process is standardised under Linux, compiling from source is often preferred. To do so, download the source code data-file, then decompress the archive using this command-line:

```
tar -zxvf stride.tar.gz
```

Change into the newly created `stride` directory, then type `make`:

```
cd stride
make
```

Messages will appear on screen as the compilation process starts. Assuming success, a new executable, also called `stride`, is created. Issue the following command to execute *STRIDE*:

```
./stride
```

## 10.9 Assigning Secondary Structures

*STRIDE*, which is short for "STRuctual IDEntification", was originally created by *Dmitrij Frishman* and *Patrick Argos*. It automatically finds secondary structure elements in proteins using a set of supplied coordinates.

> **Technical Commentary:** Another commonly used algorithm in this area is *DSSP*, short for "Define Secondary Structure of Proteins", which was created by *Wolfgang*

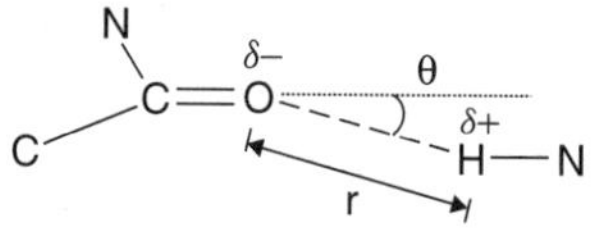

**Figure 10.6** Simplified definition of a hydrogen bond.

Kabsch and Christian Sander. DSSP is now maintained by *Elmar Krieger* at CMBI, in Nijmegen, the Netherlands[10]. It is the personal preference of the researcher as to which to use. Your authors decided to cover *STRIDE* because of its convenient downloading. Contrast the free download of *STRIDE* to *DSSP*, which requires potential users to complete licensing forms, then submit a request which must (some time later) be processed. Non-academic users may also have to pay a licence fee.

*STRIDE* works by identifying *hydrogen bonds* within the structure.

Hydrogen bonds form when a hydrogen atom attached to a donor atom is attracted by an acceptor atom because of the partial charge present on the hydrogen and the acceptor, as shown in Figure 10.6 on page 198. Despite the use of the name "bond", it is really a loose association compared to the other covalent bonds that are present in protein structures. In the protein backbone, the donors are typically carbonyl oxygen atoms, and the hydrogen is attached to amide nitrogen atoms.

With reference to Figure 10.6, the partial negative charge on the oxygen and the partial positive charge on the hydrogen result in the formation of a hydrogen bond if the distance (r) and angle $\theta$ are *realistic*.

The hydrogen bonds are diagnostic of the type of structure: in the $\alpha$ Helix they bond between successive turns of the Helix and in the "turn", there is one across the ends of the turn. In case of the $\beta$ sheet, hydrogen bonds form between two *strands*, in this case, between residues 3 and 9 and between residues 74 and 80.

First *STRIDE* searches for characteristic hydrogen bonding patterns in the protein structure, as shown in Figure 10.7 on page 199. This molecular graphic, as produced by *Open Rasmol*, shows four different types of secondary structure. For simplicity, the cartoon representation of the backbone is used with the hydrogen bond between atoms, which are shown by black rods. Residues 21 to 28 form an $\alpha$ Helix, 28 to 31 and 31 to 33 form a generic Coil and 33 to 38 form a piece of a $\beta$ sheet. For a particular amino acid K, these are:

$\alpha$ **Helices** – There is a hydrogen bond between K and K + 4 as well as one between K + 1 and K + 5.

$\beta$ **Sheets** – *STRIDE* searches for hydrogen bonds that form *bridges* between different parts of the protein structure: residues K and K + 1 must bond to at

---

[10] See http://www.cmbi.kun.nl/gv/dssp/ for more details.

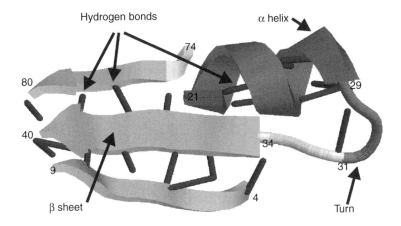

**Figure 10.7** Example of secondary structure elements in proteins.

least two consecutive amino acids somewhere else in the backbone. Consequently, the two regions can be quite distantly related in terms of amino acid sequence.

The assessment of other less common secondary structure types: $\pi$ helices, 3–10 helices or turns is performed in a similar way[11]. Any patterns of hydrogen bonds that are unrecognised are referred to as a *Coil*, which in many ways is the "catch-all" state: anything not recognised as anything else gets called *Coil* in structural biology terminology. *STRIDE* then attempts to extend the structural element along the chain.

To improve the accuracy at the end of the structural elements, *STRIDE* uses the dihedral angles $\psi$ and $\phi$ of the protein backbone, as shown in Figure 10.8 on page 200. These are specific examples of *Torsion Angles*, that measure the rotation around a particular bond with reference to four atoms.

With reference to Figure 10.8, the rotational angle about the N-C$\alpha$ bond ($\phi$) is calculated with reference to the C-N-C$\alpha$-C atoms; the rotational angle about the C$\alpha$-C bond (called $\psi$) is calculated with reference to the N-C$\alpha$-C-N atoms. The rotational angle around the C-N bond measured by the $\omega$ angle is calculated with reference to the C$\alpha$-C-N-Carbon-$\alpha$, and it varies little because of the resulting planar structure.

If the $\phi$ and $\psi$ angles found in $\alpha$ helical or beta sheets are plotted against each other, they group together in certain regions. This type of diagram is called a *Ramachandran Plot*. It is often used as a progress measure during the process of structural refinement, as certain regions correspond to more energetically favourable conformations. This also means that assessment of modern structures

---

[11] Curious readers are referred to the original *STRIDE* and *DSSP* papers for details. Both programs use broadly the same system.

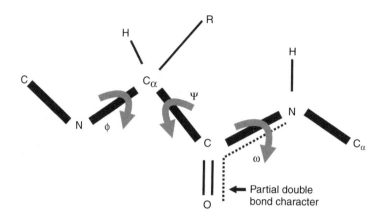

**Figure 10.8** Definition of dihedral angles in the backbone of protein structures.

using Ramachandran Plots for correctness might not be useful, as they all have good Ramachandran angles! How else would the refinement programs have proposed the structure you see[12]?

Many protein structure analysis programs, including *DeepView*, can generate these plots. Built into *STRIDE* is a probability map constructed in the same way from the observation of the dihedral angles in real proteins. This is then used in a scoring procedure to link particular combinations of $\omega$, $\phi$ and $\psi$ angles with structural states. The hydrogen bonding patterns and the Ramachandran probabilities are weighted and combined together, such that a good hydrogen bonding potential can compensate for a less than optimal geometry and vice-versa. If neither score exceeds a threshold, and none of the rules for other secondary structures indicate an alternative, then the "catch-all" designation of *Coil* is used.

### 10.9.1 Using STRIDE and parsing the output

Running *STRIDE* without any input displays the program's usage information. The message indicates that the user is expected to specify an input data-file, as follows:

```
You must specify input file

Action: secondary structure assignment
Usage: stride [Options] InputFile [ > file ]
Options:
   -fFile      Output file
   -mFile      MolScript file
```

---

[12] By ignoring the experimental evidence in favour of a nice Ramachandran Plot perhaps? Be careful to keep your training and validation datasets separate.

```
-o            Report secondary structure summary Only
-h            Report Hydrogen bonds
-rId1Id2..    Read only chains Id1, Id2 ...
-cId1Id2..    Process only Chains Id1, Id2 ...
-q[File]      Generate SeQuence file in FASTA format and die

Options are position and case insensitive
```

Executing *STRIDE* against the 1LQT structure, with the requirement that just the "A" chain be processed, is accomplished with a command like this:

```
stride  -cA  1lqt.pdb
```

The resulting output contains a number of sections.

The first section contains a header section containing instructions on how to cite the program and the methods it uses, which are identified by the REM tag. This is followed by information in a very similar form to the original PDB data-file, and it includes the name and date of the structure (HDR); the Compound (CMP), the Source (SRC) and the Authors (AUT):

```
REM  ----------------------------------------------------------------  1LQT
REM                                                                    1LQT
REM  STRIDE: Knowledge-based secondary structure assignment            1LQT
REM  Please cite: D.Frishman & P.Argos, Proteins 23, 566-579, 1995     1LQT
REM                                                                    1LQT
REM  Residue accessible surface area calculation                       1LQT
REM  Please cite: F.Eisenhaber & P.Argos, J.Comp.Chem. 14, 1272-1280, 1993 1LQT
REM               F.Eisenhaber et al., J.Comp.Chem., 1994, submitted   1LQT
REM                                                                    1LQT
REM  ---------------------- General information ---------------------- 1LQT
REM                                                                    1LQT
HDR  OXIDOREDUCTASE                          13-MAY-02   1LQT          1LQT
CMP  ...
SRC  ...
AUT  ...
```

The next section contains a summary of the secondary structure allocation. Each CHN line marks the start of a new summary for a particular chain. Each part of the summary is split across pairs of lines. The first SEQ tag identifies the amino acids. The second STR tag line uses a one-letter code to indicate the corresponding structural state, of which the most common are "H" for $\alpha$ helix, "E" for extended (one strand of a sheet), "T" for turn, "G" for 3-10 helix and a space character for *Coil*:

```
REM  -------------------- Secondary structure summary ---------------- 1LQT
REM                                                                    1LQT
CHN  ../exp_st A                                                       1LQT
REM                                                                    1LQT
REM                  .         .         .         .         .         1LQT
SEQ  1    RPYYIAIVGSGPSAFFAAASLLKAADTTEDLDMAVDMLEMLPTPWGLVRS     50    1LQT
STR       EEEEEE   HHHHHHHHHHHHHHHHHTTTT EEEEEE          HHHH          1LQT
REM                                                                    1LQT
```

```
REM              .         .         .         .         .             1LQT
SEQ  51  GVAPDHPKIKSISKQFEKTAEDPRFRFFGNVVVGEHVQPGELSERYDAVI   100       1LQT
STR      H TTTTTGGGGGGGHHHHHHHTTTEEEEETTTTTTTTTHHHHHHHTTEEE              1LQT
```

Each structural element in this section is then listed in LOC tagged lines. These correspond to those residues displayed by Figure 10.8. There are some noticeable differences between the assignments in the PDB data-file, as created by the depositors, and those produced by *STRIDE*. This is due to slightly different definitions of the secondary structure used, especially at the ends of the elements. However, all the items are present, and the variation is not great:

```
LOC  AlphaHelix   PRO   13 A    THR   28 A                              1LQT
LOC  310Helix     LYS   59 A    LYS   65 A                              1LQT
LOC  Strand       TYR    4 A    VAL    9 A                              1LQT
LOC  TurnII       THR   29 A    LEU   32 A                              1LQT
LOC  Strand       MET   34 A    LEU   39 A                              1LQT
LOC  Strand       PHE   76 A    GLY   80 A                              1LQT
```

The third and final section provides a detailed description of each residue, providing an easy-to-parse space-delimited format with actual space characters between the columns (unlike the PDB data-files the data was derived from). The remark line should (hopefully) explain what most of the fields contain. The 4th and 5th fields need further explanation. The 4th is the residue *number* as reported in the PDB data-file. The 5th is an ordinal number, which starts at one and increments by one per residue processed, and is created by *STRIDE*. The 10th "Area" field identifies the area of the amino acid exposed to the solvent:

```
REM   --------------- Detailed secondary structure assignment-------------  1LQT
REM                                                                         1LQT
REM    |---Residue---|   |--Structure--|   |-Phi-|   |-Psi-|   |-Area-|     1LQT
ASG   ARG A    2    1   C          Coil     360.00    156.52    121.3       1LQT
ASG   PRO A    3    2   C          Coil     -75.72    161.36     35.7       1LQT
ASG   TYR A    4    3   E          Strand   -71.26    145.24     21.2       1LQT
         .
         .
         .
ASG   GLY A   12   11   C          Coil     -83.55   -168.87      8.7       1LQT
ASG   PRO A   13   12   H       AlphaHelix  -53.20    -47.88     16.0       1LQT
ASG   SER A   14   13   H       AlphaHelix  -63.16    -38.61      8.3       1LQT
```

This format is so straightforward that to extract data from it using a bespoke Perl program seems excessive. As an alternative, the gawk utility can be used from the command-line to quickly parse the *STRIDE* data-file and create a custom Ramachandran Plot, as shown here:

```
gawk   '/^ASG/ {print $8 " " $9}'   1lqt.A.stride
```

The gawk utility detects the ASG tag at the start of the line[13] and prints out the 8th and 9th fields. The surrounding single-quote marks are required to prevent the

---

[13] Note the use of a regular expression.

operating system's *shell* from incorrectly interpreting the gawk program options. Here is an extract of the results produced by the execution of gawk:

```
 360.00 156.52
 -75.72 161.36
 -71.26 145.24
-111.08 119.10
-118.65 131.78
   .
   .
   .
```

This example can be extended to extract a subset of the data. To extract just those residues involved in "Strand" or "AlphaHelix" states, use command-lines like these:

```
gawk   '(/^ASG/ && /Strand/) {print $8 " " $9}'   1lqt.A.stride

gawk   '(/^ASG/ && /AlphaHelix/) {print $8 " " $9}'   1lqt.A.stride
```

Figure 10.9 on page 203 shows the grouping of the close to 450 amino acids. The figure shows a distinct grouping into certain regions. Those angles resulting from "AlphaHelix" are surround by a circle and those from "Strand" are shown surrounded by diamonds. This gives an impressive demonstration of the grouping of the dihedral angles in a two-dimensional virtual space and the power of *derived data*.

Almost any protein structural analysis program will create Ramachandran Plots, as they are a fundamental diagnostic test used to determine if a protein

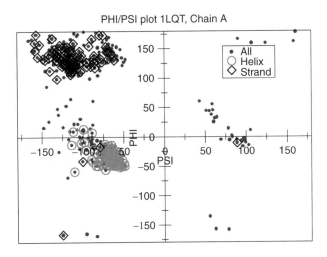

**Figure 10.9** Ramachandran plot of dihedral angles of chain A from structure 1LQT.

structure is correct. In this context, a Ramachandran Plot may reveal parts of the structure that are outside of the low-energy areas. This can be taken as an indication of a region of poor structure or that there is a good reason for the conformation to be as it is: for example, it is structurally important to the function of the protein.

## 10.9.2 Extracting amino acid sequences using STRIDE

Another common use of *STRIDE* is to extract the primary structure, that is, the amino acid sequence, from a PDB data-file. This is also straightforward, and is invoked using the -q command-line switch. The results are produced in FASTA format[14]. The following command-line:

```
stride -q 1lqt.pdb
```

produces this output:

```
>1lqt.pdb A    452    1.050
RPYYIAIVGSGPSAFFAAASLLKAADTTEDLDMAVDMLEMLPTPWGLVRSGVAPDHPKIK
   .
   .
   .
>1lqt.pdb B    454    1.050
RPYYIAIVGSGPSAFFAAASLLKAADTTEDLDMAVDMLEMLPTPWGLVRSGVAPDHPKIK
   .
   .
   .
```

If a particular chain is to be reported on, the -c option is used to specify which chain is required, as follows:

```
stride -cA -q 1lqt.pdb
```

resulting in the following output:

```
>1lqt.pdb A    452    1.050
RPYYIAIVGSGPSAFFAAASLLKAADTTEDLDMAVDMLEMLPTPWGLVRSGVAPDHPKIK
   .
   .
   .
```

This simplicity makes STRIDE the preferred method for obtaining the amino acid sequence from a PDB data-file.

---

[14] Described elsewhere in *Bioinformatics, Biocomputing and Perl*.

## 0.10 Introducing the mmCIF Protein Format

The mmCIF data format is intended to be a replacement for the legacy PDB data-file format. Designed using modern data management techniques, the contents of an mmCIF data-file are expressed in a series of *key-value pairings*. The meaning of these pairings is stored in a separate data-file called a "Dictionary", which, in essence, allows mmCIF to store data[15].

mmCIF is designed to store structures created as a result of crystallography investigations. While the storage of atom locations is the same as those in the NMR structures, the experimental details associated with the production of the structures differ. Even today, the additions needed for the mmCIF dictionaries to support this extra NMR data are still the subject of much debate. If mmCIF data-files need to be accessed directly, software libraries to process them do exist, although these are still under active development.

The mmCIF data-file format is not designed to be "easily" read by humans; computer programs are the main target audience. A side effect of this decision is that the strict format definition makes it easy to "unwind" the data-file into a software data structure. From a quality control perspective, this makes the absence or addition of data easy to verify. The mmCIF data-file format is also that which is the 'cleaned-up" version of the PDB from the *PDB Uniformity Project* is being made available.

The decision as to which format to use, either mmCIF or the legacy PDB data-file format, depends on the requirements of the researcher. For now, a good rule of thumb is to use the legacy PDB data-files, unless a very specialist application demands otherwise. One interesting caveat relates to accessing the 3D coordinate positions of atoms in the structure, together with their type/chain/residue designations. Unlike the PDB data-file, the mmCIF structure has fields that are space delimited with absent information marked by a special "spacer" character (such as "."). Therefore, the simple Perl statement

```
@Fields = split( ".", $_ );
```

will split, for example, the "ATOM" line into separate entries in the @Fields array. It is then possible to use a statement like this:

```
$X_Coordinate = $Field[ 7 ];
```

---

[15] This is similar to XML's Document Type Definition (DTD), upon which mmCIF is based.

to access the coordinates. For applications in which data is used or needs to be converted *en masse* to another data format, the precision inherent in the mmCIF structures can be helpful.

### 10.10.1 Converting mmCIF to PDB

There are a number of programs that convert between mmCIF and the PDB data-file format. Two of the most common are:

1. CIFTr – The RCSB distributes the CIFTr program, which can be used to convert from the mmCIF structure to the PDB data-file format[16].

2. pdb2cif – Again from the RCSB, the pdb2cif program can convert from the PDB data-file format to mmCIF[17].

In an ideal world, these two tools should be capable of processing each other's output, forming a *closed cycle*. That is, an mmCIF data-file can be converted into its PDB equivalent by CIFTr. The resulting PDB data-file should then be capable of being converted back to mmCIF using the pdb2cif, resulting in the original data-file. Although a reasonable assumption, this is in fact overoptimistic, but – hey – few things in life are perfect!

### 10.10.2 Converting mmCIFs to PDB with CIFTr

The installation of the CIFTr program is straightforward. A pre-compiled binary version of the program will suffice for most purposes. Browse to the CIFTr website, download the binary distribution for Linux, then unpack the downloaded file into its own directory. The RCSBROOT environment variable is then set:

```
cd
tar -zxvf ciftr-v2.0-linux.tar.gz
cd ciftr-v2.0-linux/
setenv RCSBROOT ~/ciftr-v2.0-linux
export RCSBROOT = ~/ciftr-v2.0-linux
```

To convert a mmCIF data-file to its PDB equivalent, supply the data-file name on the command-line as follows:

```
./CIFTr -i 1lqt.cif
```

---

[16] CIFTr is written by *Zukang Feng* and *John Westbrook*, and is available on-line at http://pdb.rutgers.edu/mmcif/CIFTr/index.html.

[17] pdb2cif is written by *P. E. Bourne, H. J. Bernstein* and *F. C. Bernstein*, and is available on-line at http://www.bernstein-plus-sons.com/software.

Note the use of the "-i" command-line switch (where "i" stands for "input"). If successful, a data-file called 1lqt.cif.pdb is created in the current directory. Let's try this with an example mmCIF data-file, which initially looks like this:

```
data_1LQT
#
loop_
_audit_author.name
'Bossi, R.T.'
'Aliverti, A.'
'Raimondi, D.'
'Fischer, F.'
'Zanetti, G.'
'Ferrari, D.'
'Tahallah, N.'
'Maier, C.S.'
'Heck, A.J.R.'
'Rizzi, M.'
'Mattevi, A.'
#
_pdbx_database_status.status_code              REL
_pdbx_database_status.entry_id                 1LQT
_pdbx_database_status.recvd_deposit_form       N
_pdbx_database_status.date_deposition_form     ?
_pdbx_database_status.recvd_coordinates        Y
      .
      .
      .
```

Using an appropriately formed command-line, the above entry is converted into its PDB equivalent data-file, which looks like this:

```
HEADER    OXIDOREDUCTASE                          13-MAY-02   XXXX
TITLE     A COVALENT MODIFICATION OF NADP+ REVEALED BY THE ATOMIC
TITLE    2 RESOLUTION STRUCTURE OF FPRA, A MYCOBACTERIUM TUBERCULOSIS
TITLE    3 OXIDOREDUCTASE
CAVEAT     1LQT    CHIRALITY ERROR AT THE CA CENTER OF ASP A 31.
COMPND    MOL_ID: 1;
COMPND   2 MOLECULE: FPRA;
COMPND   3 CHAIN: A, B;
COMPND   4 SYNONYM: FERREDOXIN NADP REDUCTASE;
COMPND   5 ENGINEERED: YES
SOURCE    MOL_ID: 1;
SOURCE   2 ORGANISM_SCIENTIFIC: MYCOBACTERIUM TUBERCULOSIS;
SOURCE   3 ORGANISM_COMMON: BACTERIA;
SOURCE   4 EXPRESSION_SYSTEM: ESCHERICHIA COLI;
SOURCE   5 EXPRESSION_SYSTEM_COMMON: BACTERIA
KEYWDS    NADP+ DERIVATIVE, TUBERCULOSIS, OXIDOREDUCTASE
     .
     .
     .
REMARK   4 1LQT COMPLIES WITH FORMAT V. 2.3, 09-JULY-1998
     .
     .
     .
```

### 10.10.3 Problems with the CIFTr conversion

On the whole, CIFTr works well. However, it does have a number of problems:

1. The PDB identifier is missing from the HEADER line of the resultant PDB data-file, having been replaced by "XXXX" instead. As a result, the statement in REMARK tag, type 4 (above) is invalid.

2. Each time CIFTr executes, it creates a temporary disk-file in the /tmp directory. This temporary disk-file is not removed when CIFTr exits. As the temporary disk-file is uniquely named, it remains on the hard disk until deleted manually. As a result, during large automated conversion runs, the amount of available disk space can be dramatically reduced[18].

3. The CIFTr error log is appended to each time the program is executed. There is no maximum limit set on the size of this error log. As with the previous point, if not regularly deleted, the error log grows until it occupies a large amount of disk space. Be sure to prune it regularly.

While preparing *Bioinformatics, Biocomputing and Perl*, the authors downloaded the entire PDB from the *Data Uniformity Project's* FTP site. This collection of data-files is in mmCIF format, and CIFTr converted all of them. All of the resultant PDB data-files conformed to the v2.3 standard, apart from the omission of the PDB identifier code as described above.

### 10.10.4 Some advice on using mmCIF

As suggested earlier, as a general guideline, plan to work with the PDB data-files, unless you have good reason to do otherwise. Experience has shown that working with the legacy PDB data-files is straightforward and well established. If the most up-to-date set of "super-standardised" PDBs are required, download the mmCIF versions from the *PDB Uniformity Project* and convert them using CIFTr.

### 10.10.5 Automated conversion of mmCIF to PDB

The convert_pdb program demonstrates how to use Perl to control the CIFTr program to expand a complete directory of compressed mmCIF files. Here's the source code to convert_pdb:

```
#! /usr/bin/perl

# convert_pdb - Convert PDB script.  Uses ciftr V.2.0 to convert
```

---

[18] Michael recently purchased a bigger hard disk after noticing that his current one was full. It was only after purchasing the new one that Michael noticed a large collection of temporary disk-files in his old hard-disk's /tmp directory. After deleting these, the amount of free disk-disk increased dramatically, negating the need for the new hard disk. As chance would have it, Michael's new hard-disk failed under warranty, and he returned it for a full refund!

```perl
#                   mirrored PDB from mmCIF into Legacy PDB format.  The
#                   CIFTr program should be installed at the indicated path.

use strict;

my $CIFTr_path  = "~/ciftr-v2.0-linux";
my $PDB_Path    = "~/structures/pdb-select/pdbs";
my $mmCIF_Path  = "~/structures/pdb-select/structures";

$ENV{ RCSBROOT }  = $CIFTr_path;

opendir( INPUT_DIR, "$mmCIF_Path" )
    or die "Error: Cannot read from mmCIF directory: '$mmCIF_Path'\n";

my @mmCIFdir = readdir( INPUT_DIR );

close INPUT_DIR;

open( OUTPUT_DIR, $PDB_Path )
    or die "Error: Cannot read from PDB directory: '$PDB_Path'\n";

foreach my $Current_mmCIF_file ( @mmCIFdir )
{
    if ( !( $Current_mmCIF_file =~ m/cif/i ) )
    {
        next;
    }

    my $PDB_ID    = ( $Current_mmCIF_file ) =~ m/(\d\w\w\w).cif/;
    my $PDB_name  = $PDB_ID . ".pdb";

    print "Now Processing '$Current_mmCIF_file' ";
    print "into pdb file: '$PDB_name'\n";

    my @CP_return =
        `cp $mmCIF_Path/$Current_mmCIF_file .`;
    my @Ciftr_run =
        `$CIFTr_path/bin/CIFTr -uncompress gzip -i ./$Current_mmCIF_file`;

    chomp( @Ciftr_run );

    print join ",", @Ciftr_run,"\n";

    if ( -e "./$PDB_ID.cif.pdb" )
    {
        my @Move_Result = `mv $PDB_ID.cif.pdb $PDB_Path/$PDB_name`;
    }
    else
    {
        die "ERROR: PDB file '$PDB_Path/$PDB_name' was not created!\n";
    }

    system "rm cif2pdb.err";

    system "rm /tmp/file* > /dev/null";
}
```

The result of executing this program is a list of converted PDBs in the specified directory. It is left as an extended exercise for the reader to work through this program and determine how it works[19]. Note that the *backticks* surrounding the `cp`, `CIFTr` and `mv` invocations cause `perl` to execute the specified command at the operating system level, returning any results to this program. It is very similar in operation to Perl's `system` subroutine, which is also used here.

## Where to from Here

This chapter introduced the Protein Databank, commonly referred to as the PDB. Both the legacy PDB data format and the modern replacement data format, `mmCIF`, were described, and a number of programs – some custom, bespoke and others available for download as utilities – were used to learn about the PDB and the data it holds. In the chapters that follow, the theme of Bioinformatics data and its usage is continued.

## The Maxims Repeated

Here's a list of the maxims introduced in this chapter.

- *Beware of anything in the PDB Header Section.*
- *It is often easier and desirable to regenerate database annotation than trawl through entries reconstituting the annotation using custom code.*

---

[19] This is not a cop-out on the part of your authors. You will often be presented with not much more than the source code to a program that requires amending. Learning to "read" another programmer's source code is a skill worth developing.

# 11
# Non-redundant Datasets

*The importance of non-redundant data.*

## 11.1 Introducing Non-redundant Datasets

This chapter discusses the need for, the problems associated with and the practical aspects of using *non-redundant datasets*. The focus of this chapter is on the PDB, as this is where the redundancy problems are most acute, because of the limitations of some of the processes used to determine protein structures. The fundamental concepts described here apply in a wider context.

### 11.1.1 Reasons for redundancy

There may be many reasons for redundancy in a dataset. With specific reference to the PDB, these include the following:

1. **Scientific** – It is often advantageous to study molecules with similar structures. This is a classic scientific investigative methodology: change a small part, then identify the change in structure or function to form hypotheses about the reasons for the change. Consequently, researchers are encouraged to study similar molecules to those studied previously.

2. **Technological limitations** – In X-Ray Crystallography, it is easier to obtain the structure of a molecule that is similar to one that is already known, as

molecules with similar conformations are likely to have similar crystallisation conditions. This, conveniently, allows two of the most difficult aspects of using X-Ray Crystallography to be dealt with.

## 11.1.2 Reduction of redundancy

There are two reasons for supporting the reduction of a database:

1. Conceptually, to remove bias within the database. The statistical analysis based upon the non-redundant dataset will be more representative of all the items in the database, rather than just the largest dominant group. In the PDB, the classic example of this activity is the removal of the many (several hundred) similar Lysozyme structures.

2. As a practical measure, to reduce the computational requirements caused by analysing examples that are unnecessary. For example, the *PDB-Select* structural non-redundant dataset (described below) contains approximately 1600 protein structures, whereas the entire PDB contained approximately 18,000. This ten-fold reduction in size is particularly welcome should an "all-against-all" dataset comparison be undertaken. For 1600 items, there are $1.2792^6$ (calculated as comparisons $1,600 \times 1,599/2$), whereas for 18,000 examples, there are $161.991^6$ (calculated as $18,000 \times 17,999/2$). The full comparison takes approximately 126 times *longer* than the reduced redundancy set.

## 11.1.3 Non-redundancy and non-representative

It is important to realise that a "non-redundant" dataset can contain a subset of *only* the parent dataset from which it was produced. Information absent from the parent database will remain in the dataset, and is *not* "magically" created by the removal of other repeated items. Although this may seem obvious, it is an important point that is easy to forget.

> **Maxim 11.1** *A non-redundant dataset is a subset of its parent dataset.*

It is unwise to claim that the conclusions drawn from a non-redundant PDB dataset is directly applicable to *all* proteins. Consider the case of membrane-associated proteins: these exist in close proximity to a lipid environment (inside it for some sections), which is radically different to the aqueous solution conditions of most proteins in the PDB. Membrane-bound proteins form less than 1% (about 10 structures) by proportion of the PDB, compared to the 15–30% expected from genomic prediction studies of trans-membrane helices. However, let's be optimistic. Despite what may be missing from a non-redundant version of the PDB, it still contains a lot of useful information. It is important to acknowledge that it is actually representative of globular proteins that have structures. If you avoid over-extending any conclusions in which they are inappropriate, then any conclusions drawn should be valid.

## 11.2 Non-redundant Protein Structures

The two most widely used algorithms to prepare non-redundant protein datasets are *PDB-Select* and *CD-HIT/CD-HI*. Both algorithms work in a similar, three-step way as follows:

1. Calculation of the similarity of the proteins in the PDB based upon their sequence similarity to each other.

2. Stipulation of a threshold over which proteins are deemed "similar" and below which they are not.

3. Grouping – or clustering – the proteins and selecting a representative protein from each group.

*PDB-Select* lists are prepared every six months[1] and are widely used for testing and development of protein structure prediction algorithms. The advanced search form on the PDB web-sites allows the results returned to be filtered by lists of different similarity levels, as prepared by the *CD-HIT* algorithm. This preparation occurs on a weekly basis. The *PDB-Select* list at 25% similarity from April 2002 is used as an example. The list itself can be downloaded via FTP from the *PDB-Select* site:

    ftp://ftp.embl-heidelberg.de/pub/databases/protein_extras/pdb_select

There is nothing fundamentally *wrong* with the *CD-HIT* lists, it is your author's personal preference that they are not used.

The above definition of "25% similar" is widely known as the *practical limit* at which the commonly used pairwise protein sequence comparison algorithms can identify related proteins on the basis of amino acid sequence alone. This removes many copies of the most similar proteins reducing the number of included examples to something more computationally manageable[2], while still giving a wide spread of structures. Extracts from the downloaded data file are shown in Figure 11.1 on page 214.

The meanings of the columns are given in the README data file, stored in the same directory as the downloaded dataset. The first seven columns are of most interest, and are described as follows:

---

[1] And have been since 1992.

[2] Or at least manageable in terms of *bandwidth*, especially if you have to download them over a slow connection!

```
25% threshold list: 1771 chains with 297372 residues
thrsh    ID   naa   Res  Rfac Methd n_sid n_bck n_naa n_hlx n_bta  cmpnd
25    1KBFA   49  -1.00  0.00   N     49    49    0     0    11   kinase suppressor of ras
25    1KBHA   47  -1.00  0.00   N     47    47    0    29     0   nuclear receptor coactivator
25    1KBHB   59  -1.00  0.00   N     59    59    0    32     0   nuclear receptor coactivator
25    2A93A   32  -1.00  0.00   N     32    32    0    27     0   c-myc-max heterodimeric ...
25    1C9QA  117  -1.00  0.00   N    117   117    0    34    13   apoptosis inhibitor iap ...
25    1C9FA   87  -1.00  0.00   N     87    87    0    11    22   caspase-activated dnase
25    1G84A  105  -1.00  0.00   N    105   105    0     5    61   immunoglobulin e
25    2ADX_   40  -1.00  0.00   N     40    40    0     0     0   thrombomodulin
25    2AF8_   86  -1.00  0.00   N     86    86    0    43     0   actinorhodin polyketide ...
25    1C6WA   33  -1.00  0.00   N     33    33    0     3     7   maurocalcin
      ...
thrsh    ID   naa   Res  Rfac Methd n_sid n_bck n_naa n_hlx n_bta  cmpnd
25    1C53_   79   1.80  0.19   X      0     0    0     0     0   Cytochrome c553
      ...
thrsh    ID   naa   Res  Rfac Methd n_sid n_bck n_naa n_hlx n_bta  cmpnd
25    1TIA_  271   2.00  0.19   X      3     3    0     0     0   Lipase (triacylglycerol acylhydrolase)
      ...
```

**Figure 11.1** Extracted non-redundant dataset.

1. The threshold value used to prepare the list (25% identity).

2. The PDB identifier of the protein. The first four characters are the PDB identifier code, the last is the chain identifier with the structure (or "_", if it is the only chain).

3. The number of amino acids in the structure.

4. The *Resolution* of the structure for crystallographic structures (or $-1.00$ to signal "not applicable" for NMR structures).

5. The *R-Factor* of the structure for crystallographic structures (or $-1.00$ to signal "not applicable" for NMR structures).

6. The number of residues that have *backbone atoms* reported.

7. The number of residues that have *side chain atoms* reported.

In the *PDB-Select* lists, protein chains in the same PDB data file are treated independently. Whether this is a tolerable or an unacceptable assumption depends on what is being studied or researched. For instance, this assumption may well be a problem when studying the interaction of protein sub-units. Some structures, such as 1C53 contain only carbon-alpha positions, as shown in the bottom-half of Figure 11.1 on page 214, while others (1TIA) have a small number of side chains reported.

The 1TIA structure is particular intriguing: it is an alpha-carbon trace, except for three of the 271 amino acids that have full side chains. The title of this PDB is "An unusual buried polar cluster in a family of fungal lipases", indicating that this structure was created for a very specific purpose. While very suitable for this specific study, its use in more general studies is probably limited. It can be excluded from the study by requiring that over 70% of the side chains be reported. The program that follows, called `select_filter`, filters the *PDB-Select* list in this way:

```
#! /usr/bin/perl

# select_filter: designed to filter the PDB-Select list
#                of alpha-carbon traces.

use strict;

while ( <> )
{
    if ( !/^   25/ )
    {
        next;
    }

    my @Fields = split( " ", $_ ,8 );
```

```perl
    if ( $Fields[ 7 ] / $Fields [ 2 ] > 0.7 )
    {
        my $ID    = substr( $Fields[ 1 ], 0, 4 );
        my $Chain = substr( $Fields[ 1 ], 4, 1 );

        printf( "%3s,%1s: %4i\n", $ID, $Chain, $Fields[ 7 ] );
    }
    else
    {
        print "Excluded: ", $Fields[ 1 ], "\n";
    }
}
```

This program processes all lines that start with three spaces and "25", skipping those lines that do not. The `split` subroutine, provided by Perl, breaks the line into a collection of scalars that are then assigned to the @Fields array. A test is then performed to see if the ratio of the number of amino acids in the structure (field 2) relative to the number with side chains (field 7) exceeds 0.7 (or 70%). If they do, then the chain and ID are split from each other. Note that the combined code/ID is contained in field 1. An appropriately formatted message is then displayed on STDOUT.

The `select_filter` program produces a list that contains the PDB identifier code, the chain within the data file[3] and the number of amino acids in that chain. Here is some sample output:

```
1KBF,A:   49
1KBH,A:   47
1KBH,B:   59
2A93,A:   32
1C9Q,A:  117
1C9F,A:   87
1G84,A:  105
2ADX,_:   40
2AF8,_:   86
      .
      .
      .
Excluded: 1C53_
Excluded: 1TIA_
      .
      .
      .
```

Of interest is the fact that, as this program executes, it catches other structures such as 1TIA that has 271 amino acids but only 3 complete side chains. The lines created by the excluded structures/chains can be removed by *piping* the output

---

[3] Remember: PDB data files may contain one or more protein chains.

of `select_filter` through the `grep` utility. This command-line excludes, rather than reports, all lines that contain the pattern "Excluded", thanks to the use of the "-v" switch:

```
./select_filter 2002_Apr.25 | grep -v 'Excluded'
```

Similarly, select only the excluded structures by removing the "-v" switch, as follows:

```
./select_filter 2002_Apr.25 | grep 'Excluded'
```

It is always a good idea to inspect which structures are being excluded to make sure the filters are not too stringent.

> **Maxim 11.2** *Be sure to double-check the list of excluded structures.*

The 34 structures failing the test criterion are

1C53, 2AT2(A), 1BAX, 1JQ1(A), 2ILA, 2MAD(H), 1JQS(B), 1JQS(C), 1QCR(F), 1QCR(D), 1QCR(K), 1QCR(G), 1QCR(C), 1EFM, 1TIA, 1FFK(S), 1FFK(U), 1FFK(W), 1FFK(V), 1FFK(J), 1DPI, 1FFK(B), 1FFK(G), 1FFK(F), 1FFK(D), 1FFK(C), 1ILT(A), 3HTC(I), 1AAT, 1IAN, 1JEW(2), 1JEW(4), 1JEW(3), 2DTR.

Ten are in structure 1FFK (the Large Ribosome Sub-unit from "HALOARCULA MARISMORTUI") and contain only carbon-alpha atoms. With the list of required non-redundant protein structures at hand, it is now possible to download them from the Internet.

## Where to from Here

This chapter presented the idea of non-redundant datasets, with specific reference to the PDB. Of course, storing Bioinformatics data in PDB data files is not the only option. There are other data formats. Some of these are described in the next chapter, which also presents a tutorial introduction to an important data technology: *relational database management systems*.

## The Maxims Repeated

Here's a list of the maxims introduced in this chapter.

- *A non-redundant dataset is a subset of its parent dataset.*
- *Be sure to double-check the list of excluded structures.*

# 12
# Databases

*Learning to talk database.*

## 12.1 Introducing Databases

Many modern computer systems store vast amounts of structured data. Typically, this data is held in a *database system*. Before defining what's meant by the term *database system*, let's begin with the term "database":

> A **database** is a collection of one or more related **tables**.

The use of the word "related" is important here, as is "table". The significance of "related" will soon become clear. For now, let's define "table":

> A **table** is a collection of one or more **rows of data**.

The rows of data are arranged in columns, with each intersection of a row and column containing a data item. Therefore, the definition of "row" is:

> A **row** is a collection of one or more data items, arranged in **columns**.

Within a row, the columns conform to a *structure*. For example, if the first column in a row holds a date, then every first column in every row must *also* hold a date. If the second column holds a name, then every second column must *also* hold a name, and so on.

The following data corresponds to the *structure* just identified, in that there are two columns, the first holding a date, the second holding a name:

```
1960-12-21    P. Barry
1954-6-14     M. Moorhouse
```

To further cement the notion of structure, each column is given a *descriptive name*. Here's an expanded table of data, with descriptive column names indicated (additional rows of data have also been added):

```
--------------   ---------
Discovery_Date   Scientist
--------------   ---------
1960-12-21       P. Barry
1954-6-14        M. Moorhouse
1970-3-4         J. Blow
2001-12-27       J. Doe
```

In addition to naming each column, the *structure* requires that each data item held in a column be of a specific *type*. Here's the *type information* for the data in the above table:

```
-----------      ----------------
Column name      Type restriction
-----------      ----------------
Discovery_Date   a valid Date
Scientist        a String no longer than 64 characters
```

This type information generally goes by one of two names: *metadata* or *schema*. Think of the word *structure* as a synonym for both metadata and schema. Note that the structure *restricts* the type of data that can be stored in each column, in addition to – for some columns – specifying the maximum length of the data item.

Consequently, with the example structure presented above, it is *not* possible to store a name in a column that's expecting a date. Additionally, when a name is stored in the correct column, it cannot be longer than 64 characters.

In addition to assigning descriptive names and type restrictions to each column, the entire table is given a name. Let's call this table `Discoveries`. It is now possible to answer a question like *"What is the structure of Discoveries?"*.

## 12.1.1 Relating tables

Let's extend the `Discoveries` table to include details of the discovery. An additional column is needed to hold the data, as follows:

```
--------------   ---------      ---------
Discovery_Date   Scientist      Discovery
--------------   ---------      ---------
1960-12-21       P. Barry       Flying car
1954-6-14        M. Moorhouse   Telepathic sunglasses
1970-3-4         J. Blow        Self cleaning child
2001-12-27       J. Doe         Time travel
```

The inclusion of this new column requires an update to the structure of the table. Here is a revised structure for Discoveries:

```
----------                 ----------------
Column name                Type restriction
----------                 ----------------
Discovery_Date             a valid Date
Scientist                  a String no longer than 64 characters
Discovery                  a String no longer than 128 characters
```

Now, let's assume there is a requirement to maintain additional data about each scientist, specifically their date of birth and telephone number. This data is referred to as the scientist's *identification information*, and guards against the problems that can occur when two scientists have the same name. For instance, if there are two (mad) scientists called "M. Moorhouse", the original table structure cannot distinguish between them. However, as it is highly unlikely that both scientists share the same date of birth *and* telephone number, these data items can be added to the table to guard against incorrect identifications. The structure of the Discoveries table is now:

```
----------                 ----------------
Column name                Type restriction
----------                 ----------------
Discovery_Date             a valid Date
Scientist                  a String no longer than 64 characters
Discovery                  a String no longer than 128 characters
Date_of_birth              a valid Date
Telephone_number           a String no longer than 16 characters
```

The data in the table now looks like this[1]:

```
--------------  ----------  ---------               -------------  ----------------
Discovery_Date  Scientist   Discovery               Date_of_birth  Telephone_number
--------------  ----------  ---------               -------------  ----------------
1960-12-21      P. Barry    Flying car              1966-11-18     353-503-555-91910
1954-6-14       M. Moorhouse Telepathic sunglasses  1970-3-24      00-44-81-555-3232
1970-3-4        J. Blow     Self cleaning child     1955-8-17      555-2837
2001-12-27      J. Doe      Time travel             1962-12-1      -
1974-3-17       M. Moorhouse Memory swapping toupee 1970-3-24      00-44-81-555-3232
1999-12-31      M. Moorhouse Twenty six hour clock  1958-7-12      416-555-2000
```

This version of the Discoveries table contains three rows for scientist M. Moorhouse. By examining the Date_of_birth column, it is clear that there are *two different scientists*.

---

[1] If you're thinking these dates correspond to your author's actual birth dates, think again. This data is obviously fictitious. Of course, if anyone has managed to invent a self-cleaning child, please let Paul know.

### 12.1.2 The problem with single-table databases

Although the above table structure solves the problem of uniquely identifying each scientist, it introduces some other problems, including:

1. If a scientist is responsible for a large number of discoveries, their identification information has to be entered into every row of data that refers to them. This is time-consuming and wasteful.

2. Every time identification information is added to a row for a particular scientist, it has to be entered in exactly the same way as the identification information added already. Despite the best of efforts, this level of accuracy is often difficult to achieve. In most cases, the slightly different identification information will be assumed to refer to a different scientist.

3. If a scientist changes any identification information, every row in the table that refers to the scientist's discoveries has to be changed. This is drudgery.

### 12.1.3 Solving the one-table problem

Ideally, the identification information should exist in only one place. A mechanism should provide for linking each scientist referred to in the Discoveries table to the identification information.

The problems described in the previous section are solved by breaking the all-in-one Discoveries table into two tables. Here is a new structure for Discoveries:

```
Column name            Type restriction
-----------            ----------------
Discovery_Date         a valid Date
Scientist_ID           a String no longer than 8 characters
Discovery              a String no longer than 128 characters
```

The Discoveries table reverts to three columns per row of data. The second column, originally called Scientist, is now called Scientist_ID. Its type information has also changed, from a string of up to 64 characters to one of up to 8 characters. Here is the structure for a new table, called Scientists:

```
Column name            Type restriction
-----------            ----------------
Scientist_ID           a String no longer than 8 characters
Scientist              a String no longer than 64 characters
Date_of_birth          a valid Date
Address                a String no longer than 256 characters
Telephone_number       a String no longer than 16 characters
```

The Scientists table also *critically* contains a column called Scientist_ID, which has type information identical to the corresponding column in the Discoveries table. The Scientist column from the original Discoveries table, as well as the identification information from the all-in-one Discoveries table, makes up the remainder of the columns in Scientists, together with a new column called Address.

When a new scientist comes along, a row of data is added to the Scientists table. A unique 8-character Scientist_ID is assigned to the scientist. When the same scientist discovers something, the details of the discovery are added to the Discoveries table. The unique Scientist_ID is used to link, or *relate*, the row of data in Discoveries to the scientist's identification information in Scientists. So Discoveries now contains data like this:

```
--------------  ------------  ---------
Discovery_Date  Scientist_ID  Discovery
--------------  ------------  ---------
1954-6-14       MM            Telepathic sunglasses
1960-12-21      PB            Flying car
1969-8-1        PB            A cure for bad jokes
1970-3-4        JB            Self cleaning child
1974-3-17       MM            Memory swapping toupee
1999-12-31      MM2           Twenty six hour clock
2001-12-27      JD            Time travel
```

While the Scientists table contains this data:

```
------------  ---------  -------------  -------  ----------------
Scientist_ID  Scientist  Date_of_birth  Address  Telephone_number
------------  ---------  -------------  -------  ----------------
JB            J. Blow       1955-8-17   Belfast, NI   555-2837
JD            J. Doe        1962-12-1   Sydney, AUS   -
MM            M. Moorhouse  1970-3-24   England, UK   00-44-81-555-3232
MM2           M. Moorhouse  1958-7-12   Toronto, CA   416-555-2000
PB            P. Barry      1966-11-18  Carlow, IRL   353-503-555-91910
```

Note that despite the fact that there are two scientists called M. Moorhouse, the rows of data that refer to their respective discoveries (in the Discoveries table) are easily distinguished, as "MM" identifies the English Moorhouse, whereas "MM2" identifies the Canadian.

Changes to any scientist's identification information no longer impact the Discoveries table, as only the data in the Scientists table is changed.

> **Technical Commentary:** Obviously, this last statement is not true when the change to the row in Scientists results in the value for Scientist_ID changing. In this case, every row of data in Discoveries that refers to the old Scientist_ID needs to change. Such changes, while certainly possible, are rarely justifiable.

This technique of relating data in one table to that in another forms the basis of modern database theory. It also explains why so many modern database technologies are referred to as *Relational Database Management Systems (RDBMS)*.

When a collection of tables is designed to relate to each other, as is the case with `Discoveries` and `Scientists`, they are collectively referred to as a *database*. It is usually a requirement to give the database a descriptive name.

Taking the time to think about how data relates to other data in a database is important and very worthwhile. It is so important that it warrants its very own maxim.

> **Maxim 12.1** *A little database design goes a long way.*

### 12.1.4 Database system: a definition

With the terms *database, table, row, column* and *structure* defined, it is now possible to return to the definition of "database system":

> A **database system** is a computer program (or a group of programs) that provides a mechanism to define and manipulate one or more databases.

Recall from the last section that a database contains one or more tables, and that a table contains one of more rows of columned data that conform to a defined structure. A database system allows databases, tables and columns to be created and named, and structures to be defined. It provides mechanisms to add, remove, update and interact with the data in the database. Data stored in tables can be searched, sorted, sliced, diced and cross-referenced. Reports can be generated, and calculations can be performed.

Many database systems can be extended, allowing automated interaction to occur from many programming technologies. As the next chapter will show, combining a database system with Perl is a powerful combination. But let's not get ahead of ourselves. There's additional foundation material to work through first.

After a brief survey of available database systems, the remainder of this chapter is dedicated to presenting a database case study. Bioinformatics data is used to populate a database, and then a series of interactions with the data are described. Along the way, the reader is exposed to SQL and the MySQL database system.

## 12.2 Available Database Systems

There are a large number of database systems to choose from. A simple categorisation by type of database system is as follows:

- Personal
- Enterprise
- Open source.

Which type of database system is chosen depends on a number of factors, including (but not limited to):

1. The amount of data to be stored in the database.

2. Whether the data supports a small personal project or a large collaborative one.

3. How much funds (if any) are available towards the purchase of a database system.

## 12.2.1 Personal database systems

The database systems in this category are designed to run on any personal computer, and they typically – though not exclusively – target the Microsoft Windows graphical environment. Because of their PC heritage, they are good for small personal projects, but generally *scale poorly*: as the amount of data in the database grows, the performance of these database systems degrade to the point where they become unusable. That said, most databases in the category can comfortably handle a multi-megabyte database.

Example technologies in this category include *Access*, *Paradox*, *FileMaker* and the *dBase* family of databases.

## 12.2.2 Enterprise database systems

At the other end of the spectrum, the database systems in this category are designed to support the efficient storage and retrieval of vast amounts of data. Unlike the technologies in the *Personal* category, *Enterprise* database systems can handle multi-gigabyte and increasingly multi-terabyte databases, and are designed to provide access for multiple, simultaneous users. It is possible to use an *Enterprise* database system for a personal project, but such practice is generally considered to be *overkill*. It is also possible to run the database systems in this category on standard personal computers (running operating systems such as Windows, Mac OS X and Linux), but they are designed to execute on larger enterprise-class computers such as mainframes, mini-computers or high-end servers.

Example technologies in this category include *InterBase*, *Ingres*, *SQL Server*, *Informix*, *DB2* and *Oracle*.

## 12.2.3 Open source database systems

A significant factor differentiates the database systems found in the *Personal* and *Enterprise* categories from those found in the *Open Source* category: cost. The database technologies in the *Personal* category typically cost several hundred euro, whereas those in the *Enterprise* category range in cost from several thousands to several tens of thousand euro (and sometimes more). In stark contrast, *Open Source* database systems are freely available on-line.

As a direct result of their Linux heritage, *Open Source* database technologies tend to perform equally well within a personal and an enterprise setting. Consequently, *Open Source* database technologies neatly fill the gap that exists between the personal and enterprise database worlds. Multi-megabyte and gigabyte databases can be accommodated without too much difficulty.

Example technologies in this category include *PostgreSQL* and *MySQL*.

## 12.3 SQL: The Language of Databases

If there is one technology that unites all of the technologies found in the *Personal*, *Enterprise* and *Open Source* database categories, that technology is SQL. SQL is shorthand for *Structured Query Language* and has a heritage that dates to the late 1960s.

The SQL component built into most modern database systems typically provides two facilities:

1. A Database Definition Language (DDL) and

2. A Data Manipulation Language (DML).

Prior to the arrival of SQL, every database system provided proprietary mechanisms for defining databases and then manipulating the data stored within them. To use these database systems efficiently, some knowledge of how the data was stored within the database system was required, and the effort required to acquire this specialised knowledge was considerable. Moving data from one database system to another was possible, but rarely considered, as the learning curve associated with the transition to an alternative database technology was often considerable. The skills acquired when working with one particular database system were generally not transferable to another.

The introduction, promotion and subsequent adoption of SQL as an integrated database system component changed all this.

Not only did SQL provide a *standard* mechanism for defining and manipulating data but also removed the requirement to understand the way in which the data within the database system was stored. This was a huge advantage. As the majority of database systems adopted SQL, users acquired a transferable skill that no longer bound them to a single database system (or database vendor).

### 12.3.1 Defining data with SQL

The *data definition* component of SQL provides a mechanism whereby databases can be created. Within a created database, SQL allows tables to be defined as rows of columned data conforming to a structure. Table structures can be changed, and tables can be renamed or deleted.

### 2.3.2 Manipulating data with SQL

The *data manipulation* component of SQL provides a mechanism to work with data in tables. Mechanisms exist to add data to tables a single row at a time, or in bulk (more than one row at a time). Rows can also be removed from tables.

SQL provides a powerful mechanism to search data in tables and extract row data. It is possible to extract an entire row of columned data or to specify that only certain columns are to be included in the extract. Critically, the SQL search-and-extract mechanism can be used to relate data in one table to that in another.

> **Technical Commentary:** A common question centres around the correct pronunciation of "SQL". Typically – although not exclusively – persons of a European persuasion tend to pronounce each letter individually: "S Q L". North Americans tend to pronounce SQL as "sequel", in honour of one of the earliest database technologies that provided SQL as an integrated component. It does not really matter which pronunciation is used, just so long as the use is consistent.

## 12.4 A Database Case Study: MER

The *Swiss Institute of Bioinformatics* maintains SWISS-PROT, an annotated protein sequence database. Unlike the example database from the start of this chapter, the SWISS-PROT database is not maintained as a collection of tables. Instead, the SWISS-PROT database uses what's commonly referred to as a *flat-file (or text-based)* format to represent protein structures. The protein data is stored as text in data files.

The SWISS-PROT database does define a specific structure for the contents of the data file, which can contain one or more protein structures. Each structure is referred to as an *entry*. The SWISS-PROT data format is described in detail in the SWISS-PROT manual, which is available at:

   http://www.expasy.org/sprot/userman.html

Obviously, the definition of "database" as it relates to SWISS-PROT is somewhat different to the definition from earlier in this chapter. Although the SWISS-PROT definition of "database" may be confusing to some readers, it is acceptable as the meaning of the word "database" can vary depending on context. When working with RDBMSs, "database" is defined as a collection of one or more related tables. When working with SWISS-PROT data files, "database" is defined as a collection of similarly formatted flat files.

The SWISS-PROT structure is designed to be highly compatible with that of the EMBL Nucleotide Sequence Database. The EMBL database is maintained by the *EMBL Outstation* at the *European Bioinformatics Institute.* Unlike SWISS-PROT, which stores data on protein structures, the EMBL database stores DNA sequence data. Like SWISS-PROT, the data in the EMBL database is a collection of similarly formatted, text-based data files.

With modern database systems now in widespread use, one might be forgiven for asking why the data in these two important databases is provided in its current form. Why not use an RDBMS? This answer is taken from the on-line documentation to the EMBL database:

> *"An attempt has been made to make the collected data as easily accessible as possible without restricting their usefulness to a particular type of computing environment. For this reason, the simplest possible organisation ('flat file') has been chosen."*

The entire EMBL manual is available at this web-site:

http://www.ebi.ac.uk/embl/Documentation/User_manual/home.html

By choosing a simple, open format, the SWISS-PROT and EMBL databases can be put to many different uses. Here's an example SWISS-PROT entry:

```
ID   MERT_ACICA     STANDARD;      PRT;    116 AA.
AC   Q52106;
DT   01-NOV-1997 (Rel. 35, Created)
DT   01-NOV-1997 (Rel. 35, Last sequence update)
DT   15-JUN-2002 (Rel. 41, Last annotation update)
DE   Mercuric transport protein (Mercury ion transport protein).
GN   MERT.
OS   Acinetobacter calcoaceticus.
OG   Plasmid pKLH2.
OC   Bacteria; Proteobacteria; Gammaproteobacteria; Pseudomonadales;
OC   Moraxellaceae; Acinetobacter.
OX   NCBI_TaxID=471;
RN   [1]
RP   SEQUENCE FROM N.A.
RX   MEDLINE=94134837; PubMed=8302940;
RA   Kholodii G.Y., Lomovskaya O.L., Gorlenko Z.M., Mindlin S.Z.,
RA   Yurieva O.V., Nikiforov V.G.;
RT   "Molecular characterization of an aberrant mercury resistance
RT   transposable element from an environmental Acinetobacter strain.";
RL   Plasmid 30:303-308(1993).
CC   -!- FUNCTION: INVOLVED IN MERCURIC TRANSPORT. PASSES A HG(2+) ION
CC       FROM THE PERIPLASMIC MERP PROTEIN TO THE MERCURIC REDUCTASE
CC       (MERA).
CC   -!- SUBCELLULAR LOCATION: INTEGRAL MEMBRANE PROTEIN. INNER MEMBRANE
CC       (BY SIMILARITY).
CC   -------------------------------------------------------------------
CC   This SWISS-PROT entry is copyright. It is produced through a collaboration
CC   between  the Swiss Institute of Bioinformatics  and the  EMBL outstation -
CC   the European Bioinformatics Institute.  There are no  restrictions on  its
CC   use by  non-profit  institutions as long  as its content  is  in  no  way
CC   modified and this statement is not removed.  Usage  by  and for commercial
CC   entities requires a license agreement (See http://www.isb-sib.ch/announce/
CC   or send an email to license@isb-sib.ch).
CC   -------------------------------------------------------------------
```

```
DR   EMBL; AF213017; AAA19679.1; -.
DR   InterPro; IPR003457; Transprt_MerT.
DR   Pfam; PF02411; MerT; 1.
KW   Transport; Mercuric resistance; Inner membrane; Mercury; Plasmid;
KW   Transmembrane.
FT   TRANSMEM     16     36       POTENTIAL.
FT   TRANSMEM     46     66       POTENTIAL.
FT   TRANSMEM     94    114       POTENTIAL.
FT   METAL        24     24       HG(2+) (BY SIMILARITY).
FT   METAL        25     25       HG(2+) (BY SIMILARITY).
FT   METAL        76     76       HG(2+) (BY SIMILARITY).
FT   METAL        82     82       HG(2+) (BY SIMILARITY).
SQ   SEQUENCE   116 AA;  12510 MW;  2930A92CF88EB10F CRC64;
     MSEPQNGRGA LFAGGLAAIL ASACCLGPLV LIALGFSGAW IGNLTVLEPY RPIFIGAALV
     ALFFAWRRIV RPTAACKPGE VCAIPQVRTT YKLIFWFVAV LVLVALGFPY VMPFFY
//
```

The exact meaning of each *line type* in this SWISS-PROT entry is described in the SWISS-PROT manual. Although convenient for humans (the entry *is* easy to read), processing the entry by computer is complicated by a number of factors, including the following:

1. Not all SWISS-PROT line types are required. A number of the line types are optional. The above entry, for example, does not contain the optional RC line type, which refers to a Reference Comment.

2. Some of the line types can extend over any number of lines. For example, the CC line type, which refers to a Comment, extends over 14 lines in the above entry.

3. Other line types, for example the RN line type, which refers to a citation Reference Number, contain a block of line types, some of which are optional and some of which can extend over a number of lines. Note that the RA line type, which refers to the Reference Author(s), and the RT line type, which refers to the Reference Title, both extend over two lines in the above entry. As described in point 1, the RC line type is not used in this RN entry.

These factors make processing SWISS-PROT entries a challenge. Luckily, Perl is on our side and, as will be demonstrated shortly, Perl is a natural at working with this type of data.

The EMBL data format is similar to SWISS-PROT[2]. Here's an abridged example EMBL entry:

```
ID   PPMERR     standard; DNA; UNC; 2923 BP.
XX
AC   M24940;
XX
SV   M24940.1
XX
DT   02-FEB-1990 (Rel. 22, Created)
```

---

[2] In fact, the SWISS-PROT format was designed to be highly complementary to the EMBL format, which predates the SWISS-PROT database by a number of years.

```
DT   06-JUL-1999 (Rel. 60, Last updated, Version 3)
XX
DE   Plasmid pDU1358 (from Serratia marcescens) mercury resistance protein genes
DE   merR, merP and merT, complete cds, and merA gene, 5' end.
XX
KW   merA gene; mercury resistance protein; merP gene; merR gene; merT gene.
XX
OS   Plasmid pDU1358
OC   plasmids.
OG   Plasmid pDU1358
XX
RN   [1]
RP   1-2923
RX   MEDLINE; 89327136.
RA   Nucifora G., Chu L., Silver S., Misra T.K.;
RT   "Mercury operon regulation by the merR gene of the organomercurial
RT   resistance system of plasmid pDU1358";
RL   J. Bacteriol. 171(8):4241-4247(1989).
XX
DR   GOA; P08662; P08662.
DR   GOA; P13111; P13111.
DR   GOA; P13112; P13112.
DR   GOA; P13113; P13113.
DR   SWISS-PROT; P08662; MERA_SERMA.
DR   SWISS-PROT; P13111; MERR_SERMA.
DR   SWISS-PROT; P13112; MERT_SERMA.
DR   SWISS-PROT; P13113; MERP_SERMA.
XX
CC   Draft entry and computer-readable sequence for [1] kindly provided
CC   by Nucifora,G. 13-JUN-1989.
XX
FH   Key             Location/Qualifiers
FH
FT   source          1..2923
FT                   /db_xref="taxon:2547"
FT                   /organism="Plasmid pDU1358"
FT                   /plasmid="pDU1358"
FT                   /specific_host="Serratia marcescens"
FT   gene            complement(677..1111)
FT                   /gene="merR"
FT   CDS             complement(677..1111)
FT                   /codon_start=1
FT                   /db_xref="GOA:P13111"
FT                   /db_xref="SWISS-PROT:P13111"
FT                   /transl_table=11
FT                   /gene="merR"
FT                   /product="mercury resistance protein"
FT                   /protein_id="AAA98221.1"
FT                   /translation="MEKNLENLTIGVFAKAAGVNVETIRFYQRKGLLPEPDKPYGSIRR
FT                   YGEADVTRVRFVKSAQRLGFSLDEIAELLRLDDGTHCEEASSLAEHKLQDVREKMTDLA
FT                   RMETVLSELVFACHARQGNVSCPLIASLQGEKEPRGADAV"
          .
          .
          .
XX
SQ   Sequence 2923 BP; 617 A; 882 C; 820 G; 604 T; 0 other;
     ttaatctgct caacaagata gtgataatgc tgttgtaatt tagcaataac tggctaggta        60
     aagaggcaaa ctattatcct caagaatggt actcagtcgg ctaataacgg cagctcctcg       120
     gggaacgcta atgccaaatt ccagcagaaa agcatgcatt tgattggttg ttttcaccct       180
     atcctgaacc agggattcac ggacacgatg cagagcccgc attgcctgct gagattccgt       240
          .
          .
          .
```

```
           acccgtccat cggcgaggcc gtcacagccg ctttccgtgc cgaagggatc aaggtactgg        2880
           aacacacgca agccagccag gtcgcgcatg tgaacggcga att                          2923
   //
```

Processing EMBL entries is complicated by similar factors to those discussed above in relation to the SWISS-PROT entries. However, the inclusion of the XX line type, which refers to a separator line, can help when processing EMBLs.

## 2.4.1 The requirement for the MER database

A small collection of SWISS-PROT and EMBL entries are taken from the *Mer Operon*, a bacterial gene cluster that is found in many bacteria for the detoxification of Mercury $Hg2+$ ions. These provide the *raw* data to a database, which is called *MER*. The MER database contains four tables:

proteins – A table of protein structure details, extracted from a collection of SWISS-PROT entries.

dnas – A table of DNA sequence details, extracted from a collection of EMBL entries.

crossrefs – A table that links the extracted protein structures to the extracted DNA sequences.

citations – A table of literature citations extracted from both the SWISS-PROT and EMBL DNA entries.

Once the raw data is in the database, SQL can be used to answer questions about the data, for instance:

1. How many protein structures in the database are longer than 200 amino acids in length?
2. How many DNA sequences in the database are longer than 4000 bases in length?
3. What's the largest DNA sequence in the database?
4. Which protein structures are cross-referenced with which DNA sequences?
5. Which literature citations reference the results from the previous question?

Of course, it is possible to determine answers to these questions manually, as follows:

- Print out all the SWISS-PROT and EMBL entries of interest.
- Sift through the printouts visually, noting the data of interest.

**232** *Databases*

which is probably (depending on the number of entries examined) no more than a few hours' work. A computer program could be written to automate the collection of the interesting pieces of data, which would probably reduce the amount of time required from hours to tens of minutes, depending on how complicated the computer programs are and whether they have to be written from scratch. Compare tens of minutes and a few hours to the length of time it takes an SQL-capable database system to answer each of these questions: *no more than a few seconds.*

### 12.4.2 Installing a database system

MySQL is a modern, capable and SQL-enabled database system. It is *Open Source* and freely available for download from the MySQL web-site:

    http://www.mysql.com

Especially well-written documentation, in the form of the *MySQL Manual*, is also available for download in a number of formats, including HTML and PDF. At over 800 pages, this may be all the documentation most MySQL users ever need. However, a good collection of third-party texts also exist (see Appendix D, *Suggestions for Further Reading*, for some recommendations).

MySQL is so popular that it comes as a standard, installable component of most Linux distributions, and is the database system of choice within *Bioinformatics, Biocomputing and Perl*. However, as far as is possible, the material presented in this chapter is database system neutral: most commands should work unaltered with any modern database system, not just with MySQL.

The following commands switch on MySQL on RedHat and RedHat-like Linux distributions[3]:

    chkconfig  --add  mysqld
    chkconfig  mysqld  on

If the first `chkconfig`[4] command produces an error messages like this:

    error reading information on service mysqld: No such file or directory

This means that MySQL is not installed and the second command will also fail. Check the CD-ROMs that came with the Linux distribution to see if the required software is available, or download MySQL from its web-site. Be careful to read and follow the installation instructions as described on the MySQL web-site and in the MySQL Manual.

---
[3] But may not work on your Linux distribution: check your documentation.
[4] To learn about `chkconfig`, type "`man chkconfig`" at the Linux prompt.

Once MySQL is installed, it needs to be configured. The first requirement is to assign a password to the MySQL superuser, known as "root". The `mysqladmin` program does this, as follows:

```
mysqladmin -u root password 'passwordhere'
```

It is now possible to securely access the *MySQL Monitor* command-line utility with the following command, providing the correct password when prompted:

```
mysql -u root -p
```

The *MySQL Monitor* is an interactive, command-line tool that comes with MySQL. It can be used to issue SQL queries[5] to the MySQL database system. If the password was entered successfully, the *MySQL Monitor* command-prompt appears. It looks like this:

```
mysql>
```

Certain actions can be performed only within MySQL when operating as the superuser. These actions include the ability to create a new database.

### 2.4.3 Creating the MER database

SQL queries can be entered directly at the *MySQL Monitor* prompt. Let's use an SQL DDL query, CREATE DATABASE, to create a database called MER. Note that throughout this section, the typed query is shown in *an italic font*:

```
mysql> create database MER;
```

This, if successful, should produce the following – rather cryptic – message from MySQL:

```
Query OK, 1 row affected (0.36 sec)
```

MySQL confirms that the entered SQL query was OK. Be advised that interactive SQL queries are terminated by a semicolon, the ";" character. The *MySQL Monitor* does not execute the query until it sees the semicolon, so be careful to always include it at the end of each query.

The "1 row affected" part of the message refers to the fact that MER has been added as a database within the system. As might be expected, MySQL uses internal tables to store this "system information". To view the list of databases in the system, use the SHOW DATABASES query. Here's the query, together with the results returned from the *MySQL Monitor*:

---

[5] In SQL-speak, the word "query" has the same meaning as "command".

```
mysql> show databases;
+------------+
| Databases  |
+------------+
| MER        |
| test       |
| mysql      |
+------------+
3 rows in set (0.00 sec)
```

A list of databases is returned by MySQL. This particular list was produced on Paul's laptop, which is running a default installation of MySQL (version 3.23). There are three identified databases:

**MER** – The just-created database that will store details on the extracted protein structures, DNA sequences, cross references and literature citations.

**test** – A small test database that is used by MySQL and other technologies to test the integrity of the MySQL installation.

**mysql** – The database that stores the internal "system information" used by the MySQL database system.

> **Technical Commentary:** Throughout this chapter, when a snippet of SQL or an SQL query is described in the text, it is always shown in UPPERCASE. As can be seen from the interactive examples, SQL queries can be entered into the *MySQL Monitor* in lowercase. It makes no difference to MySQL whether uppercase or lowercase is used when entering SQL. With some other database systems, case *is* important, so be sure to check the documentation.

It is possible to use the MySQL superuser to create tables within the MER database. However, it is better practice to create a user within the database system to have authority over the database, and then perform all operations on the MER database as this user. The queries to do this are entered at the *MySQL Monitor* prompt. Here are the queries and the messages returned:

```
mysql> use mysql;
Database changed

mysql> grant all on MER.* to bbp identified by 'passwordhere';
Query OK.  0 rows affected (0.00 sec)
```

The first query tells MySQL that any subsequent queries are to be applied to the named database, which in this case is the `mysql` database. The message returned confirms this. The second query does three things:

1. It creates a new MySQL user called "bbp".

2. It assigns a password with the value of "passwordhere" to user "bbp".

3. It grants every available privilege relating to the MER database to "bbp".

Note that all of the above queries are terminated by the required semicolon. With the "bbp" user created, and appropriate rights granted, the *MySQL Monitor* can be exited by typing QUIT at the prompt:

```
mysql> quit
Bye
```

The MER database and the "bbp" user now exist within the MySQL database system. The next task is to create the required tables within the database.

## 2.4.4 Adding tables to the MER database

The ability to interactively type an SQL query into the *MySQL Monitor* and execute it can be very convenient, especially when the amount to type is small. For larger tasks (which involve more typing), it is often better to put the SQL query in a text file and then "feed" it to the *MySQL Monitor* from the Linux command-line. For instance, assume a text file called create_proteins.sql contains the SQL DDL queries to create an appropriately structured proteins table. This text file can be fed to the *MySQL Monitor* with the following command:

```
mysql -u bbp -p MER < create_proteins.sql
```

The "<" character *redirects* the contents of the create_proteins.sql text file and sends it to the *MySQL monitor* as standard input. Note how the database to use, MER, is specified on the command-line.

> **Technical Commentary:** If a message similar to *"ERROR 1045: Access denied for user: 'bbp@localhost' (Using password: YES)"* appears at this stage, don't fret. MySQL is supplied with an *anonymous* user enabled, and this user can sometimes cause problems. Remove the anonymous user from the MySQL system by issuing these commands as *root*: "use mysql;", "delete from user where User = '';" and "flush privileges;". That should fix the problem.

The create_proteins.sql text file contains a valid CREATE TABLE query. This is the SQL DDL query that defines the structure for, and creates a new table in, a database.

Before examining the contents of create_proteins.sql, let's return to the SWISS-PROT entry from earlier and highlight the data that populates the proteins table. However, before proceeding, let's have another maxim.

> **Maxim 12.2** *Understand the data before designing the tables.*

As the intention is to show the correct process to go through when designing, creating and populating a table, only a subset of the available line types are extracted from each SWISS-PROT entry for eventual inclusion in the table.

Referring back to the SWISS-PROT entry on page 228, the line types to be extracted are as follows:

- ID – The identification tag, specifically the mnemonic code and species sub-parts of the ID tag.
- AC – The accession number.
- DT – The date, specifically the last of the three dates provided.
- DE – The description.
- SQ – The sequence header, with a specific extraction of the sequence length.
- The actual sequence data (which has a blank line type).

For reasons that will become clear later, the order of these line types in the proteins table is AC, ID (code sub-part), ID (species sub-part), DT, DE, SQ, sequence length and then the sequence data.

That's a total of eight columns per row of data in the table. With this in mind, let's examine the contents of the create_proteins.sql text file (which has been formatted with plenty of whitespace to make it easy to read):

```
create table proteins
(
    accession_number varchar (6)   not null,
    code             varchar (4)   not null,
    species          varchar (5)   not null,
    last_date        date          not null,
    description      text          not null,
    sequence_header  varchar (75)  not null,
    sequence_length  int           not null,
    sequence_data    text          not null
)
```

A single CREATE TABLE query defines and creates the proteins table. Note the format used on each column specification line:

**Column name** – This is also referred to as the *field name*, and it uniquely identifies the columned data item within the table row.

**Column type** – This restricts the *type of data* that can be stored in the column. There are a number of different column types supported by MySQL, and the proteins table uses four of them:

1. The *varchar* type restriction is a string, which can vary in length from 0 to 255 characters.

2. The *date* type restriction is a valid date in YYYY-MM-DD format.

3. The *text* type restriction is a sequence of text from 0 to 65,535 characters in size.

4. The *int* type restriction is a number in the range −2,147,483,648 to 2,147,483,647.

**Column length** - An optional *maximum length* for the data item stored in the column. For the `proteins` table, each of the *varchar* columns indicates the largest string that can be stored in the data item, by including the maximum length value in parentheses.

**Column specifier** - An optional *null specifier* that can be used to indicate that data must be entered into the column. All of the columns in `proteins` are specified to be *NOT NULL*, which means that data must be provided for every column in the row.

Each column specification (expect for the last) ends with a comma, ",", and the entire list of fields is enclosed in parentheses, "(" and ")". Note that when fed to the *MySQL Monitor* from the command-line, this SQL DDL query *does not* need to end with a semicolon, as the use of the semicolon is implied.

With the table created, use the *MySQL Monitor* to access the MER database and issue a SHOW TABLES query against the database, as follows:

```
mysql -u bbp -p MER

mysql> show tables;
+----------------+
| Tables_in_MER  |
+----------------+
| proteins       |
+----------------+
1 row in set (0.00 sec)
```

It is also possible to ask MySQL to provide details on the structure of the `proteins` table using the DESCRIBE query, as follows:

```
mysql> describe proteins;
+------------------+-------------+------+-----+------------+-------+
| Field            | Type        | Null | Key | Default    | Extra |
+------------------+-------------+------+-----+------------+-------+
| accession_number | varchar(6)  |      |     |            |       |
| code             | varchar(4)  |      |     |            |       |
| species          | varchar(5)  |      |     |            |       |
| last_date        | date        |      |     | 0000-00-00 |       |
| description      | text        |      |     |            |       |
| sequence_header  | varchar(75) |      |     |            |       |
| sequence_length  | int(11)     |      |     | 0          |       |
| sequence_data    | text        |      |     |            |       |
+------------------+-------------+------+-----+------------+-------+
8 rows in set (0.04 sec)
```

**238**  *Databases*

Additional information is provided on each column, over and above the name and type restriction. Refer to the *MySQL Manual* for more details and the meaning of these additional data.

With the table ready, the next task is to populate it with data derived from a collection of SWISS-PROT entries.

### 12.4.5  Preparing SWISS-PROT data for importation

There are two common techniques for populating a table with data. The first is to use an SQL DML query, INSERT, to add data one row at a time, via *MySQL Monitor*. This is effective only when a small amount of row data is to be added to a table. When a large amount of data is to be added, most modern database systems provide a mechanism to import data in bulk from a correctly formed data file. Such a mechanism is provided by MySQL.

MySQL expects data files that contain importable data to be *tab-delimited*, with each row of data on its own line. This means that each piece of columned data in a row is separated from the next by the *tab* character (often written as "\t"), and each row is separated from the next by the *newline* character (often written as "\n").

Obviously, this is a very different format to that used by SWISS-PROT. What's required is a mechanism to convert the SWISS-PROT entry into a tab-delimited line of data that MySQL can import into the `proteins` table. As described on page 229, this is complicated by a number of factors. However, with Perl, these complicating factors can be dealt with without too much difficulty. The conversion strategy is as follows:

1. Process the SWISS-PROT data file by examining each entry one line at a time.
2. For each line, perform a series of pattern matches against the line in order to determine the line type.
3. When a line type of interest is matched, extract any interesting data from the line, and use the extracted data to construct the tab-delimited line of data.
4. When an entire tab-delimited line has been constructed, print it to the screen.
5. Process the next SWISS-PROT entry by returning to point 2 above, and iterate.
6. Finish when there are no more lines of input.

The tab-delimited line must conform to the following format in order to allow for bulk-importation into the `proteins` table:

- An accession number, extracted from the AC line.
- A *tab* character.

## A Database Case Study: MER 239

- The mnemonic code of the protein name, extracted from the ID line.
- A *tab* character.
- The mnemonic species identification code, extracted from the ID line.
- A *tab* character.
- A date (in YYYY-MM-DD format) extracted and converted from the last DT line. Note: the SWISS-PROT date format is DD-MMM-YYYY.
- A *tab* character.
- A description copied from any DT lines.
- A *tab* character.
- A sequence header copied from the SQ line.
- A *tab* character.
- The sequence length, extracted from the SQ line.
- A *tab* character.
- The sequence data copied from any sequence data lines.
- A *newline* character.

With the strategy determined, let's examine the get_proteins program, which takes any collection of SWISS-PROT entries and converts them into correctly formatted tab-delimited lines:

```perl
#! /usr/bin/perl -w

# get_proteins - given a list of SWISS-PROT files, extract data
# from them in preparation for importation into a database system.
#
# Note that the results produced are TAB-delimited.

BEGIN {
    push @INC, "$ENV{'HOME'}/bbp/";
}
use UsefulUtils qw( biodb2mysql );

use strict;

my ( $table_line, $code, $species );

while ( <> )
{
    if ( /^ID   (.+)_(.+?) / )
    {
        ( $code, $species ) = ( $1, $2 );
```

```perl
    }
    if ( /^AC   (.+?);/ )
    {
        $table_line = $1 . "\t" . $code . "\t" . $species . "\t";

        while ( <> )
        {
            last unless /^AC/;
        }
    }
    if ( /^DT/ )
    {
        my $date_line = $_;

        while ( <> )
        {
            last unless /^DT/;
            $date_line = $_;
        }
        $date_line =~ /^DT   (.+?) /;
        $table_line = $table_line . biodb2mysql( $1 ) . "\t";
    }
    if ( /^DE   (.+)/ )
    {
        my $descr_lines = $1;

        while ( <> )
        {
            last unless /^DE   (.+)/;
            $descr_lines = $descr_lines . ' ' . $1
        }
        $table_line = $table_line . $descr_lines . "\t";
    }
    if ( /^SQ   (.+)/ )
    {
        my $header = $1;

        $header =~ /(\d+)/;

        $table_line = $table_line . $header . "\t" . $1 . "\t";
    }
    if ( /^   (.+)/ )
    {
        my $sequence_lines = $1;

        while ( <> )
        {
            if ( m[^//] )
            {
                last;
            }
```

```
            else
            {
                /^    (.+)/;
                $sequence_lines = $sequence_lines . $1;
            }
        }
        $table_line = $table_line . $sequence_lines;
    }
    if ( m[^//] )
    {
        print "$table_line\n";
        $table_line = '';
    }
}
```

Let's describe the workings of this program in detail. Take a moment to print out a SWISS-PROT entry so that it can be referred to while working through the description of this program.

After the standard first line and a comment, a BEGIN block pushes the location of the *Bioinformatics, Biocomputing and Perl* shared code directory onto the @INC array. This allows the program to find the utilities module developed in the *Getting Organised* chapter[6]. A use of the utilities module comes immediately after the BEGIN block. Note the explicit mention of the biodb2mysql subroutine, which is used to convert the SWISS-PROT formatted date into a date format that is acceptable to MySQL:

```
BEGIN {
    push @INC, "$ENV{'HOME'}/bbp/";
}
use UsefulUtils qw( biodb2mysql );
```

Strictness is switched on, then three scalar variables are declared:

```
use strict;

my ( $table_line, $code, $species );
```

The $table_line scalar holds the (soon to be constructed) tab-delimited line, whereas the $code and $species scalars hold the extracted mnemonic protein code and species values, respectively.

A loop is started that continues to execute while there are lines of data arriving from standard input:

```
while ( <> )
{
```

---

[6] We could also use a use lib statement here (as described in *Getting Organised*), but we wished to show the other popular technique for including "local" modules.

The current line of data is assigned to the Perl's default scalar variable, $_. Once assigned, the line is matched against a series of patterns. The first of these patterns looks for the ID line type:

```
if ( /^ID   (.+)_(.+?) / )
{
    ( $code, $species ) = ( $1, $2 );
}
```

Specifically, the pattern attempts to match against a line that starts with the letter "I", followed by the letter "D" and three space characters. After the space characters, the match looks for two series of one or more characters (the ".+" pattern), separated from each other by an underscore character, and followed by a single-space character. If the pattern matches, the program knows it has found an identification line type within the SWISS-PROT entry.

Note that the second series of characters near the end of the pattern is *non-greedy* because of the use of the "?" qualifier. This stops the second ".+" pattern from attempting to match as much of the line as possible by forcing the pattern to match *as soon as possible*.

The parentheses "(" and ")" that surround the two ".+" patterns arrange for perl to remember the matched values in the $1 and $2 scalars. These values correspond to the mnemonic code for the protein and its associated species, and they are used within the if block to initialise the $code and $species scalars.

The second pattern looks for the AC line type, and upon a match, the program starts to construct the tab-delimited line. The matched accession number, together with the code and species values, with each data value separated from the next by a *tab* character, is assigned to the $table_line scalar[7]:

```
if ( /^AC   (.+?);/ )
{
    $table_line = $1 . "\t" . $code . "\t" . $species . "\t";

    while ( <> )
    {
        last unless /^AC/;
    }
}
```

Processing the AC line type is complicated by the fact that a SWISS-PROT entry can have more than one AC line type. Additionally, there can be more than one accession number on each AC line. Only the first accession number is of interest, so the pattern *non-greedily* matches against the first, which is a series of characters immediately followed by a semicolon, which is matched by the *non-greedy* pattern "(.+?);".

---

[7] Remember that "." is the Perl *concatenation operator*.

# A Database Case Study: MER

The second `while` loop within the `if` block (often referred to as an *inner loop*) reads and discards any additional lines that match the letters "AC" at the start of the line. In this way, any additional AC line types are ignored. Note the use of `last`, which when invoked ensures that the inner loop ends as soon as a line that starts with anything other than "AC" is encountered.

When it comes to extracting the last date from any DT lines, the program first needs to find the last date line. Once found, it matches against the date part of the line, then calls the `biodb2mysql` subroutine to convert the SWISS-PROT date into a format that is acceptable to MySQL. The converted date is then added to the `$table_line` scalar, together with a *tab* character:

```
if ( /^DT/ )
{
    my $date_line = $_;

    while ( <> )
    {
        last unless /^DT/;
        $date_line = $_;
    }
    $date_line =~ /^DT    (.+?) /;
    $table_line = $table_line . biodb2mysql( $1 ) . "\t";
}
```

Note that unlike the AC line type, in which the requirement was to extract the first accession number from the first AC line than ignore the rest, this `if` block ignores all but the last DT line. As each DT line is read, the current line is temporarily stored in the `$date_line` scalar, then the pattern match is applied to `$date_line` once there are no more DT lines to process. Again, the use of *non-greedy* pattern qualifiers ensure that only the required information is matched and remembered in the `$1` scalar.

The DE line type contains the description of the protein structure. As there can be more than one DE line type, the `if` block matches a pattern against the description text, remembers the description in the `$descr_lines` scalar, then processes any remaining DE line types, concatenating the matched description to the description already in `$descr_lines`:

```
if ( /^DE   (.+)/ )
{
    my $descr_lines = $1;

    while ( <> )
    {
        last unless /^DE   (.+)/;
        $descr_lines = $descr_lines . ' ' . $1
    }
    $table_line = $table_line . $descr_lines . "\t";
}
```

With all the description lines determined, they are added to $table_line, together with a *tab* character.

The SQ line type provides sequence header details for the SWISS-PROT entry. The first number in this line is the sequence length, and it is extracted from the sequence header in order to import it into the `proteins` table as a separate data item. The `if` block starts by remembering the sequence header in a scalar called $header. A second pattern match is then performed against the value in $header to determine the first number, which is matched against the "\d+" pattern:

```
if ( /^SQ   (.+)/ )
{
    my $header = $1;

    $header =~ /(\d+)/;

    $table_line = $table_line . $header . "\t" . $1 . "\t";
}
```

The sequence header (in $header) and the determined sequence length (in $1) are then added to the $table_line scalar, separated from each other by the required *tab* character.

The actual data associated with the protein structure is in the sequence data line type, which does not have a two-letter line tag (unlike the line types ID, AC, DT, DE and SQ). As with the DE line type, there can be more than one line of data in the sequence. The strategy for determining the entire sequence is similar to that used to determine the entire description. The sequence data is located immediately before the end of the SWISS-PROT entry, which is indicated by a double slash (//) at the start of a line on its own. The `if` block looks for this pattern, and when it is found, it uses `last` to break out of the inner loop. Note the use of the square brackets as delimiters around the "//" pattern, because the forward-leaning slash character is the default pattern-matching delimiter:

```
if ( /^    (.+)/ )
{
    my $sequence_lines = $1;

    while ( <> )
    {
        if ( m[^//] )
        {
            last;
        }
        else
        {
            /^    (.+)/;
            $sequence_lines = $sequence_lines . $1;
        }
```

```
        }
        $table_line = $table_line . $sequence_lines;
    }
```

When all the sequence data lines are in the $sequence_lines scalar, they are added to the $table_line scalar. As the sequence data is at the end of a row of data within the proteins table, there is no need to add a *tab* character. Instead, the line of data will be terminated by the *newline* character.

The final pattern match in get_proteins checks for the end of entry double slash. When it is found, the if block prints out the value of $table_line with the required *newline*. Once printed, the value of $table_line is reset to the empty string, in preparation for processing the next SWISS-PROT entry (if there is one):

```
        if ( m[^//] )
        {
            print "$table_line\n";
            $table_line = '';
        }
    }
```

When provided with the names of a collection of data files containing one or more SWISS-PROT entries, the get_proteins program converts all the entries in each of the data files into individual tab-delimited lines of data, one line per entry. The line of data is then printed to standard output. Assume that a collection of SWISS-PROT data files are named as follows:

```
acica_ADPT.swp.txt
serma_abdprt.swp.txt
shilf_seq_ACDP.swp.txt
```

The following invocation of the get_proteins program takes these files, performs the conversion on each entry and then writes the output to a data file called proteins.input:

```
./get_proteins  *swp*  >  proteins.input
```

The ">" character on the command-line *redirects* the output away from standard output and towards the named file.

## 2.4.6 Importing tab-delimited data into proteins

There now exists a collection of tab-delimited rows of data in proteins.input. Importing this data into the proteins table is straightforward:

```
mysql  -u  bbp  -p  MER

mysql> load data local infile "proteins.input" into table proteins;

Query OK, 14 rows affected (0.07sec)
Records: 14  Deleted: 0, Skipped: 0, Warnings: 0
```

246  Databases

After logging-in to the MER database as the "bbp" user, a LOAD DATA query is issued to import the data in the file `proteins.input` into the `proteins` table. MySQL responds by stating that the query was OK, and indicates that 14 records were affected. 14 rows of data have been successfully added to the `proteins` table. Of note is the fact that the addition of the 14 rows of data took all of 0.07 seconds.

### 12.4.7 Working with the data in `proteins`

The SQL DML query, SELECT, allows data in a table to be displayed[8]. The basic form of the SELECT query involves specifying the names of the columns to display, together with the table name. Here is a SELECT query that displays the `accession_number` and `sequence_length` values for all the rows in the `proteins` table:

```
mysql> select accession_number, sequence_length
    -> from proteins;
+------------------+-----------------+
| accession_number | sequence_length |
+------------------+-----------------+
| Q52109           |             561 |
| Q52110           |             121 |
| Q52107           |              91 |
| Q52106           |             116 |
| P08662           |             460 |
| P08664           |             212 |
| P08654           |             121 |
| P13113           |              91 |
| P13111           |             144 |
| P13112           |             116 |
| P08332           |             564 |
| P04337           |              60 |
| P20102           |             120 |
| P04129           |              91 |
+------------------+-----------------+
14 rows in set (0.06 sec)
```

The SELECT query *extracts* the columns from the `proteins` table and displays the data in the form of a table. As expected, this new (temporary) table has two columns and 14 rows of data. The word FROM has special meaning when used with SELECT: it identifies the table against which to execute the query.

Note how this query is entered into the *MySQL Monitor* over two lines. If, when entering a query, the *Enter* key is pressed, the `MySQL Monitor` prompts for an additional line of input with the "->" symbol. Remember, the query is

---
[8] It may be helpful to refer to the description of the `proteins` table on page 237 while working through this section.

not executed until the required semicolon is encountered. In this query, the semicolon appears at the end of the second line. The *MySQL Monitor* treats the two lines as one single query.

SELECT queries can be qualified in a number of ways[9]. The ORDER BY qualifier sorts the results on the basis of a column name. In this next query, the results from the query are sorted by accession_number:

```
mysql> select accession_number, sequence_length
    -> from proteins
    -> order by accession_number;
+------------------+-----------------+
| accession_number | sequence_length |
+------------------+-----------------+
| P04129           |              91 |
| P04337           |              60 |
| P08332           |             564 |
| P08654           |             121 |
| P08662           |             460 |
| P08664           |             212 |
| P13111           |             144 |
| P13112           |             116 |
| P13113           |              91 |
| P20102           |             120 |
| Q52106           |             116 |
| Q52107           |              91 |
| Q52109           |             561 |
| Q52110           |             121 |
+------------------+-----------------+
14 rows in set (0.01 sec)
```

A further qualifier, WHERE, filters the results from the query on the basis of a condition. In this next example query, only those results in which the length of the sequence is greater than 200 are displayed:

```
mysql> select accession_number, sequence_length
    -> from proteins
    -> where sequence_length > 200
    -> order by sequence_length;
+------------------+-----------------+
| accession_number | sequence_length |
+------------------+-----------------+
| P08664           |             212 |
| P08662           |             460 |
| Q52109           |             561 |
| P08332           |             564 |
+------------------+-----------------+
4 rows in set (0.04 sec)
```

---

[9] Refer to the *MySQL Manual* for the full list of qualifiers.

And with this query, Question 1 from page 231 is answered: *How many protein structures in the database are longer than 200 amino acids in length?*. The answer is 4. Note that the results from this query are sorted by `sequence_length`, as opposed to `accession_number` (as they were with the last query).

## 12.4.8 Adding another table to the MER database

More data is required to answer the rest of the questions on page 231. Another table needs to be created in the MER database to accommodate this additional data. Specifically, this table holds DNA sequences extracted from a series of EMBL entries.

The `create_dnas.sql` text file contains a CREATE TABLE query that defines the structure for a new table, called `dnas`:

```
create table dnas
(
    accession_number varchar (8)   not null,
    entry_name       varchar (9)   not null,
    sequence_version varchar (16)  not null,
    last_date        date          not null,
    description      text          not null,
    sequence_header  varchar (75)  not null,
    sequence_length  int           not null,
    sequence_data    text          not null
)
```

This table structure is not unlike that for the `proteins` table (on page 237). However, it is different. The `accession_number` in the `dnas` table can be 8 characters long, whereas the similarly named column in `proteins` is restricted to a maximum of 6 characters. Also, the second and third columns in this table hold data on the EMBL entries name and the version number, respectively. Recall that columns 2 and 3 in the `proteins` table hold data on the mnemonic code and species for a protein structure.

As with the creation of the `proteins` table, the `create_dnas.sql` text file can be fed to the *MySQL Monitor* to create the `dnas` table. The *MySQL Monitor* is then used to issue a SHOW TABLES and DESCRIBE query to confirm that the `dnas` table exists within the MER database, as follows:

```
mysql  -u  bbp  -p  MER  <  create_dnas.sql

mysql  -u  bbp  -p  MER

mysql> show tables;
+-----------------+
| Tables_in_MER   |
+-----------------+
| dnas            |
```

```
| proteins       |
+----------------+
2 rows in set (0.00 sec)

mysql> describe dnas;
+------------------+--------------+------+-----+------------+-------+
| Field            | Type         | Null | Key | Default    | Extra |
+------------------+--------------+------+-----+------------+-------+
| accession_number | varchar(8)   |      |     |            |       |
| entry_name       | varchar(9)   |      |     |            |       |
| sequence_version | varchar(16)  |      |     |            |       |
| last_date        | date         |      |     | 0000-00-00 |       |
| description      | text         |      |     |            |       |
| sequence_header  | varchar(75)  |      |     |            |       |
| sequence_length  | int(11)      |      |     | 0          |       |
| sequence_data    | text         |      |     |            |       |
+------------------+--------------+------+-----+------------+-------+
8 rows in set (0.00 sec)
```

## 2.4.9 Preparing EMBL data for importation

The strategy for populating the dnas table with data is very similar to that used with proteins. A program called get_dnas (which is based on get_proteins) processes any number of EMBL entries and converts each entry into an appropriately formatted tab-delimited line of data. Here is the get_dnas program:

```
#! /usr/bin/perl -w

# get_dnas - given a list of EMBL files, extract data
# from them in preparation for importation into a database system.
#
# Note that the results produced are TAB-delimited.

BEGIN {
    push @INC, "$ENV{'HOME'}/bbp/";
}
use UsefulUtils qw( biodb2mysql );

use strict;

my ( $table_line, $name );

while ( <> )
{
    if ( /^ID       (.+?) / )
    {
        $name = $1;
    }
    if ( /^AC       (.+?);/ )
    {
```

```perl
            $table_line = $1 . "\t" . $name . "\t";

            while ( <> )
            {
                last unless /^AC/;
            }
        }
        if ( /^SV    (.+)/ )
        {
            $table_line = $table_line . $1 . "\t";
        }
        if ( /^DT/ )
        {
            my $date_line = $_;

            while ( <> )
            {
                last unless /^DT/;
                $date_line = $_;
            }
            $date_line =~ /^DT   (.+?) /;
            $table_line = $table_line . biodb2mysql( $1 ) . "\t";
        }
        if ( /^DE   (.+)/ )
        {
            my $descr_lines = $1;

            while ( <> )
            {
                last unless /^DE   (.+)/;
                $descr_lines = $descr_lines . ' ' . $1
            }
            $table_line = $table_line . $descr_lines . "\t";
        }
        if ( /^SQ   (.+)/ )
        {
            my $header = $1;

            $header =~ /(\d+)/;

            $table_line = $table_line . $header . "\t" . $1 . "\t";
        }
        if ( /^     (.+?)\s+\d+/ )
        {
            my $sequence_lines = $1;

            while ( <> )
            {
                if ( m[^//] )
                {
                    last;
```

```
            }
            else
            {
                /^        (.+?)\s+\d+$/;
                $sequence_lines = $sequence_lines . ' ' . $1;
            }
        }
        $table_line = $table_line . $sequence_lines;
    }
    if ( m[^//] )
    {
        print "$table_line\n";
        $table_line = '';
    }
}
```

Rather than describe the workings of this program in detail (it is very similar to get_proteins, after all), let's examine the differences between this program and the get_proteins program.

The ID line in the EMBL entry is easy to process because there is no mnemonic code nor species sub-parts to extract, as there was with the SWISS-PROT entry. The identification of the EMBL entry is *non-greedily* matched against the line and assigned to the $name scalar:

```
if ( /^ID     (.+?) / )
{
    $name = $1;
}
```

Note that the value of $name is added to the tab-delimited line during the processing of the AC line type.

The SV line is not found within SWISS-PROT entries, so the get_dnas program adds a pattern match to first find, and then extract, the EMBL sequence version and add it to the tab-delimited line of data:

```
if ( /^SV     (.+)/ )
{
    $table_line = $table_line . $1 . "\t";
}
```

Refer back to the sample EMBL entry on page 229, and note the format of the sequence data. Unlike the sequence data within a SWISS-PROT entry, each line of EMBL sequence data ends with a number. These numbers indicate the number of the last base in each line of sequence data, and are included to allow readers to quickly locate a particular region of interest. There is no requirement to include these numbers in the dnas table, so the pattern match used within the if block ensures that the numbers are not concatenated with the list of bases:

```
         if ( /^     (.+?)\s+\d+/ )
         {
             my $sequence_lines = $1;

             while ( <> )
             {
                 if ( m[^//] )
                 {
                     last;
                 }
                 else
                 {
                     /^     (.+?)\s+\d+$/;
                     $sequence_lines = $sequence_lines . ' ' . $1;
                 }
             }
             $table_line = $table_line . $sequence_lines;
         }
```

The pattern used to extract these bases are:

```
    /^     (.+?)\s+\d+/
```

This matches five space characters at the start of a line, "^", followed by a collection of one or more characters, ".+?", followed by one or more space characters, "\s+", followed by one or more digits, "\d+", positioned at the end of a line, "$". The collection of characters is remembered in the $1 scalar, then used to construct the line of sequence data. When all the lines that contain sequence data are exhausted, the list of bases is added to the tab-delimited line.

The rest of get_dnas is as per the description of the get_proteins program. Let's assume a small series of EMBL entries is contained in a collection of data files with the following names:

```
AF213017.EMBL.txt
J01730.embl.txt
M15049.embl.txt
M24940.embl.txt
```

The following invocation of the get_dnas program takes these files, performs the conversion on each entry, and then writes any output to a data file called dnas.input:

```
./get_dnas   *EMBL*   *embl*   >   dnas.input
```

Remember that the ">" character on the command-line *redirects* the output away from standard output and towards the named file.

## 4.10 Importing tab-delimited data into dnas

There now exists a collection of tab-delimited rows of data in dnas.input. Importing this data into the dnas table is accomplished by logging-in to the MER database (using *MySQL Monitor*), and issuing the following LOAD DATA query:

```
mysql> load data local infile "dnas.input" into table dnas;

Query OK, 4 rows affected (0.01sec)
Records: 4   Deleted: 0, Skipped: 0, Warnings: 0
```

MySQL responds by stating that the query is OK, and indicates that 4 rows of data were added to the dnas table.

## 4.11 Working with the data in dnas

Answering Question 2 from page 231 is easy, as the SQL DML query is based on the query used to answer Question 1 from the last section. Here's the SELECT query and the results returned from MySQL:

```
mysql> select accession_number, sequence_length
    -> from dnas
    -> where sequence_length > 4000
    -> order by sequence_length;
+------------------+-----------------+
| accession_number | sequence_length |
+------------------+-----------------+
| J01730           |            5747 |
| AF213017         |            6838 |
+------------------+-----------------+
2 rows in set (0.00 sec)
```

Which answers the question: *How many DNA sequences in the database are longer than 4000 bases in length?*

Answering Question 3, *What's the largest DNA sequence in the database?*, is complicated by the fact that MySQL, version 3, does not yet support a technology called *sub-select*. This is the ability of SQL to take the results of one SELECT query and use them as part of another. For instance, this SELECT query returns the largest sequence_length value from the dnas table:

```
select max( sequence_length ) from dnas;
```

It would be convenient to embed the result from this query into another SELECT query, and then extract a list of columns, like this:

```
select accession_number, entry_name, sequence_length
from dnas
where sequence_length = ( select max( sequence_length ) from dnas );
```

That is, the sub-select determines the largest `sequence_length` value, which is then used to extract the `accession_number`, `entry_name` and `sequence_length` columns from the `dnas` table for the row that contains a value equal to the maximum. This would be nice, if only MySQL supported this feature[10].

Other than using *another* database system, this MySQL limitation can be worked around using a number of techniques. One is to simply order the results by `sequence_length`, and arrange to display the list in descending order. That way, the row (or rows) with the largest `sequence_length` appear at the top of the results. Here, again, is the query that answered Question 2, this time with the ORDER BY clause qualified by the word DESC, which orders the results in descending order:

```
mysql> select accession_number, sequence_length
    -> from dnas
    -> where sequence_length > 4000
    -> order by sequence_length desc;
+------------------+-----------------+
| accession_number | sequence_length |
+------------------+-----------------+
| AF213017         |            6838 |
| J01730           |            5747 |
+------------------+-----------------+
2 rows in set (0.00 sec)
```

Another technique is to arrange to display only a single row from the results. The LIMIT query qualifier does just this:

```
mysql> select accession_number, entry_name, sequence_length
    -> from dnas
    -> order by sequence_length desc
    -> limit 1;
+------------------+------------+-----------------+
| accession_number | entry_name | sequence_length |
+------------------+------------+-----------------+
| AF213017         | AF213017   |            6838 |
+------------------+------------+-----------------+
1 row in set (0.01 sec)
```

And there it is, the answer to Question 3: *What's the largest DNA sequence in the database?*

## 12.4.12 Relating data in one table to that in another

The real power of a database system comes from its ability to relate the data in one table to that in another. As they stand, the `proteins` and `dnas` tables are

---

[10] Version 4.1 of MySQL is under active development as this book is being written. Current plans call for the inclusion of *sub-select*.

independent of one another. Although the table structures are similar in that, for instance, they both contain a column called `accession_number`, this alone does not allow the tables to be related to each other. The AC values in `proteins` are unique to the SWISS-PROT database, just as those in `dnas` are unique to the EMBL database.

Both the SWISS-PROT and EMBL entries contain an optional DR line type, which contains a list of database cross references for the entry. Here are the DR lines from the sample SWISS-PROT entry (page 228):

```
DR   EMBL; AF213017; AAA19679.1; -.
DR   InterPro; IPR003457; Transprt_MerT.
DR   Pfam; PF02411; MerT; 1.
```

Note the EMBL line, which cross-references this SWISS-PROT entry to an identified EMBL entry. It is this information that can be used to relate the data in the `proteins` table to that in `dnas`. Here are the DR lines from the sample EMBL entry (page 229):

```
DR   GOA; P08662; P08662.
DR   GOA; P13111; P13111.
DR   GOA; P13112; P13112.
DR   GOA; P13113; P13113.
DR   SWISS-PROT; P08662; MERA_SERMA.
DR   SWISS-PROT; P13111; MERR_SERMA.
DR   SWISS-PROT; P13112; MERT_SERMA.
DR   SWISS-PROT; P13113; MERP_SERMA.
```

Again, notice that there are DR lines that cross-reference this EMBL entry to a small collection of SWISS-PROT entries. It is this information that can be used to relate the data in the `dnas` table to that in `proteins`.

## 2.4.13 Adding the `crossrefs` table to the MER database

The `create_crossrefs.sql` text file contains a CREATE TABLE query that defines the structure for a new table, called `crossrefs`:

```
create table crossrefs (
    ac_protein varchar (6) not null,
    ac_dna     varchar (8) not null
)
```

The `crossrefs` table contains two columns. The first, `ac_protein`, holds the accession number extracted from a SWISS-PROT entry, while the second, `ac_dna`, holds the accession number extracted from an EMBL entry. This table is added to the MER database with the now familiar commands:

```
mysql -u bbp -p MER < create_crossrefs.sql

mysql -u bbp -p MER

mysql> show tables;
+-----------------+
| Tables_in_MER   |
+-----------------+
| crossrefs       |
| dnas            |
| proteins        |
+-----------------+
3 rows in set (0.00 sec)

mysql> describe crossrefs;
+------------+------------+------+-----+---------+-------+
| Field      | Type       | Null | Key | Default | Extra |
+------------+------------+------+-----+---------+-------+
| ac_protein | varchar(6) |      |     |         |       |
| ac_dna     | varchar(8) |      |     |         |       |
+------------+------------+------+-----+---------+-------+
2 rows in set (0.00 sec)
```

The create_crossrefs.sql text file is fed to the *MySQL Monitor*, then the database is logged into by the "bbp" user. The SHOW TABLES query confirms that the database now contains three tables, and a DESCRIBE query issued against the crossrefs table provides details on the structure of crossrefs.

## 12.4.14 Preparing cross references for importation

The strategy for determining cross-reference data from both the SWISS-PROT and EMBL entries is the same. Each entry is processed one line at a time in order to determine the AC line type. When this is found, the accession number is remembered in a scalar variable container called $ac. A pattern then matches against the DR line type.

For SWISS-PROT entries that cross-reference the EMBL database, the DR line begins with "DR EMBL;", followed by the accession number of the cross-referenced EMBL entry. If a match is found on this pattern, the current SWISS-PROT accession number (stored in $ac), a *tab* character and the EMBL accession number (stored in $1) are printed to standard output.

Here is a small program, called get_protein_crossrefs, that implements this algorithm for any collection of SWISS-PROT entries:

```
#! /usr/bin/perl -w

# get_protein_crossrefs - given a list of SWISS-PROT files, extract
# data in preparation for importation into a database system.
# The AC number is extracted, together with any EMBL AC's.
```

```perl
#
# Note that the results produced are TAB-delimited.

use strict;

my ( $ac );

while ( <> )
{
    if ( /^AC   (.+?);/ )
    {
        $ac = $1;

        while ( <> )
        {
            last unless /^AC/;
        }
    }
    if ( /^DR   EMBL; (.+?); /)
    {
        print "$ac\t$1\n";
    }
}
```

Similarly, for EMBL entries that cross-reference the SWISS-PROT database, the DR line begins with "DR SWISS-PROT;", followed by the accession number of the cross-referenced SWISS-PROT entry. If a match is found on this pattern, the current SWISS-PROT accession number (stored in $1), a *tab* character and the EMBL accession number (stored in $ac) are printed to standard output.

Here is a program, called get_dna_crossrefs, that implements this algorithm for any collection of EMBL entries:

```perl
#! /usr/bin/perl -w

# get_dna_crossrefs - given a list of EMBL files, extract data
# from them in preparation for importation into a database system.
# The AC number is extracted, together with any SWISS-PROT AC's.
#
# Note that the results produced are TAB-delimited.

use strict;

my ( $ac );

while ( <> )
{
    if ( /^AC   (.+?);/ )
    {
        $ac = $1;
```

```
            while ( <> )
            {
                last unless /^AC/;
            }
        }
        if ( /^DR   SWISS-PROT; (.+?); / )
        {
            print "$1\t$ac\n";
        }
    }
```

Note that both programs produce a list of cross references, one cross reference per line, in SWISS-PROT, EMBL order.

The following invocations of both programs produce two cross-referenced lists from the same collection of SWISS-PROT and EMBL entries used earlier in this chapter:

```
./get_protein_crossrefs  *swp*               > protein_crossrefs
./get_dna_crossrefs      *embl*  *EMBL*      > dna_crossrefs
```

Two lists of cross references now exist. It is possible to load each of these lists into the crossrefs table. However, as there is a high likelihood that the combination of the two lists will result in some duplicate cross references, it is prudent to remove the duplicates *before* loading the data into the database.

Another small program, called unique_crossrefs, does just this. Using a very popular Perl *programming idiom*, it reads any number of cross references and inserts them into a hash called %unique. The name part of %unique is set to the cross-reference value, while the value part is set to 42 (for want of a better value[11]). The unique_crossrefs program ignores the value part of the hash, and takes advantage of the fact that the name parts must be unique:

```
#! /usr/bin/perl -w

# unique_crossrefs - read the cross reference files produced by
# get_dna_crossrefs and get_protein_crossrefs and produce a unique
# list by removing duplicates.

use strict;

my %unique;

while ( <> )
{
    chomp;
    $unique{ $_ } = 42;
```

---

[11] It does not really matter which value we set the value part to, as the value is never used. However, the use of 42 on this occasion may have something to do with *Douglas Adams*.

```
}
foreach my $crossref ( keys %unique )
{
    print "$crossref\n";
}
```

The cross references are read one line at a time and added to the hash. Note the use of chomp to remove the *newline* character from the end of each line of input. Once the list of cross references is exhausted, a foreach statement extracts the name parts from the %unique hash using keys, then prints them to standard output (one at a time). The following command-line takes the data files produced by protein_crossrefs and dna_crossrefs and runs the unique_crossrefs program against them. The results are written to a new data file, called unique.input:

```
./unique_crossrefs  protein_crossrefs  dna_crossrefs  >  unique.input
```

## 2.4.15 Importing tab-delimited data into crossrefs

Importing the unique.input data into the crossrefs table is accomplished by logging-in to the MER database (using *MySQL Monitor*), and issuing the following LOAD DATA query:

```
mysql> load data local infile "unique.input" into table crossrefs;
Query OK, 22 rows affected (0.04 sec)
Records: 22  Deleted: 0  Skipped: 0  Warnings: 0
```

A total of 22 distinct cross references now exist in the database.

## 2.4.16 Working with the data in crossrefs

A quick way to view all the data in a table is to use the wildcard version of the SELECT query. Use this SELECT query to view every row and column in the crossrefs table:

```
mysql> select * from crossrefs;
+------------+----------+
| ac_protein | ac_dna   |
+------------+----------+
|   P04336   | J01730   |
|   P08332   | J01730   |
|   P08654   | M15049   |
|   P20102   | X03405   |
|   Q52107   | AF213017 |
|   P03827   | J01730   |
|   P13113   | M24940   |
|   P04129   | J01730   |
```

```
| P13112      | M24940    |
| P04337      | J01730    |
| P04129      | K03089    |
| P08662      | M24940    |
| P08662      | M15049    |
| P13111      | M24940    |
| P08332      | K03089    |
| P20102      | L29404    |
| P03830      | J01730    |
| Q52109      | AF213017  |
| P20102      | J01730    |
| Q52106      | AF213017  |
| P04337      | K03089    |
| P08664      | M15049    |
+-------------+-----------+
22 rows in set (0.02 sec)
```

The `crossrefs` table provides the needed link to relate the `proteins` table to `dnas`. Specifically, the SWISS-PROT accession number stored in the `proteins` table can be related to the SWISS-PROT accession number in `crossrefs`. The EMBL accession number cross-referenced with the same SWISS-PROT accession number in `crossrefs` can be used to relate the EMBL accession number in `crossrefs` with the EMBL accession number stored in the `dnas` table.

Here's a SELECT query to extract data from the `proteins` and `dnas` tables on the basis of the existence of a cross reference:

```
mysql> select proteins.sequence_header, dnas.sequence_header
    -> from proteins, dnas, crossrefs
    -> where proteins.accession_number = crossrefs.ac_protein
    -> and dnas.accession_number = crossrefs.ac_dna
    -> order by proteins.sequence_header;
```

The `sequence_header` columns from both tables are explicitly identified by prefixing each column name with its associated table name. Unlike the SELECT queries from earlier, this query extracts its data from three tables: `proteins`, `dnas` and `crossrefs`, and these tables are identified as part of the FROM clause.

The WHERE qualifier relates the data in all three tables to each other. If the SWISS-PROT accession number in `crossrefs` is identical to the SWISS-PROT accession number in `proteins`, in addition to the EMBL accession number in the same row in `crossrefs` being identical to that in `dnas`, a link can be established between the protein structure and the DNA sequence.

The ORDER BY qualifier arranges to display the results sorted by SWISS-PROT sequence header. The results from this query are shown in Figure 12.1 on page 261.

```
+------------------------------------------------------------+------------------------------------------------------------+
| sequence_header                                            | sequence_header                                            |
+------------------------------------------------------------+------------------------------------------------------------+
| SEQUENCE 116 AA; 12510 MW; 2930A92CF88EB10F CRC64;         | Sequence 6838 BP; 1540 A; 1858 C; 1790 G; 1650 T; 0 other; |
| SEQUENCE 116 AA; 12511 MW; 4BE13C204E21A79E CRC64;         | Sequence 2923 BP;  617 A;  882 C;  820 G;  604 T; 0 other; |
| SEQUENCE 120 AA; 12723 MW; EB5716CA4138EB4D CRC64;         | Sequence 5747 BP; 1105 A; 1775 C; 1771 G; 1096 T; 0 other; |
| SEQUENCE 121 AA; 12954 MW; 5FF0C68DDE1F4103 CRC64;         | Sequence 2153 BP;  391 A;  641 C;  680 G;  441 T; 0 other; |
| SEQUENCE 144 AA; 16033 MW; 05FBF5224B89C052 CRC64;         | Sequence 2923 BP;  617 A;  882 C;  820 G;  604 T; 0 other; |
| SEQUENCE 212 AA; 23078 MW; 165E6640E11950E2 CRC64;         | Sequence 2153 BP;  391 A;  641 C;  680 G;  441 T; 0 other; |
| SEQUENCE 460 AA; 48587 MW; 628466EF3F653F05 CRC64;         | Sequence 2923 BP;  617 A;  882 C;  820 G;  604 T; 0 other; |
| SEQUENCE 460 AA; 48587 MW; 628466EF3F653F05 CRC64;         | Sequence 6838 BP; 1540 A; 1858 C; 1790 G; 1650 T; 0 other; |
| SEQUENCE 561 AA; 58558 MW; 111E02A702C157D6 CRC64;         | Sequence 5747 BP; 1105 A; 1775 C; 1771 G; 1096 T; 0 other; |
| SEQUENCE 564 AA; 58974 MW; 582E74333BDA88EF CRC64;         | Sequence 5747 BP; 1105 A; 1775 C; 1771 G; 1096 T; 0 other; |
| SEQUENCE  60 AA;  6456 MW; 66501654D90DB45  CRC64;         | Sequence 5747 BP; 1105 A; 1775 C; 1771 G; 1096 T; 0 other; |
| SEQUENCE  91 AA;  9414 MW; 822183AC32303lA5 CRC64;         | Sequence 2923 BP;  617 A;  882 C;  820 G;  604 T; 0 other; |
| SEQUENCE  91 AA;  9548 MW; 21EB0D79E9795069 CRC64;         | Sequence 2923 BP;  617 A;  882 C;  820 G;  604 T; 0 other; |
| SEQUENCE  91 AA;  9636 MW; 84321669C5305F6E CRC64;         | Sequence 6838 BP; 1540 A; 1858 C; 1790 G; 1650 T; 0 other; |
+------------------------------------------------------------+------------------------------------------------------------+
14 rows in set (0.28 sec)
```

**Figure 12.1** The cross-referenced sequence headers from the proteins and dnas tables.

A variation on the last SELECT query may produce more meaningful results. This query extracts the `code` and `species` values from the `proteins` table, together with any associated DNA `entry_name` for all cross references:

```
mysql> select proteins.code, proteins.species, dnas.entry_name
    -> from proteins, dnas, crossrefs
    -> where proteins.accession_number = crossrefs.ac_protein
    -> and dnas.accession_number = crossrefs.ac_dna;
+------+---------+------------+
| code | species | entry_name |
+------+---------+------------+
| MERA | SHIFL   | EC4        |
| MERD | SERMA   | PPMER      |
| MERP | ACICA   | AF213017   |
| MERP | SERMA   | PPMERR     |
| MERP | SHIFL   | EC4        |
| MERT | SERMA   | PPMERR     |
| MERC | SHIFL   | EC4        |
| MERA | SERMA   | PPMERR     |
| MERA | SERMA   | PPMER      |
| MERR | SERMA   | PPMERR     |
| MERA | ACICA   | AF213017   |
| MERD | SHIFL   | EC4        |
| MERT | ACICA   | AF213017   |
| MERB | SERMA   | PPMER      |
+------+---------+------------+
14 rows in set (0.05 sec)
```

The presentation of these results can be improved. Specifically, sorting the results by SWISS-PROT mnemonic code improves readability. Also, the ability to provide more descriptive names for each column of results also helps. Here's a variation on the last query that implements both improvements:

```
mysql> select
    -> proteins.code as 'Protein Code',
    -> proteins.species as 'Protein Species',
    -> dnas.entry_name as 'DNA Entry Name'
    -> from proteins, dnas, crossrefs
    -> where proteins.accession_number = crossrefs.ac_protein
    -> and dnas.accession_number = crossrefs.ac_dna
    -> order by proteins.code;
+--------------+-----------------+----------------+
| Protein Code | Protein Species | DNA Entry Name |
+--------------+-----------------+----------------+
| MERA         | SERMA           | PPMERR         |
| MERA         | SERMA           | PPMER          |
| MERA         | ACICA           | AF213017       |
| MERA         | SHIFL           | EC4            |
| MERB         | SERMA           | PPMER          |
| MERC         | SHIFL           | EC4            |
```

```
| MERD      | SERMA        | PPMER        |
| MERD      | SHIFL        | EC4          |
| MERP      | ACICA        | AF213017     |
| MERP      | SERMA        | PPMERR       |
| MERP      | SHIFL        | EC4          |
| MERR      | SERMA        | PPMERR       |
| MERT      | ACICA        | AF213017     |
| MERT      | SERMA        | PPMERR       |
+-----------+--------------+--------------+
14 rows in set (0.08 sec)
```

The use of the "as" keyword allows for the renaming of each column in the results table, and an appropriate ORDER BY qualifier sorts the results by SWISS-PROT mnemonic code. And with that query, Question 4 is answered: *Which protein structures are cross-referenced with which DNA sequences?*

## 4.17  Adding the `citations` table to the MER database

To answer Question 5, *Which literature citations reference the results from the previous question?* (that is, Question 4), more data is required than currently exists in the MER database. A new table, called `citations`, stores data on the citation information extracted from a collection of SWISS-PROT and EMBL entries. Here is the content of the `create_citations.sql` text file:

```
create table citations (
    accession_number varchar (8)  not null,
    number           int          not null,
    author           text         not null,
    title            text         not null,
    location         text         not null,
    annotation       text
)
```

The `citations` table is populated with data from any reference lines that exist in either type of entry. These are easily identified: simply look for a series of lines that start with the RN line type. Let's refer to this series of lines as a *reference record*. Here is the reference record from the sample SWISS-PROT entry on page 228:

```
RN   [1]
RP   SEQUENCE FROM N.A.
RX   MEDLINE=94134837; PubMed=8302940;
RA   Kholodii G.Y., Lomovskaya O.L., Gorlenko Z.M., Mindlin S.Z.,
RA   Yurieva O.V., Nikiforov V.G.;
RT   "Molecular characterization of an aberrant mercury resistance
RT   transposable element from an environmental Acinetobacter strain.";
RL   Plasmid 30:303-308(1993).
```

And here is the reference record from the sample EMBL entry on page 229:

```
RN   [1]
RP   1-2923
RX   MEDLINE; 89327136.
RA   Nucifora G., Chu L., Silver S., Misra T.K.;
RT   "Mercury operon regulation by the merR gene of the organomercurial
RT   resistance system of plasmid pDU1358";
RL   J. Bacteriol. 171(8):4241-4247(1989).
XX
```

Notice how both reference records are similar in that each has the same sequence of line types, presented in the following order: RN, RP, RX, RA, RT and RL. However, not all of these line types are required within the reference record and, to make matters slightly more complicated, the SWISS-PROT manual identifies a different set of mandatory and optional line types for its reference records than does the EMBL manual.

The `citations` table, as defined above, provides columns to hold the reference number (RN), author (RA), title (RT) and location (RL). The other columns store an accession number (extracted from the AC line type) and an optional annotation.

The usual sequence of commands is used to create the `citations` table, check to see that the table has been added to the database (using SHOW TABLES) and display the structure of the newly created table (using DESCRIBE):

```
mysql  -u  bbp  -p  MER  <  create_citations.sql

mysql  -u  bbp  -p  MER

mysql> show tables;
+---------------+
| Tables_in_MER |
+---------------+
| citations     |
| crossrefs     |
| dnas          |
| proteins      |
+---------------+
4 rows in set (0.00 sec)

mysql> describe citations;
+------------------+------------+------+-----+---------+-------+
| Field            | Type       | Null | Key | Default | Extra |
+------------------+------------+------+-----+---------+-------+
| accession_number | varchar(8) |      |     |         |       |
| number           | int(11)    |      |     | 0       |       |
| author           | text       |      |     |         |       |
| title            | text       |      |     |         |       |
| location         | text       |      |     |         |       |
```

```
| annotation       | text        | YES  |     | NULL     |       |
+------------------+-------------+------+-----+----------+-------+
6 rows in set (0.00 sec)
```

Note that the annotation column accepts NULL values, as it is defined as optional.

## 4.18 Preparing citation information for importation

The get_citations program processes any collection of SWISS-PROT and EMBL entries, extracting any found reference records. Both types of entry can contain zero, one or more reference records, and the get_citations program needs to accommodate this. Additionally, the RT line type, which contains the reference title, is – somewhat surprisingly – optional within SWISS-PROT entries, but not within an EMBL entry. This also has to be taken into consideration. Here's the entire get_citations program:

```perl
#! /usr/bin/perl -w

# get_citations - given a list of SWISS-PROT and EMBL files, extract
# data in preparation for importation into a database system.
# Specifically, extract the RN citation information from the files.
#
# Note that the results produced are TAB-delimited.

use strict;

my ( $table_line, $ac, $title_lines );

while ( <> )
{
    if ( /^AC   (.+?);/ )
    {
        $ac = $1;

        while ( <> )
        {
            last unless /^AC/;
        }
    }
    if ( /^RN   \[(\d+)\]/ )
    {
        print "$table_line\t\n" if defined $table_line;

        $table_line = $ac . "\t" . $1 . "\t";

        while ( <> )
        {
            if ( /^RA   (.+)/ )
```

```perl
            {
                my $author_lines = $1;

                while ( <> )
                {
                    last unless /^RA   (.+)/;
                    $author_lines = $author_lines . ' ' . $1
                }
                $table_line = $table_line . $author_lines . "\t";
            }
            if ( /^RT   (.+)/ )
            {
                $title_lines = $1;

                while ( <> )
                {
                    last unless /^RT   (.+)/;
                    $title_lines = $title_lines . ' ' . $1
                }
                $table_line = $table_line . $title_lines . "\t";
            }
            if ( /^RL   (.+)/ )
            {
                my $location_lines = $1;

                if ( !defined( $title_lines ) )
                {
                    $table_line = $table_line . '(no title)' . "\t";
                }
                $title_lines = undef;

                while ( <> )
                {
                    last unless /^RL   (.+)/;
                    $location_lines = $location_lines . ' ' . $1
                }
                $table_line = $table_line . $location_lines;

                if ( /^RN   \[(\d+)\]/ )
                {
                    print "$table_line\t\n" if defined $table_line;

                    $table_line = $ac . "\t" . $1 . "\t";

                    redo;
                }
                else
                {
                    last;
                }
            }
```

```
            }
        }
    }
    print "$table_line\t\n" if defined $table_line;
```

The accession number is extracted from the AC line type in the usual way, and stored in the $ac scalar. A pattern match then looks for "RN" at the start of a line. If this is not found, the entry has no reference records and the get_citations program ends, producing no results. This explains the use of the if defined statement qualifier appended to each of the print statements. That is, if there's no $table_line to print, don't print it.

If the pattern *is* found, the program processes the reference record. Recall that more than one reference record can exist in either entry. With SWISS-PROT entries, the first, second (and subsequent) reference records are positioned immediately after each other in the entry. With EMBL entries, the first, second (and subsequent) reference records are separated from each other by a XX line type. This helps explain the inclusion of a pattern match for "RN" at the start of the line within the inner loop, as follows:

```
if ( /^RN    \[(\d+)\]/ )
{
    print "$table_line\t\n" if defined $table_line;

    $table_line = $ac . "\t" . $1 . "\t";

    redo;
}
else
{
    last;
}
```

If another RN line type is encountered within the inner loop (that is, while already processing a reference record), it is highly likely that a SWISS-PROT entry is being processed. The if block prints the current $table_line, starts another $table_line and then invokes Perl's redo subroutine. This causes the current (inner) loop to restart *without* re-evaluating the loop condition. As the program has determined that a new reference record is starting, and as the program has already read the first line of the record (the RN line type), this is the most sensible thing to do at this stage.

If, having reached the end of a reference record and having read a line that starts with something other than "RN", the program can assume that it is reading an EMBL entry or that it has reached the end of the reference records in the SWISS-PROT file. Either way, the invocation of last within the else block ensures that the inner loop ends.

The following command-line executes the get_citations program against all the files in the current directory[12]. The results are written to a new data file, called citations.input:

```
./get_citations  *  >  citations.input
```

### 12.4.19 Importing tab-delimited data into citations

Importing citations.input into the crossrefs table is accomplished by logging-in to the MER database (using *MySQL Monitor*), and issuing the following LOAD DATA query:

```
mysql> load data local infile "citations.input" into table citations;
Query OK, 34 rows affected (0.08 sec)
Records: 34  Deleted: 0  Skipped: 0  Warnings: 0
```

Thirty-four citations are now stored in the table.

### 12.4.20 Working with the data in citations

It is possible to exploit the fact that the citations table includes a column of data that contains a mix of SWISS-PROT and EMBL accession numbers. Specifically, the accession_number column in citations can be related to the similarly named column in both proteins and dnas, as well as the ac_protein and ac_dna columns in crossrefs.

Here is a SELECT query that answers Question 5, *Which literature citations reference the results from the previous question?*:

```
mysql> select
    -> proteins.code as 'Protein Code',
    -> proteins.species as 'Protein Species',
    -> dnas.entry_name as 'DNA Entry Name',
    -> citations.location as 'Citation Location'
    -> from proteins, dnas, crossrefs, citations
    -> where proteins.accession_number = crossrefs.ac_protein
    -> and dnas.accession_number = crossrefs.ac_dna
    -> and dnas.accession_number = citations.accession_number
    -> order by proteins.code;
```

This query is the longest in this chapter. Despite this, it is not too difficult to understand. In essence, it is the same query that answered Question 4, with the main difference being that the accession_number column in the dnas table is

---

[12] The assumption is that the current directory contains the collection of SWISS-PROT and EMBL data files.

also related to the `accession_number` column in the `citations` table. The FROM clause includes the `citations` table in its list, and the `location` column of data (from `citations`) is included in the results for this query as the "Citation Location" column.

The abridged results from this query are shown in Figure 12.2 on page 270.

## Where to from Here

A lot of ground has been covered in this chapter. Despite this, there is much more to databases – this chapter is merely an introduction. No consideration has been given to important database topics such as primary/secondary keys, indices and normalisation. Nevertheless, the simple technique described in this chapter can be applied to many situations. The mechanism is as follows:

- Design the table structures.
- Prepare the data for importation.
- Import the data.
- Process the data.

In the next chapter, the emphasis shifts from interacting with MySQL manually (using the *MySQL Monitor*) to interacting automatically with the Perl programming language. However, before moving on, take a moment to consider one more maxim.

> **Maxim 12.3** *The SELECT query can do no harm.*

All SELECT can do is extract data from a collection of database tables. SELECT cannot be used to insert, delete, replace or update data, which has the effect of making SELECT a relatively safe database query to work with. Readers are encouraged to do just that: experiment with SELECT, safe in the knowledge that it can do no harm.

## The Maxims Repeated

Here's a list of the maxims introduced in this chapter.

- *A little database design goes a long way.*
- *Understand the data before designing the tables.*
- *The SELECT query can do no harm.*

```
+--------------+-----------------+------------------+----------------------------------------------------------+
| Protein Code | Protein Species | DNA Entry Name   | Citation Location                                        |
+--------------+-----------------+------------------+----------------------------------------------------------+
| MERA         | ACICA           | AF213017         | Plasmid 30(3):303-308(1993).                             |
| MERA         | ACICA           | AF213017         | Gene 269(1-2):121-130(2001).                             |
| .            | .               | .                | .                                                        |
| .            | .               | .                | .                                                        |
| MERC         | SHIFL           | EC4              | Unpublished.                                             |
| MERD         | SHIFL           | EC4              | Proc. Natl. Acad. Sci. U.S.A. 75(2):615-619(1978).       |
| MERD         | SHIFL           | EC4              | Nature 292(5824):640-643(1981).                          |
| MERD         | SHIFL           | EC4              | Proc. Natl. Acad. Sci. U.S.A. 81(19):5975-5979(1984).    |
| MERD         | SHIFL           | EC4              | Gene 34(2-3):253-262(1985).                              |
| MERD         | SERMA           | PPMER            | Unpublished.                                             |
| MERD         | ACICA           | AF213017         | Proc. Natl. Acad. Sci. U.S.A. 84(10):3112-3116(1987).    |
| MERP         |                 |                  | Plasmid 30(3):303-308(1993).                             |
| .            | .               | .                | .                                                        |
| .            | .               | .                | .                                                        |
| MERR         | SERMA           | PPMER            | J. Bacteriol. 171(8):4241-4247(1989).                    |
| MERT         | ACICA           | AF213017         | Plasmid 30(3):303-308(1993).                             |
| MERT         | ACICA           | AF213017         | Gene 269(1-2):121-130(2001).                             |
| MERT         | ACICA           | AF213017         | Unpublished.                                             |
| MERT         | ACICA           | AF213017         | Submitted (09-DEC-1999) to the EMBL/GenBank/DDBJ databases ... |
| MERT         | SERMA           | PPMER            | J. Bacteriol. 171(8):4241-4247(1989).                    |
+--------------+-----------------+------------------+----------------------------------------------------------+
39 rows in set (0.01 sec)
```

**Figure 12.2** The results of the citation cross reference between the proteins and dna tables.

# Exercises

1. Create a new MySQL database called *MAD*. Within MAD, create two tables, one called Discoveries and another called Scientists. Model the structure of these tables on the descriptions from the start of this chapter. Use the *MySQL Monitor* to add ten rows of data to each table.

2. If you have not already done so, create the MER database and its four tables. Populate the tables with your own selection of SWISS-PROT and EMBL entries.

# 13
# Databases and Perl

*Using Perl to talk to databases.*

## 13.1 Why Program Databases?

Why would anybody want to program a database? *Paul Dubios*, a highly respected member of the MySQL community[1], suggests four reasons:

1. **Customised output handling** – The standard output produced by most database systems (including MySQL) is often bland, as demonstrated by the examples in the last chapter. To "fancy things up a little", programs can be written to post-process the results of any SQL query and display them in any number of preferred formats.

2. **Customised input handling** – This can provide a more intuitive (and easier to use) mechanism for getting data into a database and for issuing queries. Users of customised input handling programs do not need to know anything about SQL – all they need to know and understand is their data.

3. **Extending SQL** – Some tasks that are difficult or impossible to do with SQL can be programmed more easily.

4. **Integrating MySQL into custom applications** – Having the power of MySQL as a component of an application can be very powerful. This is especially true of web-based applications (which is covered in Part III, *Working with the Web*).

[1] He is also the author of some *excellent* books on MySQL.

This chapter presents a collection of Perl programs written to communicate directly with a database system (specifically, with MySQL).

## 13.2 Perl Database Technologies

A number of third-party CPAN modules provide access to MySQL from within a Perl program. One such module is `Net::MySQL` by *Hiroyuki Oyama*, which, despite being at release 0.08, provides a stable programming interface to MySQL functionality. In fact, nearly every database system provides a specific technology for programmers to use when programming their particular database. This technology is referred to as an API, an *application programming interface*.

Unfortunately, the effort expended in learning how to use `Net::MySQL` is of little use when a program has to be written to interface with Oracle or Sybase (or any other database system, for that matter). Imagine spending six months designing and writing a large series of programs to work with MySQL (utilising `Net::MySQL`) only to be informed that a decision has been made in your organisation to store all data in Oracle, not MySQL.

In order to avoid such situations while working with Perl, *Tim Bunce* developed a module called `DBI`. The `DBI` module provides a *database independent interface* for Perl. By providing a generalised API, programmers can program at a "higher level" than the API provided by the database system, in effect insulating programs from changes to the database system. To connect the high-level `DBI` technology to a particular database system, a special *driver* converts the general `DBI` API into the database system-specific API. These drivers are implemented as CPAN modules.

*Alligator Descartes* wrote the first version of the `DBD::mysql` module to allow `DBI` programs to interface with MySQL. Today, *Jochen Wiedmann* is primarily responsible for the ongoing development of `DBD::mysql`. DBD stands for *database driver*.

The theory is that if a program is written to use `DBI` and then connected to a particular database system using the appropriate driver module, then the database system can be changed *at any time* without severely impacting the program[2]. All that is required is to change the program to use the driver module for the newly selected database system. This typically involves changing only a single line of code. *Everything else stays the same.*

In practice, each driver includes a series of "enhancements" that provide a mechanism to access database system-specific functionality. Although this can be very convenient, it is best avoided, as accessing database system-specific functionality defeats the whole purpose of using `DBI` in the first place.

> **Maxim 13.1** *If at all possible,*
> *avoid the use of database driver "enhancements".*

---

[2] A similar theory underlies the *ODBC* technology from *Microsoft*. However, the use of ODBC requires more work from the programmer than does `DBI`.

## 13.3 Preparing Perl

Depending on the version of Linux installed, the `DBI` module and any associated DBD drivers may already be installed[3]. Paul was pleasantly surprised to find that the installation of MySQL from the `up2date` service provided by *RedHat Linux* included the downloading, installation and configuration of the `DBI` and `DBD::mysql` modules for Perl.

The manual installation of the `DBI` and `DBD::mysql` modules conforms to the standard method discussed in the *Getting Organised* chapter in Part I. To check if `DBI` and `DBD::mysql` are installed on a computer, try these commands to view the associated documentation:

```
man   DBI
man   DBD::mysql
```

Of course, it is possible that the modules are installed, but that the documentation is missing or incorrectly installed. If the above commands produce "*No manual entry for ...*" messages, use these command-lines to search your computer for installations of the `DBI` and `DBD::mysql` modules:

```
find  `perl -Te 'print "@INC"' ` -name '*.pm' -print | grep 'DBI.pm'
find  `perl -Te 'print "@INC"' ` -name '*.pm' -print | grep 'mysql.pm'
```

or use the `locate` utility provided by most Linux distributions:

```
locate   DBI.pm
locate   mysql.pm
```

If these commands produce no output, then the missing modules need to be installed from CPAN[4].

### 13.3.1 Checking the DBI installation

A simple program, called `check_drivers`, checks the status of the DBI installation, listing any installed database drivers. Here it is:

```
#! /usr/bin/perl -w

# check_drivers - check which drivers are installed with DBI.
```

---

[3] *Microsoft Windows* users will find *ActivePerl* from *ActiveState* provides excellent support to the DBI programmer in the form of PPM modules.

[4] Note that a similar `find` command can be used to locate any installed Perl module. However, custom modules cannot be found by `find` in this way. Instead, they can often be found with `locate`. Unfortunately, not all Linux distributions include, nor enable, the `locate` utility.

```
use strict;

use DBI;

my @drivers = DBI->available_drivers;

foreach my $driver ( @drivers )
{
    print "Driver: $driver installed.\n";
}
```

After enabling strictness, the check_drivers program uses the DBI module. A call is then made to the available_drivers subroutine included with the DBI module. This returns a list of installed database drivers, which are assigned to an array called @drivers. This array is then iterated over using foreach to display a formatted list of installed drivers. Here's the output produced on Paul's computer[5]:

```
Driver: ExampleP installed.
Driver: Pg installed.
Driver: Proxy installed.
Driver: mysql installed.
```

The first three drivers are included with DBI. The final driver, referred to as mysql, is the DBD::mysql driver. So, Paul's database programming environment is ready to go!

## 13.4 Programming Databases with DBI

Let's start with a simple example. A program, called show_tables, connects to the MER database (created during the last chapter) and determines the list of tables in the database. Obviously, this can easily be achieved using the *MySQL Monitor*, as it is just a matter of logging in to MySQL, using the MER database and issuing a SHOW TABLES query, effectively negating the need for a custom program. But bear with us, as all this example is designed to do is get things going. Here is the entire show_tables program:

```
#! /usr/bin/perl -w

# show_tables - list the tables within the MER database.
#               Uses "DBI::dump_results" to display results.

use strict;
```

---

[5] This is running release 9 of RedHat Linux, version 3.23.56 of MySQL, version 1.32 of DBI and version 2.1021 of DBD::mysql.

```perl
    use DBI qw( :utils );

    use constant DATABASE => "DBI:mysql:MER";
    use constant DB_USER  => "bbp";
    use constant DB_PASS  => "passwordhere";

    my $dbh = DBI->connect( DATABASE, DB_USER, DB_PASS )
        or die "Connect failed: ", $DBI::errstr, ".\n";

    my $sql = "show tables";

    my $sth = $dbh->prepare( $sql );

    $sth->execute;

    print dump_results( $sth ), "\n";

    $sth->finish;

    $dbh->disconnect;
```

Let's take a look at what's going on.

After the usual first line, a comment and the switching on of strictness, the DBI module is used. Note the inclusion of the ":utils" tag, which brings in a collection of DBI database utility routines. Three constants are defined:

**DATABASE** - Identifies the *data source* to use. In show_tables, the data source is identified as DBI, the mysql driver, and the MER database.

**DB_USER** - Identifies the username to use when connecting to the data source.

**DB_PASS** - Identifies the password to use when authenticating to the data source.

An invocation of the connect subroutine (included with DBI) establishes a database connection (or *session*) between the show_tables program and MySQL:

```perl
    my $dbh = DBI->connect( DATABASE, DB_USER, DB_PASS )
        or die "Connect failed: ", $DBI::errstr, ".\n";
```

Note the specification of the three constants as parameters to connect.

If the connection cannot be established, connect returns undef and the show_tables program dies with an appropriate error message, which includes a message from DBI (included in the $DBI::errstr scalar). If the connection succeeds, connect returns a *database handle*, which is assigned to the $dbh scalar. Database handles have specific DBI functionality associated with them. The functionality is accessed through subroutine calls[6].

---

[6] Which are referred to as *methods* within the DBI documentation. For our purposes, *method* and *subroutine* mean the same thing.

Bearing in mind that it is always a good idea to use good, descriptive names for variables, why use $dbh instead of the more descriptive $database_handle? In this case, the earlier maxim is considered, but ignored, because of the convention within the DBI programming community to use specifically named variables for certain purposes. When established conventions exist within a programming community, it is often better to follow them since nearly ever published work on DBI uses $dbh as the database handle variable name. Another example of a standard DBI variable name is $sth, which is used with *statement handles*.

> **Maxim 13.2** *Be sure to adhere to any established naming conventions within a programming community.*

With a connection established to the database, the show_tables program assigns the SQL query to a scalar called $sql. This scalar is then used in a call to the database handle's prepare subroutine that gets the SQL query ready for use. Note that the requirement to terminate the SQL query with a semicolon (as is the case with the *MySQL Monitor*) is typically not required when working with DBI. The prepared SQL query is assigned to a *statement handle* called $sth.

As with database handles, statement handles also have functionality associated with them (in the form of invokable subroutines). The execute subroutine takes the prepared SQL query and asks the database system to execute it. Any results are returned to the show_tables program and stored within the statement handle identified by $sth.

To access the results returned from the database, the show_tables program invokes the DBI utility dump_results, made available to the program as a result of the :utils tag. The dump_results subroutine displays the results of the executed query in a relatively raw, unformatted way.

The program concludes by calling the finish subroutine on the statement handle, which tells the database system that the query session is over, and disconnects from the database by calling disconnect on the database handle.

Use the following command-lines to turn the show_tables program into an executable, then invoke it:

```
chmod +x show_tables
./show_tables
```

The following results are displayed on STDOUT:

```
'citations'
'crossrefs'
'dnas'
'proteins'
4 rows
4
```

This is the default (raw) format produced by the call to dump_results, and it provides the names of the four tables currently included within the MER database. It also confirms that four rows of data were processed. As shown, it is just a *dump* of the *results*, including any extra messages produced by the database (which helps explain the last two lines of output).

### 3.4.1 Developing a database utility module

The three constants at the top of the show_tables program, together with the call to connect, are used by every other program in this chapter. Rather than cut 'n' paste these five lines of code into every program, a small database utilities module is created to store them in a central location. As well as eliminating all that cutting 'n' pasting, this strategy avoids the situation whereby a large collection of programs include the username and password. Generally, it is not a good idea to litter disk-files with username/password combinations (for obvious security reasons). Also, by specifying the username and password in a single location, they can easily be changed as required.

> **Maxim 13.3** *Avoid littering programs with username/password combinations.*

Here's another custom module, called DbUtilsMER, which is stored in the same directory as the UsefulUtils module (introduced in the *Getting Organised* chapter from Part I):

```
package DbUtilsMER;

# DbUtilsMER.pm - the database utilities module from "Bioinformatics,
#                 Biocomputing and Perl".
#
# Version 0.01:   module created to hold MERconnectDB.

require Exporter;

our @ISA           = qw( Exporter );

our @EXPORT        = qw( MERconnectDB );
our @EXPORT_OK     = qw();
our %EXPORT_TAGS   = ();

our $VERSION       = 0.01;

use constant DATABASE => "DBI:mysql:MER";
use constant DB_USER  => "bbp";
use constant DB_PASS  => "passwordhere";

sub MERconnectDB {
    #
    # Given:  nothing.
```

## 280  Databases and Perl

```
        # Return: a "connected" database handle to the MER database or
        #         if no connection is possible, return "undef".
        #
        return DBI->connect( DATABASE, DB_USER, DB_PASS );
}

1;
```

In addition to defining the three constants, this module also defines a small subroutine called MERconnectDB, which either successfully establishes a connection to the MER database using the specified username/password pairing, or returns undef if the connection cannot be established. With the DbUtilsMER module available, each program that wishes to communicate with the database includes the following code:

```
use lib "$ENV{'HOME'}/bbp/";
use DbUtilsMER;

my $dbh = MERconnectDB
    or die "Connect failed: ", $DBI::errstr, ".\n";
```

as opposed to this code:

```
use constant DATABASE => "DBI:mysql:MER";
use constant DB_USER  => "bbp";
use constant DB_PASS  => "passwordhere";

my $dbh = DBI->connect( DATABASE, DB_USER, DB_PASS )
    or die "Connect failed: ", $DBI::errstr, ".\n";
```

which despite "saving" only one line, does significantly improve the maintainability of any program using it.

### 13.4.2 Improving upon dump_results

A second version of show_tables, called show_tables2, uses the DbUtilsMER module and the MERconnectDB subroutine. It also processes the results from the query within a loop and, as such, reformats the output to be more presentable. Here's the code:

```
#! /usr/bin/perl -w

# show_tables2 - list the tables within the MER database.
#                Uses "fetchrow_array" to display results.

use strict;

use DBI;
```

```
use lib "$ENV{'HOME'}/bbp/";
use DbUtilsMER;

my $dbh = MERconnectDB
    or die "Connect failed: ", $DBI::errstr, ".\n";

my $sql = "show tables";

my $sth = $dbh->prepare( $sql );

$sth->execute;

print "The MER database contains the following tables:\n\n";

while ( my @row = $sth->fetchrow_array )
{
    foreach my $column_value ( @row )
    {
        print "\t$column_value\n";
    }
}

$sth->finish;

$dbh->disconnect;
```

When executed, show_tables2 produces the following on STDOUT:

```
The MER database contains the following tables:

        citations
        crossrefs
        dnas
        proteins
```

The interesting (and new) lines of code are these:

```
print "The MER database contains the following tables:\n\n";

while ( my @row = $sth->fetchrow_array )
{
    foreach my $column_value ( @row )
    {
        print "\t$column_value\n";
    }
}
```

A simple `print` statement displays a friendly message, then a `while` loop displays the results from the database. With each iteration, the `fetchrow_array` subroutine (which is part of DBI) is invoked against the statement handle[7]. This subroutine returns a single row from the results as an array. When used within a loop, as is the case within show_tables2, each iteration results in fetchrow_array returning the next row of data from the results (until there are no more rows).

Each row is assigned to an array called @row. The body of the while loop contains another loop, this time a foreach that iterates over the @row array and prints the values contained therein. In this case, there's only one value per row (that is, one column in the table). Each value is printed to STDOUT, with the tab (\t) and newline (\n) characters helping out with the required formatting.

Unlike the output produced by show_tables, which used the dump_results utility, the show_tables2 program does not see the "4 rows" and "4" messages from earlier. The use of fetchrow_array produces only the results that are likely to be of interest to the user, as opposed to all the messages generated by the database system.

## 13.5 Customising Output

There are a number of ways to process the results returned to a statement handle from a database system. To demonstrate the most common techniques, the crossrefs table from the MER database is used in the examples in this section. A very simple query is issued against the database to retrieve every row of data from the table:

```
select * from crossrefs
```

The which_crossrefs program processes the results from the query and displays them in a nice, human-friendly format. Here's what the output looks like:

```
There are 22 cross references in the database.

The protein P04336 is cross referenced with J01730.
The protein P08332 is cross referenced with J01730.
The protein P08654 is cross referenced with M15049.
The protein P20102 is cross referenced with X03405.
The protein Q52107 is cross referenced with AF213017.
The protein P03827 is cross referenced with J01730.
The protein P13113 is cross referenced with M24940.
The protein P04129 is cross referenced with J01730.
The protein P13112 is cross referenced with M24940.
The protein P04337 is cross referenced with J01730.
The protein P04129 is cross referenced with K03089.
The protein P08662 is cross referenced with M24940.
```

---

[7] Remember: any results from the database are *stored* in the statement handle.

```
    The protein P08662 is cross referenced with M15049.
    The protein P13111 is cross referenced with M24940.
    The protein P08332 is cross referenced with K03089.
    The protein P20102 is cross referenced with L29404.
    The protein P03830 is cross referenced with J01730.
    The protein Q52109 is cross referenced with AF213017.
    The protein P20102 is cross referenced with J01730.
    The protein Q52106 is cross referenced with AF213017.
    The protein P04337 is cross referenced with K03089.
    The protein P08664 is cross referenced with M15049.
```

A message indicates the number of cross references in the database[8]. After producing a blank line, a series of messages is produced, one for each cross reference. Here's the source code to the which_crossrefs program, which produces the above output:

```perl
#! /usr/bin/perl -w

# which_crossrefs - nicely displayed list of protein->dna
#                   cross references.

use strict;

use DBI;

use lib "$ENV{'HOME'}/bbp/";
use DbUtilsMER;

my $dbh = MERconnectDB
    or die "Connect failed: ", $DBI::errstr, ".\n";

my $sql = "select * from crossrefs";

my $sth = $dbh->prepare( $sql );

$sth->execute;

print "There are ", $sth->rows, " cross references in the database.\n\n";

while ( my @row = $sth->fetchrow_array )
{
    print "The protein $row[0] is cross referenced with $row[1].\n";
}

$sth->finish;

$dbh->disconnect;
```

As can be seen, other than the fact that the $sql scalar has a different SQL query assigned to it, this program is very similar to the show_tables2 program. The lines of interest are these:

---

[8] More correctly, the number of cross references in the crossrefs table *within* the database.

```
print "There are ", $sth->rows, " cross references in the database.\n\n";

while ( my @row = $sth->fetchrow_array )
{
    print "The protein $row[0] is cross referenced with $row[1].\n";
}
```

The rows subroutine, invoked through the statement handle, returns the number of rows contained in the results, and the value is used within the opening message.

The while loop processes the results, employing fetchrow_array to return each row of data and assign it to @row. The crossrefs table has two columns, so knowing this, each column's data can be accessed using standard array indexing notation. The first column is therefore referred to as $row[0] and the second column is referred to as $row[1]. The which_crossrefs program exploits this when producing the message for each cross reference.

An alternative to using array indices is to assign the array returned from fetchrow_array to a list of named scalars. This technique is implemented in which_crossrefs2 and involves changing the condition-part of the loop, as well as the print statement within the loop block:

```
while ( my ( $protein, $dna ) = $sth->fetchrow_array )
{
    print "The protein $protein is cross referenced with $dna.\n";
}
```

This loop produces the same output as the loop used in which_crossrefs.

The third technique uses the names of the columns to reference the values contained in them. Rather than invoke fetchrow_array, the which_crossrefs3 program employs fetchrow_hashref to return a *reference to* a hash. The name-parts of this referenced hash are set to the names of the columns in the table, while the value-parts in the referenced hash are set to the values associated with each individual row. The *hash reference* is assigned to the $row scalar, then the values are accessed as follows:

```
while ( my $row = $sth->fetchrow_hashref )
{
    print "The protein $row->{ ac_protein } is cross referenced ";
    print "with $row->{ ac_dna }.\n";
}
```

The hash name-parts are identical to the column names as defined in the crossrefs table within the MER database.

The output produced by which_crossrefs3 is the same as that produced by both which_crossrefs and which_crossrefs2. Which technique is used often depends on personal preference, as they all work. The final technique, using a hash reference, guards against changes to the ordering of the columns within a table. For example, the ordering of the columns within the crossrefs table is ac_protein followed by ac_dna. If for some reason this ordering was to be changed to ac_dna, then ac_protein, of the three cross-referencing programs, only which_crossrefs3 would continue to produce correct output[9]. The which_crossrefs and which_crossrefs2 programs assume the ordering of the columns within the table *never* changes.

> **Maxim 13.4** *Use* fetchrow_hashref *to guard against changes to the structure of a database table.*

## 13.6 Customising Input

The ability to process the results produced by the execution of a query allows a programmer to customise them for any particular purpose. It is also possible to customise the input to a query.

The specific_crossref program provides a mechanism whereby a user of the program can check the crossrefs table for a specific protein cross reference. Here's a captured usage session, showing the messages produced and the input provided by the user (which is shown in *italics*):

```
Provide a protein accession number to cross reference ('quit' to end): p03377
Not found: there is no cross reference for that protein in the database.

Provide a protein accession number to cross reference ('quit' to end): p04337
Found: P04337 is cross referenced with J01730.

Provide a protein accession number to cross reference ('quit' to end): q52109
Found: Q52109 is cross referenced with AF213017.

Provide a protein accession number to cross reference ('quit' to end): x6587
Not found: there is no cross reference for that protein in the database.

Provide a protein accession number to cross reference ('quit' to end): quit
```

The user is prompted (with an appropriate message) to provide the accession number of a protein to cross reference. Once entered, the cross reference is searched for within the table and either a "found" or "not found" message is generated. Of interest is the fact that any input provided by the user is converted to uppercase, in order to match the data in the table (which is stored in all uppercase). Critically, the use of the specific_crossrefs program does not require the user to know anything about SQL syntax nor queries (which is not the case when using *The MySQL Monitor*). All the user of this program needs

---

[9] Although, frankly, it is hard to imagine a situation whereby such a reordering would be justified. Then again, you never know.

to understand is the meaning of the data contained in the table, that is, that it contains protein to DNA cross references. In effect, the `specific_crossrefs` program *shields* its users from the technical details of SQL, MySQL and database systems. Before discussing how the `specific_crossrefs` program works in detail, consider its source code:

```perl
#! /usr/bin/perl -w

# specific_crossref - allow for the "interactive" checking
#                     of crossrefs from the command-line.
#                     Keep going until the user enters "quit".

use strict;

use DBI;

use lib "$ENV{'HOME'}/bbp/";
use DbUtilsMER;

use constant TRUE => 1;

my $dbh = MERconnectDB
    or die "Connect failed: ", $DBI::errstr, ".\n";

my $sql = qq/ select ac_dna from crossrefs where ac_protein = ? /;

my $sth = $dbh->prepare( $sql );

while ( TRUE )
{
    print "\nProvide a protein accession number to cross ";
    print "reference ('quit' to end): ";

    my $protein2find = <>;

    chomp $protein2find;

    $protein2find = uc $protein2find;

    if ( $protein2find eq 'QUIT' )
    {
        last;
    }

    $sth->execute( $protein2find );

    my $dna = $sth->fetchrow_array;

    $sth->finish;

    if ( !$dna )
    {
        print "Not found: there is no cross reference for that protein ";
        print "in the database.\n";
    }
```

```
    else
    {
        print "Found: $protein2find is cross referenced with $dna.\n";
    }
}

$dbh->disconnect;
```

Despite the fact that specific_crossrefs is longer than the programs discussed so far in this chapter, it conforms to the same design. Strictness is enabled, the DBI and DbUtilsMER modules are used and a connection is established with the database system. The SQL query (assigned to $sql) is a little more complicated:

```
my $sql = qq/ select ac_dna from crossrefs where ac_protein = ? /;
```

The above line uses Perl's qq *generalised quote operator*, delimiting the SQL query with slash characters. This has the effect of double-quoting the string and removing the need to escape any otherwise escapable characters within the SQL query. Note the use of the "?" character. This is referred to as a *placeholder* within DBI. A placeholder (which is always identified by the "?" character) indicates a place where a value in the SQL query will be provided later in the program, *after* the statement has been prepared but *before* it is executed.

An infinite while loop provides a mechanism within which the user of the program is repeatedly asked to supply the accession number of a protein to cross reference. Entering "quit" results in the program terminating. The input from the user is assigned to a scalar called $protein2find, then processed:

```
my $protein2find = <>;

chomp $protein2find;

$protein2find = uc $protein2find;

if ( $protein2find eq 'QUIT' )
{
    last;
}
```

Any trailing newline character is removed from the user's input using chomp, then Perl's in-built uc subroutine is used to convert the input to uppercase. If the value in $protein2find is now "QUIT", the loop is exited by calling last. If the value in $protein2find is anything else, the loop continues.

The execute subroutine associated with the statement handle is then executed, and the $protein2find scalar is passed as a parameter to execute:

```
$sth->execute( $protein2find );
```

The above statement has the effect of taking the value of `$protein2find` and using it within the SQL query at the position indicated by the placeholder. So, if the value of `$protein2find` is entered as "p04377", the SQL query that starts out looking like this:

```
select ac_dna from crossrefs where ac_protein = ?
```

is transformed into the following as a result of the invocation of `execute`:

```
select ac_dna from crossrefs where ac_protein = P04377
```

So, each time a different protein accession number is provided by the user, it is used to execute a slightly different SQL query against the database. As before, the `fetchrow_array` subroutine is used to retrieve the results returned to the program and stored within the statement handle. Rather than assign the row returned by `fetchrow_array` to an array, the `specific_crossrefs` program assigns any results to a scalar:

```
my $dna = $sth->fetchrow_array;
```

At first, this may appear to be a strange thing to do. However, when used in this way, `fetchrow_array` does not return an array of row values; instead it provides a single value equal to the first (or last) column value in the row. As it is known from the query that only a single column is retrieved (the column holding the `ac_dna` values), the `specific_crossref` program exploits the scalar context behaviour of `fetchrow_array` and assigns the value returned to the `$dna` scalar. If no cross reference is found, the `$dna` scalar is assigned `undef`.

The remaining code in the loop checks the value of `$dna`, printing an appropriate message to `STDOUT` depending on whether the SQL query resulted in a found cross reference:

```
if ( !$dna )
{
    print "Not found: there is no cross reference for that protein ";
    print "in the database.\n";
}
else
{
    print "Found: $protein2find is cross referenced with $dna.\n";
}
```

As with all DBI programs, the `specific_crossrefs` program concludes by closing the connection, invoking the `disconnect` subroutine on the database handle.

## 13.7 Extending SQL

The ability to integrate the features of a relational database with the features of a programming language can leverage the best of both, leading to powerful programming solutions.

Recall the `match_embl` program from page 144, which provides a mechanism to interactively match a small DNA sequence against a sequence entry from the EMBL database, reporting successful matches and failures. The problem with `match_embl` is that it works only with a single EMBL entry. It would be nice if the program could be extended to look for matches in more than one EMBL entry. It would be *great* if the program could be extended to look for matches in the EMBL entries stored within the `dnas` table, which is part of the MER database. Achieving this is straightforward with Perl and DBI.

The `db_match_embl` program extends `match_embl` to communicate with the MER database and attempts to match a user-supplied sequence against any of the EMBL sequences in the `dnas` table. Here's the source code to `db_match_embl`:

```perl
#! /usr/bin/perl -w

# The 'db_match_embl' program - check a sequence against each EMBL
#                               database entry stored in the dnas
#                               table within the MER database.

use strict;

use DBI;

use lib "$ENV{'HOME'}/bbp/";
use DbUtilsMER;

use constant TRUE  => 1;
use constant FALSE => 0;

my $dbh = MERconnectDB
    or die "Connect failed: ", $DBI::errstr, ".\n";

my $sql = qq/ select accession_number, sequence_data,
              sequence_length from dnas /;

my $sth = $dbh->prepare( $sql );

while ( TRUE )
{
    my $sequence_found = FALSE;

    print "Please enter a sequence to check ('quit' to end): ";

    my $to_check = <>;

    chomp( $to_check );
    $to_check = lc $to_check;
```

```perl
        if ( $to_check =~ /^quit$/ )
        {
            last;
        }

        $sth->execute;

        while ( my ( $ac, $sequence, $sequence_length ) = $sth->fetchrow_array )
        {
            $sequence =~ s/\s*//g;

            if ( $sequence =~ /$to_check/ )
            {
                $sequence_found = TRUE;

                print "The EMBL entry in the database: ",
                        $ac,
                            " contains: $to_check.\n";
                print "[Lengths: ",
                        length $sequence,
                            "/$sequence_length]\n\n";
            }
        }

        if ( !$sequence_found )
        {
            print "No match found in database for: $to_check.\n\n";
        }

        $sth->finish;
    }

    $dbh->disconnect;
```

Before describing the inner workings of the db_match_embl program, let's take a look at the program in action. Here's a captured usage session, showing the messages produced and the input provided by the user (again shown in *italics*):

```
Please enter a sequence to check ('quit' to end): aattgc
The EMBL entry in the database: AF213017 contains: aattgc.
[Lengths: 6838/6838]

Please enter a sequence to check ('quit' to end): aatttc
The EMBL entry in the database: AF213017 contains: aatttc.
[Lengths: 6838/6838]

The EMBL entry in the database: J01730 contains: aatttc.
[Lengths: 5747/5747]

Please enter a sequence to check ('quit' to end): accttaaatttgtacgtg
No match found in database for: accttaaatttgtacgtg.

Please enter a sequence to check ('quit' to end): tcatgcacctgatgaacgtgcaaaaccacagtca
The EMBL entry in the database: AF213017 contains: tcatgcacctgatgaacgtgcaaaaccacagtca.
[Lengths: 6838/6838]

Please enter a sequence to check ('quit' to end): aatgc
The EMBL entry in the database: AF213017 contains: aatgc.
```

```
[Lengths: 6838/6838]

The EMBL entry in the database: J01730 contains: aatgc.
[Lengths: 5747/5747]

The EMBL entry in the database: M15049 contains: aatgc.
[Lengths: 2153/2153]

The EMBL entry in the database: M24940 contains: aatgc.
[Lengths: 2923/2923]

Please enter a sequence to check ('quit' to end): quit
```

The user is asked to enter a sequence, which is then checked against the dnas table. For each successful match found, the db_match_embl program displays the accession number of the EMBL entries that do match. Alternatively, a "no match found" message is printed. Note how the "aatttc" and "aatgc" sequences match more than one EMBL entry in the table.

The db_match_embl program begins by using the appropriate modules, defines constants for *true* and *false*, connects to the database and then assigns the following SQL query to the $sql scalar:

```
select accession_number, sequence_data, sequence_length from dnas
```

A while loop (which is *initially* infinite) repeatedly asks the user to enter a sequence to check. Note this line of code:

```
my $sequence_found = FALSE;
```

The $sequence_found scalar is set to *true* whenever a match is made, and is used later in the program to decide whether to display the "*no match found*" message. Borrowing code directly from the match_embl program, the db_match_embl program reads a sequence to check from STDIN, assigns it to the $to_check scalar, converts it to lowercase and checks to see if the value entered by the user was "quit". If it was, the loop terminates (by invoking last), otherwise the program continues.

The while loop contains another loop, an *inner* while loop, which fetches a row of results from the database and assigns the values from the row to three scalars: $ac, $sequence and $sequence_length. The EMBL entry sequence data is assigned to the $sequence scalar in this way, which has any space characters within it removed by a substitution regular expression. The "sanitised" value in $sequence is then bound against the $to_check scalar and, if successfully matched, the $sequence_found scalar is set to *true*:

```
while ( my ( $ac, $sequence, $sequence_length ) = $sth->fetchrow_array )
{
    $sequence =~ s/\s*//g;

    if ( $sequence =~ /$to_check/ )
    {
        $sequence_found = TRUE;
```

With a match found, two `print` statements output to `STDOUT`, confirming that the match was successful and providing a pair of length values:

```
print "The EMBL entry in the database: ",
        $ac,
            " contains: $to_check.\n";
print "[Lengths: ",
        length $sequence,
            "/$sequence_length]\n\n";
    }
}
```

The first length value is determined by invoking the in-built `length` subroutine against the sanitised value of `$sequence`. The second length value is the value for the length of the sequence retrieved from the database. This is a simple integrity check: both values should be the same. If they differ, this may indicate a problem with the data in the `dnas` table.

There is a temptation to add `last` to the end of the inner loop, resulting in the `do_match_embl` program stopping after the first successful match within the database table. Tempting maybe, but not advisable, as it defeats the whole purpose of trying to successfully match with as many EMBL entries as possible.

The program ends when the rows of data from the table have been exhausted, closing the connection with the database by invoking `disconnect` against the database handle.

## Where to from Here

The ability to program databases is important. Using `DBI`, it is possible to produce custom programs that interact with a database system on behalf of users, producing useful applications with relatively little effort. Indeed, the combination of MySQL, Perl and `DBI` is a potent one.

The `DBI` and `DBD::mysql` modules provide many more facilities to those described in this chapter. Take some time to work through the documentation provided with both modules. The use of `DBI` is revisited in the *Working with the Web* part of *Bioinformatics, Biocomputing and Perl*.

## The Maxims Repeated

Here's a list of the maxims introduced in this chapter.

- *If at all possible, avoid the use of database driver "enhancements".*
- *Be sure to adhere to any established naming conventions within a programming community.*

- *Avoid littering programs with username/password combinations.*
- *Use* `fetchrow_hashref` *to guard against changes to the structure of a database table.*

## Exercises

1. Install the latest version of the `Net::MySQL` module from CPAN, then rewrite the `db_match_embl` program to use `Net::MySQL` as opposed to `DBI`.

2. Write a program that prompts a user to supply two values: a protein code and a protein species value. Using these two values, the program is to use an appropriately formed SQL query to return a list of citations associated with the entered protein code/species combination. The program is to continue prompting for protein code/species combinations until such time as the user enters "`quit`". [Hint: review the SQL query used to answer Question 5 from the last chapter].

# Part III

# Working with the Web

# 14

# *The Sequence Retrieval System*

*The EBI's web-based sequence retrieval system.*

## 14.1 An Example of What's Possible

The *Sequence Retrieval System* (SRS) is often considered by Bioinformaticians to be the "next level up" from a relational database[1]. In fact, it is a *web-based database integration system* that allows for the querying of data contained in a multitude of databases, all through a single-user interface. This makes the individual databases appear as if they are really one big relational database, organised with different subsections: one called SWISS-PROT, one called EMBL, one called PDB, and so on. SRS makes it very easy to query the entire data set, using common search terms that work across all the different databases, regardless of what they are.

Everything contained within the SRS is "tied together" by the web-based interface. Figure 14.1 on page 298 is the database selection page from the EBI's SRS web-site, which can be navigated to from the following Internet address:

    http://srs.ebi.co.uk

[1] SRS is a trademark and the intellectual property of *Lion Bioscience*.

298    *The Sequence Retrieval System*

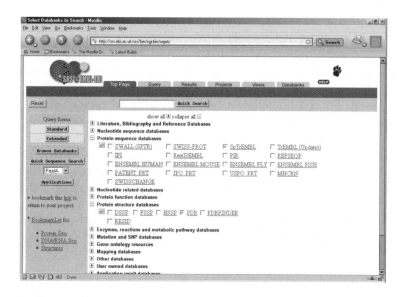

Figure 14.1   EBI's SRS database selection page.

## 14.2  Why SRS?

SRS is presented here for two reasons:

1. It is a useful and convenient service that every Bioinformatician should know about.

2. It is an excellent example of what can be created when the World Wide Web, databases and programming languages are combined.

The next two chapters cover the skills required to develop for the web. The remainder of this chapter is given over to exploring SRS.

## 14.3  Using SRS

As may be expected, some SRS queries make sense only for certain databases. For instance, trying to extract X-Ray Crystallographic resolutions is not possible from the sequence-orientated SWISS-PROT database.

On the basis of the specified query, SRS intelligently presents only those fields common to the selected databases (or datasets). Obviously, the more dissimilar the selected data types, the fewer the fields. Figure 14.1 shows the SpTrEMBL and PDB databases selected. Even when the *extended* query form is used (refer to Figure 14.2 on page 299), only a few common fields exist in each of these datasets, and these include "ID", "bibliographic details" and "date of submission".

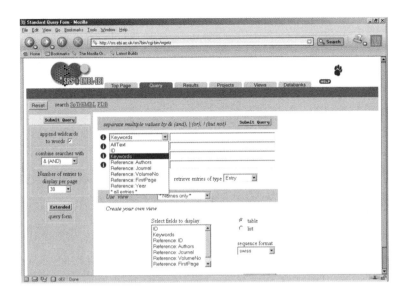

**Figure 14.2** EBI's SRS extended query page.

SRS indexes data stored in a variety of formats: flat data files, relational databases (using either *Oracle* or *MySQL*) and/or XML. Special *construction tools* allow researchers to add support to SRS for their unique data format, although a better strategy may be to reformat unique data to map onto an existing standard as this may be easier than writing a new one from the beginning.

**Maxim 14.1** *Don't create a new data format unless absolutely necessary. Use an existing format whenever possible.*

EMBL originally developed SRS, although it is now available from *Lion Bioscience*[2]. Despite being the property of a commercial organisation, usage of SRS is provided free to academic institutions.

Before showing what SRS can do, let's look at what SRS cannot do: it cannot transform information between radically different data types. However, SRS has extensive control and user interface tools, as well as an internal scripting language. This makes it straightforward to use the power of external packages if need be, such as *BLAST* or *EMBOSS*. For example, the EBI SRS implementation offers the facility to do a *protein sequence similarity search* on the basis of either a sequence pasted into a text box, as shown on Figure 14.3 on page 300, or by using the results of another SRS database search.

From an administrative standpoint, the automatic update features of SRS are very useful when maintaining copies of all the databases (of which there are currently 150 or more). In summary, SRS is a useful database integration tool

---

[2] This explains the footprint logo at the top right of each page.

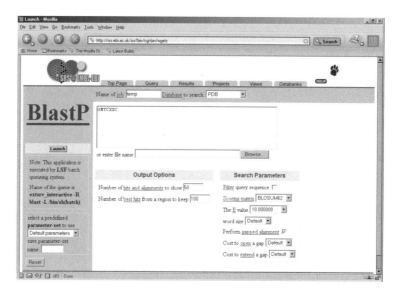

**Figure 14.3** EBI's SRS BlastP service form.

that presents a uniform search engine interface to a wide range of Bioinformatics databases.

Once selections have been made to the SRS web-based interface, the user then clicks on *Submit* or *Query* to start the search. The SRS system is then sent a message that tells it what to do. The message is sent to a *server-side program* that communicates automatically with one or more databases, extracting data of interest. Results are calculated and processed into a HTML page that is then presented back to the user. As all of this occurs within the user's web browser, it is both easy and convenient to use.

## Where to from Here

In the next chapter, the automatic creation of web pages is described. When web pages are linked to server-side programs, they become that program's user interface. The next two chapters describe the techniques used in creating this linkage, with Perl providing the programming technology that ties it all together[3].

## The Maxims Repeated

Here's the maxim introduced in this chapter.

---

[3] This helps explain why Perl is often referred to as "the duct-tape of the Internet". This description is attributed to *Sun Microsystems*, who are heavy users of Perl, despite their Marketing Department's preference for the home-grown *Java programming language*.

- Don't create a new data format unless absolutely necessary. Use an existing format whenever possible.

## Exercises

1. Skip forward to Chapter 18 and use SRS to access the entries for that chapter's DNA and protein sequences, using their associated accession codes. What information is provided by the system? What happens when the *view* of the data is changed?

2. Explore SRS's ability to pass data to external programs. Arrange for the results of an SRS search to be passed to the ClustalW program (see Chapter 17), performing a *multiple sequence analysis* of the MerP protein.

3. Experiment with the SRS "Start with Session" feature, which allows for the saving of individual searches.

4. Download and read the "Linking to SRS" document, then experiment with the wget program (introduced in Chapter 9) to download specific SWISS-PROT or EMBL database entries from SRS.

# 15
# Web Technologies

*Using the Internet to publish data and applications.*

## 15.1 The Web Development Infrastructure

The ability to publish data and applications on the Internet, in the form of custom web pages, is now considered an essential skill in many disciplines, including Biology. The development infrastructure of the World Wide Web (WWW) is well established and well understood. There is a standard set of components:

**The web server** – A program that when loaded onto a computer system, provides for the publication of data and applications (often referred to collectively as *content*). Example web-server technologies include Apache, Jigsaw and Microsoft's IIS.

**The web client** – A program that can request content from a web server and display the content (typically) within a graphical window, providing a mechanism whereby users can interact with the content. The common name for the web client is *web browser*, and there are many browsers available, including Mozilla, Netscape Navigator, Microsoft Internet Explorer, KDE Konqueror, Opera and Lynx.

**The transport protocol** – The "language" that the web client and web server use when communicating with each other. Think of this as the set of rules and regulations to which the client and server must adhere. The transport protocol employed by the WWW is called the *HyperText Transport Protocol* (HTTP). The transport protocol is the most abstract of the four components, since the web client and web server shield users from having to interact directly with it.

**The content** – The data and applications published by the web server. At its most basic level, this is textual data formatted to conform to one of the *HyperText Mark-up Language* standards (HTML)[1]. In addition to textual data, HTML can be enhanced with embedded graphics (with PNG, GIF, BMP and JPEG the most common graphic file formats). Data published in the form of HTML is often referred to as *HTML pages* or *web pages* (and both are used interchangeably in this chapter).

Much can be accomplished with these four infrastructural components, and when the WWW was invented in 1991 by *Tim Berners-Lee*, this is *all* that was available. Considerably more can be achieved today, as the standard infrastructure has been enhanced with additional components, specifically:

**Client-side programming** – A technology used to program the web client, providing a way to enhance the user's interactive experience. Common client-side programming technologies include *Java Applets*, *JavaScript* and *Macromedia Flash*.

**Server-side programming** – A technology used to program the web server, providing a mechanism to extend the services provided by the web server. Common server-side programming technologies include *Java Servlets*, *JSP*, *Python*, *ASP*, *PHP* and Perl (although in practice, just about any programming technology can be used to support server-side programming).

**Backend database technology** – A place to store the data to be published, which is accessed by the server-side programming technology. Although almost any database system can be used to provide this backend, *MySQL* is very popular on the Internet.

> **Technical Commentary:** And it is not just MySQL that is popular. The acronym *LAMP* is used to describe the favoured WWW development infrastructure of many programmers. The letters that form the acronym are taken from the words *Linux*, *Apache*, *MySQL* and Perl (although the "P" is also used to refer to *Python* and/or *PHP*). *O'Reilly & Associates* provides an excellent LAMP web-site, available on-line at http://www.onlamp.com.

These additional components turn the standard web development infrastructure into a dynamic and powerful *application development environment* that is straightforward to learn, use and exploit. One of the reasons the WWW is so popular is the fact that creating content is so straightforward. Adding a programming

---

[1] There are three versions of HTML in widespread use: 2, 3 and 4.

language into the mix allows even more to be accomplished and, as this and the next chapter demonstrate, programming the WWW is not difficult.

This chapter describes how the components introduced above are used to publish data and applications on the WWW. The examples build upon, and draw from, the material presented in earlier chapters.

## 15.2 Creating Content for the WWW

When it comes to producing content for the WWW, a number of techniques can be employed to create HTML. These include the following:

**Creating content manually** – Any text editor can be used to create HTML, since HTML is mostly text. Special *tags* within the text guide the web browser when it comes to displaying the web page on screen. The tags are also textual and any text editor can produce them.

**Creating content visually** – Special-purpose editors can create HTML pages visually, displaying the web page as it will appear in the web browser as it is edited. Programs that work in this way include *Netscape Composer*, *Microsoft FrontPage* and *Macromedia Dreamweaver*.

**Creating content dynamically** – Since HTML is text, it is also possible to create HTML from a program. The need for this technique is illustrated later in this chapter.

Each technique has advantages and disadvantages. Creating HTML manually can be time-consuming and tedious, as the creator of the page has to write the content as well as decide which tags to use and where. However, manually creating HTML web pages provides the maximum amount of flexibility as the creator has complete control over the process. It can also be advantageous to know what's going on behind the scenes, so learning HTML is highly recommended.

> **Maxim 15.1** *Take the time to learn HTML.*

Using a visual HTML editor can be very convenient, as there's really no need to know anything about HTML. The editor adds the required tags to the text that's entered by the user. Unfortunately, in your author's experience, the HTML that's produced by visual editors adds considerably more tags than typically required, resulting in the content within the page being obscured by those extra tags. Additionally, HTML pages produced with a visual editor are typically larger than a similar page produced manually[2].

---

[2] Although this is really a concern only when accessing the Internet over a slow data connection, which is becoming less of an issue within the developed world but still an issue within developing nations.

Creating HTML pages dynamically from within a program has the obvious disadvantage of requiring the web page creator to write a program to produce even the simplest of pages, which is normally not very clever. However, HTML pages produced in this way can sometimes be useful when combined with a web server that allows for server-side programming of a backend database, but more on this later.

The assumption in this chapter is that the reader is already familiar with the WWW, and that creating web pages, either manually or with a visual editor, is a skill that has already been acquired. If this is *not* the case, take some time to work through some of the on-line tutorials available on the Internet. A useful web-site in this regard is:

```
http://www.htmlprimer.com
```

Let's take a look at the simplest of web pages:

```
<HTML>
<HEAD>
    <TITLE>A Simple HTML Page</TITLE>
</HEAD>
<BODY>
This is as simple a web page as there is.
</BODY>
</HTML>
```

The header part of the HTML page contains a four word title, and the body of the HTML page contains a single ten word sentence. This web page, called `simple.html`, takes no more than a few minutes to produce, whether created manually, visually or from a program. In fact, using a HERE document to produce a HTML page is only marginally more effort than producing the same page manually.

In fact, those readers who completed the *Chapter 8* chapter exercises (at the end of Part I) already know how to produce the above HTML web page with a Perl program using a HERE document:

```
#! /usr/bin/perl -w

# produce_simple - produces the "simple.html" web page using
#                  a HERE document.

use strict;

print <<WEBPAGE;
<HTML>
<HEAD>
    <TITLE>A Simple HTML Page</TITLE>
</HEAD>
```

```
        <BODY>
        This is as simple a web page as there is.
        </BODY>
        </HTML>
        WEBPAGE
```

Of course, with Perl, there's always more than one way to do things, so here's another version of `produce_simple`, written to use Perl's standard CGI module:

```
        #! /usr/bin/perl -w

        # produce_simpleCGI - produces the "simple.html" web page using
        #                     Perl's standard CGI module.

        use strict;

        use CGI qw( :standard );

        print start_html( 'A Simple HTML Page' ),
              "This is as simple a web page as there is.",
                end_html;
```

Among other things, the `CGI` module is designed to make the production of HTML pages as convenient as possible. Written by *Lincoln D. Stein*, this module is, more than likely, the most-used module of all of those that come with Perl (after `strict`, that is). In fact, `CGI` can claim to account for Perl's huge popularity as a server-side programming technology on the WWW.

> **Technical Commentary:** An interesting aside regarding the CGI module relates to its creator: *Lincoln D. Stein*. Dr Stein is a researcher at *Cold Springs Harbor Laboratory*, using technology such as Perl to develop sophisticated Bioinformatics tools. Dr Stein is as well regarded for his contributions to the Perl programming community as he is for his contributions to, and observations of, the field of Bioinformatics. Among other things, Dr Stein has worked extensively on the AceDB database.

The `produce_simpleCGI` program uses the `CGI` module, importing a set of subroutines by specifying the `:standard` tag. Some of these subroutines are used within the program's sole `print` statement:

```
        print start_html( 'A Simple HTML Page' ),
              "This is as simple a web page as there is.",
                end_html;
```

The `print` statement contains invocation of two CGI subroutines, `start_html` and `end_html`. When invoked, the `start_html` subroutine produces the tags that appear at the start of a web page. Any string supplied as a parameter to `start_html` is used as the web page's title. The above invocation produces the following HTML:

```
        <html><head><title>A Simple HTML Page</title></head><body>
```

The end_html subroutine produces the following HTML, representing the tags that conclude a web page:

```
</body></html>
```

As the invocations of both of these subroutines occur as part of a `print` statement, they are displayed on `STDOUT`, together with the one-line message (which is the actual content). When executed, the `produce_simpleCGI` program generates the following HTML[3]:

```
<?xml version="1.0" encoding="iso-8859-1"?>
<!DOCTYPE html
        PUBLIC "-//W3C//DTD XHTML 1.0 Transitional//EN"
        "http://www.w3.org/TR/xhtml1/DTD/xhtml1-transitional.dtd">
<html xmlns="http://www.w3.org/1999/xhtml" lang="en-US" xml:lang="en-US">
<head><title>A Simple HTML Page</title>
</head><body>This is as simple a web page as there is.</body></html>
```

What's all that extra stuff at the start? The first point to make is that it is *optional*. The web page displays in any browser regardless. What these extra tags do is tell the web browser exactly which version of HTML the web page conforms to. The CGI module includes these tags for web browsers that *can* interpret the information, allowing the browser to optimise its behaviour to the version of HTML identified. Other web browsers simply ignore them.

### 15.2.1 The static creation of WWW content

The `simple.html` web page, as well as being simple, is also *static*. If the web page is put on a web server, and served up to a web browser, it always appears in exactly the same way every time it is accessed. It's static, and remains as it is until someone takes the time to change it. It rarely makes sense to create such a web page with a program unless there is some other special requirement.

> **Maxim 15.2** *Create static web pages either manually or visually.*

### 15.2.2 The dynamic creation of WWW content

When the web page includes content that is not static, it's referred to as a *dynamic* web page. An example of a dynamic web page is one that includes the current date and time. It is not possible to create a web page either manually or visually that includes dynamic content, and this is where server-side programming technologies come into their own. Here's a program, called `whattimeisit`, that creates a HTML page that includes the current date and time:

---

[3] Note that HTML tags can be in lower or uppercase.

```
#! /usr/bin/perl -wT

# whattimeisit - create a dynamic web page that includes the
#                current date/time.

use strict;

use CGI qw( :standard );

print start_html( 'What Date and Time Is It?' ),
        "The current date/time is: ", scalar localtime,
            end_html;
```

which, when executed, produces the following HTML:

```
<?xml version="1.0" encoding="iso-8859-1"?>
<!DOCTYPE html
        PUBLIC "-//W3C//DTD XHTML 1.0 Transitional//EN"
        "http://www.w3.org/TR/xhtml1/DTD/xhtml1-transitional.dtd">
<html xmlns="http://www.w3.org/1999/xhtml" lang="en-US" xml:lang="en-US">
<head><title>What Date and Time Is It?</title></head>
<body>The current date/time is: Mon Aug 25 23:21:55 2003</body></html>
```

And there it is, surrounded by the <BODY> and </BODY> tags, the date and time when the page was created. Execute the program sometime later, and the date and time change (as expected):

```
<?xml version="1.0" encoding="iso-8859-1"?>
<!DOCTYPE html
        PUBLIC "-//W3C//DTD XHTML 1.0 Transitional//EN"
        "http://www.w3.org/TR/xhtml1/DTD/xhtml1-transitional.dtd">
<html xmlns="http://www.w3.org/1999/xhtml" lang="en-US" xml:lang="en-US">
<head><title>What Date and Time Is It?</title></head>
<body>The current date/time is: Tue Aug 26 08:04:23 2003</body></html>
```

This web page, if served up by a web server, changes with each serving, as it is *dynamic*.

Note the use of the "T" command-line option at the start of `whattimeisit`. This switches on Perl's *taint mode*, which enables a set of special security checks on the behaviour of the program. Enabling these checks is particularly important when it comes to server-side programs. Be advised that although the program is created by a trusted source (namely, *you*), when executed as a result of a request from a user of a web browser on the WWW, the user may or may not be a trusted source: it could literally be *anybody*.

If a server-side program does something that could potentially be exploited and, as a consequence, pose a security threat, Perl refuses to execute the program when *taint mode* is enabled. Here's a really important maxim.

> **Maxim 15.3** *Always enable "taint mode" for server-side programs.*

## 15.3 Preparing Apache for Perl

Arranging for a web server to serve up any web page, whether static or dynamic, involves configuring the server to display the static web page or execute the server-side program as a result of a request from a web browser. Before doing this, let's ensure a web server is installed and ready to start servicing requests. As with MySQL earlier in this book, the Linux `chkconfig` command is used to add the Apache web server program and get it ready:

```
chkconfig   --add   httpd
chkconfig   httpd   on
```

On Linux systems, `httpd` is the name commonly given to the web-serving program.

The Apache web server is by far the most widespread web-server implementation, and its configuration details are maintained within a disk-file called `httpd.conf`. It is important to check (and possibly adjust) some of the settings in this disk-file. This can be accomplished only by the superuser (*root* on Linux). After becoming the superuser (or logging in as *root*), find the `httpd.conf` disk-file using the `locate` utility:

```
locate   httpd.conf
```

On Paul's computer (running RedHat Linux), the above `locate` command produces the following output:

```
/etc/httpd/conf/httpd.conf
/usr/share/apacheconf/httpd.conf.xsl
```

The first line of output reveals that the `httpd.conf` disk-file is located within the `/etc/httpd/conf/` directory[4]. Using any text editor, edit this disk-file. In *Section 2* of the `httpd.conf` disk-file, adjust the server administrator's e-mail address to something other than the default, which may look something like this:

```
ServerAdmin root@localhost
```

Whenever a problem occurs with a request on the web server, the web browser is told about the problem and given an e-mail address to which to send a "complaint message" (when appropriate). The e-mail address to use is set by the `ServerAdmin` directive.

Later in the `httpd.conf` disk-file, the `DocumentRoot` directive indicates the default directory location for static web pages:

```
DocumentRoot "/var/www/html"
```

---

[4] The second line of output refers to some other disk-file.

Any HTML page placed in the directory associated with the `DocumentRoot` directive (which is `/var/www/html` in this case) can be requested by a web browser. There's no need for the browser to specify the actual directory location of the HTML page, as its name is sufficient. For example, assume a web server called `www.example.com` and a disk-file called `index.html`. The disk-file is copied into the `/var/www/html` directory, so its name is:

`/var/www/html/index.html`

whereas it is accessed from any web browser by entering the following into the web browser's location/address bar:

`http://www.example.com/index.html`

That is, the `DocumentRoot` directive specifies the top level from where the web server starts looking for content. Unless there's a really good reason to change the value associated with the `DocumentRoot` directive, leave it as it is.

Another important Apache directive is `ScriptAlias`. This directive has two objectives: it identifies the directory location that contains any server-side programs and it provides a shorthand notation for referring to the location. Here's the default `ScriptAlias` line from the `httpd.conf` configuration disk-file:

`ScriptAlias /cgi-bin/ "/var/www/cgi-bin/"`

As with `DocumentRoot`, a server-side program copied into the `/var/www/cgi-bin/` directory is referred to by a shorthand notation. If the disk-file is:

`/var/www/cgi-bin/whattimeisit`

the server-side program (again on the `www.example.com` computer) is accessed from a web browser by entering the following into the browser's location/address bar:

`http://www.example.com/cgi-bin/whattimeisit`

One of the conditions placed on the server-side program is that it *must* be set to *executable*. More on this later.

> **Technical Commentary:** The name of the directory within which server-side programs are located is `cgi-bin`. The "bin" part is short for *binary*, which is another name for a disk-file that is set to executable, and the word is a throwback to the early days of computing. The "cgi" part is short for *Common Gateway Interface*, which is an Internet standard that describes a technique for executing server-side programs on web servers. As will be shown later, this also explains Lincoln D. Stein's choice of the name `CGI` for his module.

To test that the web server is working properly, either reboot the computer (which arranges to start the web server as the computer *boots*) or start the web-serving service using this command-line (while operating as superuser):

```
/etc/init.d/httpd start
```

With the web server up-and-running, type the following Internet address into any browser's location/address bar (assuming the browser is executing on the *same* computer as the web server):

```
http://localhost/
```

The name `localhost` is another name for the current computer being used[5]. After a short delay, the Apache "welcoming" web page should appear within the browser's window. The web page contains a message similar to this, which confirms that the web server is ready for action:

> This page is used to test the proper operation of the Apache Web server after it has been installed. If you can read this page, it means that the Apache Web server installed at this site is working properly.

> **Maxim 15.4** *Test your web-site on `localhost` prior to deployment on the Internet.*

## 15.3.1 Testing the execution of server-side programs

Let's test that the web server is capable of processing server-side programs by copying the `whattimeisit` program into the appropriate directory location, then setting the disk-file to executable. Use these commands:

```
su
cp whattimeisit /var/www/cgi-bin
chmod +x /var/www/cgi-bin/whattimeisit
<Ctrl-D>
```

Note the requirement to issue the `cp` and `chmod` commands as *root*, as it is not possible to write to the `/var/www/cgi-bin` directory as a regular user. Now, using any web browser on the same computer, surf to this web page by typing the following Internet address into your browser's location/address bar:

```
http://localhost/cgi-bin/whattimeisit
```

---

[5] That is, your computer, the one that you just started running the web server on. The `localhost` mechanism allows web-sites to be tested locally on a computer, *before* unleashing them onto the Internet.

Preparing Apache for Perl 313

**Figure 15.1** The "Server Error" web page.

This requests the execution of the `whattimeisit` program as stored in the aliased `cgi-bin` directory on the web server operating on a computer called `localhost`. Unfortunately, the web server responds with the web page as shown on Figure 15.1 on page 313.

Whoops! Something has gone wrong and the web server is not happy with the request. In fact, the web server attempted to execute the server-side program but ran into trouble, gave up and produced this error message. What is wrong is that having a program produce HTML dynamically from the web server is only part of the solution. A small amount of additional information is required to allow the web server to process the output produced by the program. Specifically, the program needs to tell the web server the *type* of information it is creating. To fix this problem, add the following line of code to the start of the `whattimeisit` program (just before the other `print` statement):

```
print header;
```

The `header` subroutine is provided by the `CGI` module and (by default) tells the web server that the output produced by the program is textual and that it is formatted as HTML. When the web page is accessed again, the web server responds with the message as shown in Figure 15.2 on page 314. Click on the browser's "reload" button, and the web page should reload with an advanced date/time value.

Note that the `whattimeisit` program sends any output to `STDOUT`. When executed as a server-side program by the web browser, this output is *captured* by the web server and *redirected* to the web-browser screen.

314     Web Technologies

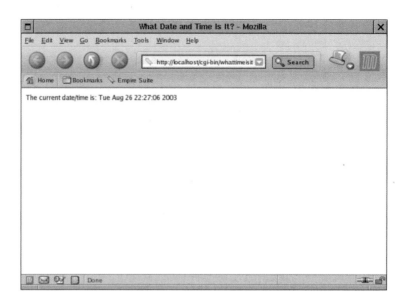

**Figure 15.2**   The "What Date and Time Is it?" web page.

Executing the server-side program via the web browser's location/address bar is good for testing the functionality of the generated web page. To execute the server-side program from another web page include its Internet address within a HTML <A> anchor tag. Here is such a tag, which is part of a HTML page called time.html:

```
<HTML>
<HEAD>
    <TITLE>Get your data and time here!</TITLE>
</HEAD>
<BODY>
Click <a href="/cgi-bin/whattimeisit">here</a> to
get the date and time.
</BODY>
</HTML>
```

The href part of the anchor tag specifies the *hypertext reference* that identifies the whattimeisit program within the cgi-bin directory on the web server. Now, when the user of a web browser requests the time.html web page from the server and clicks on the word "here", the web server executes the whattimeisit program, sending the HTML page produced by the program to the web browser screen.

This mode of executing a server-side program employs the *Common Gateway Interface*, and it is unlikely there's a web server on the planet that does not support this standard server-side programming mechanism. Programs that operate using this mechanism are referred to as *CGI programs*, *CGI scripts* or simply *CGIs*.

## 15.4 Sending Data to a Web Server

Being able to dynamically create a web page as a result of executing a server-side program is very useful and with the help of Perl and the CGI module, is not difficult. Of course, the server-side program can produce any amount of HTML, no matter how simple or complicated. Things become interesting when the server-side program accepts input from the user of the web browser.

Recall the match_embl program from page 144. By providing a short DNA sequence when prompted, match_embl looks for a match within a specific EMBL entry. Although match_embl works well, it suffers from the drawback that the user of the program has to be physically using the computer that runs match_embl. The advantage of using the WWW is that the requirement to be physically close to the content being accessed is nullified. If the match_embl program can be "moved" to the WWW, it can be accessed by any user from anywhere, which is quite an advantage.

Moving the match_embl program to the WWW is straightforward and follows a simple recipe:

- Switch on *taint mode* on the Perl command-line.
- Use the CGI module, importing (at least) the :standard set of subroutines.
- Ensure the first print statement within the program is "print header;".
- Envelope any output sent to STDOUT with calls to the start_html and end_html subroutines.
- Create a static web page to invoke the server-side program, providing input as necessary.

Except for the last step, there's not much new to learn here. Before dealing with the last step, let's look at the source code to the match_emblCGI program:

```
#! /usr/bin/perl -wT

# The 'match_emblCGI' program - check a sequence against the EMBL
#                               database entry stored in the
#                               embl.data.out data-file on the
#                               web server.

use strict;

use CGI qw/:standard/;

print header;

open EMBLENTRY, "embl.data.out"
    or die "No data-file: have you executed prepare_embl?\n";
```

```perl
    my $sequence = <EMBLENTRY>;

    close EMBLENTRY;

    print start_html( "The results of your search are in!" );

    print "Length of sequence is: <b>", length $sequence,
            "</b> characters.<p>";

    print h3( "Here is the result of your search:" );

    my $to_check = param( "shortsequence" );

    $to_check = lc $to_check;

    if ( $sequence =~ /$to_check/ )
    {
        print "Found.  The EMBL data extract contains: <b>$to_check</b>.";
    }
    else
    {
        print "Sorry.  No match found for: <b>$to_check</b>.";
    }

    print p, hr,p;

    print "Press <b>Back</b> on your browser to try another search.";

    print end_html;
```

The `match_emblCGI` program is very similar to the `match_embl` program, except for all the extra HTML-specific program code. Rather than produce straight text, a HTML web page is produced instead. Note the use of the `h3` subroutine (from `CGI`) that adds a level three HTML header to the web page. The `p` and `hr` subroutines (also from `CGI`) insert a paragraph break and horizontal rule, respectively. The critical line of code is this one:

```perl
    my $to_check = param( "shortsequence" );
```

which uses the `CGI`-supplied `param` subroutine to assign the web browser-supplied value associated with `shortsequence` to the `$to_check` scalar. But just what is `shortsequence` and when is its value set?

The `shortsequence` parameter is set within a web page, specifically within a web page that contains a form. It is a HTML *named parameter*. Here's a HTML page called `mersearch.html` that associates `shortsequence` with a HTML *textarea* component within a form:

```html
<HTML>
<HEAD>
    <TITLE>Search the Sequence for a Match</TITLE>
</HEAD>
<BODY>
```

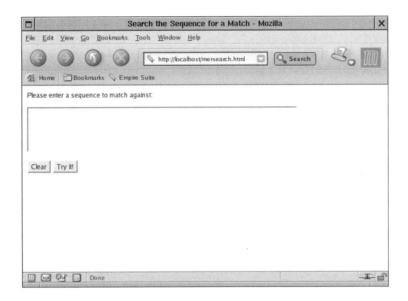

**Figure 15.3** The "Search the Sequence for a Match" web page.

```
Please enter a sequence to match against:<p>
<FORM ACTION="/cgi-bin/match_emblCGI">
    <p>
        <textarea name="shortsequence" rows="4" cols="60"></textarea>
    </p>
    <p>
        <input type="reset"  value="Clear">
        <input type="submit" value="Try it!">
    </p>
</FORM>
</BODY>
</HTML>
```

When loaded into a web browser, the mersearch.html web page should look like Figure 15.3 on page 317.

The large textarea at the top of the web page is used to enter the sequence to check[6].

The value entered by this page's user is associated with the shortsequence named parameter. When the user *submits* the web page (by clicking on the *Try it!* button), the web browser sends the named parameters *and their associated values* to the web server. The web server arranges to send the parameter/value pairings to the server-side program identified by the ACTION attribute of the form. In this example, the ACTION attribute identifies the match_emblCGI server-side program.

---

[6] Or to cut 'n' paste that little bit of sequence that you've just received via e-mail.

318     *Web Technologies*

To try out this web page and server-side program combination, use the following commands (as *root*) to copy the disk-files to the appropriate locations on the web server, noting the requirement to copy the `embl.data.out` disk-file to the `cgi-bin` directory so that the `match_emblCGI` program can find it:

```
su
cp  mersearch.html   /var/www/html
cp  match_emblCGI    /var/www/cgi-bin
chmod +x  /var/www/cgi-bin/match_embl
cp  embl.data.out    /var/www/cgi-bin
<Ctrl-D>
```

Load the `mersearch.html` web page into a browser and enter a short sequence into the textarea, then click on the *Try it!* button. If a match is found, the web browser displays a HTML page similar to Figure 15.4 on page 318. If no match is found, the web browser displays a HTML page similar to Figure 15.5 on page 319. Either way, it's the `match_emblCGI` program on the web server that decides which of the two HTML pages to return to the web browser and it decides this on the basis of whether the value associated with the `shortsequence` named parameter matches the EMBL sequence contained in the `embl.data.out` disk-file.

Note the message at the bottom of the returned web page: the user is advised to click on the *Back* button to try another search. This is a reasonable strategy to follow, but quickly becomes tiresome (for the user) when a large number of sequences have to be checked against the EMBL entry. It would be better if the results returned from the web server contained the form from the

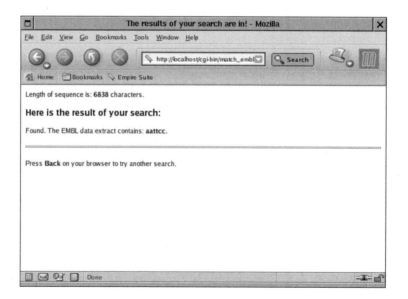

**Figure 15.4**  The "Results of your search are in!" web page.

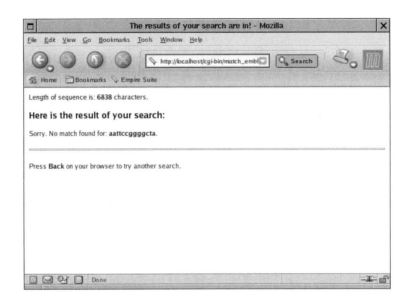

**Figure 15.5** The "Sorry! Not Found" web page.

mersearch.html web page, to allow the user to quickly check another sequence without having to click on *Back*.

The match_emblCGIbetter server-side program and its associated web-page combination implement this strategy. The only change to the web page (over mersearch.html) is to arrange to invoke the "better" CGI script from the form's ACTION attribute:

```
<FORM ACTION="/cgi-bin/match_emblCGIbetter">
```

The match_emblCGIbetter program is nearly identical to the match_emblCGI program, except that the end of page message, thus:

```
print "Press <b>Back</b> on your browser to try another search.";
```

is replaced with the following HERE document:

```
print <<MERFORM;

Please enter another sequence to match against:<p>
<FORM ACTION="/cgi-bin/match_emblCGIbetter">
    <p>
        <textarea name="shortsequence" rows="4" cols="60"></textarea>
    </p>
    <p>
        <input type="reset"  value="Clear">
        <input type="submit" value="Try it!">
    </p>
</FORM>
MERFORM
```

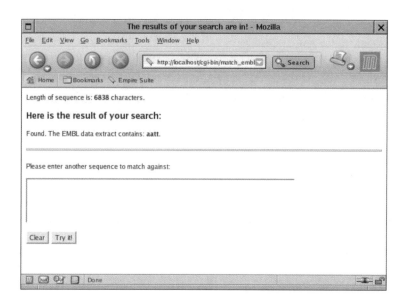

**Figure 15.6** The better version of the "Results of your search are in!" web page.

When the `mersearchbetter.html` web page is loaded into a web browser and provided with a sequence to check, clicking on the *Try it!* button results in the web server responding with a web page similar to that in Figure 15.6 on page 320.

Note the inclusion of the form at the bottom of the results HTML page. Of note, too, is the decision within the `match_emblCGIbetter` server-side program to produce the form using a HERE document. It is a straightforward matter to create a form using the subroutines provided by `CGI` but as the static web page including the form already exists, it makes sense to include the form verbatim as a HERE document. Remember: with Perl, there's more than one way to do it.

## 15.5 Web Databases

The ability to execute a server-side program as a result of a user of a web browser interacting with a HTML form opens up all types of possibilities. When a database system is added to the mix, the development environment is augmented with a powerful backend data repository. To demonstrate what's possible (while keeping everything as straightforward as possible), let's revisit the `db_match_embl` program from page 289.

The `db_match_embl` program communicates with the MER database and attempts to match a user-supplied sequence against all the EMBL entries in the `dnas` table. As with the `match_embl` program, the `db_match_embl` program requires its user to be using the same computer as that which runs the program. By moving the `db_match_embl` program to a web server and providing a web browser interface to it, the program can be accessed from anywhere on the Internet, by any user.

Here's a web page called `mersearchmulti.html` that provides the interface to the `db_match_embl` program:

```
<HTML>
<HEAD>
    <TITLE>Search the "dnas" Table for a Match</TITLE>
</HEAD>
<BODY>

Please enter a sequence to match against the database:<p>
<FORM ACTION="/cgi-bin/db_match_emblCGI">
    <p>
        <textarea name="shortsequence" rows="4" cols="60"></textarea>
    </p>
    <p>
        Include a border around the results:
        <input type="checkbox" name="printborder" value="on"
            checked="checked" />
    </p>
    <p>
        <input type="reset"  value="Clear">
        <input type="submit" value="Try it!">
    </p>
</FORM>
</BODY>
</HTML>
```

This HTML page is very similar to the `mersearch.html` web page. The page title is different, and there's a *checkbox* under the textarea that provides the user of the page with a choice as to whether the results produced by clicking on the *Try it!* button are surrounded by a border. Note that the checkbox is named `printborder`. The `ACTION` attribute within the `FORM` tag identifies `db_match_emblCGI` as the server-side program to execute when the user clicks on the *Try it!* button. Figure 15.7 on page 322 shows how the web page looks when loaded into a browser.

Note that as the checkbox is initially selected, any results produced will include the border. The results are contained within a HTML table, and Figure 15.8 on page 322 shows the results produced by entering "aatttc" into the textarea, leaving the checkbox selected and clicking on the *Try it!* button.

As can be seen in Figure 15.8, two of the four sequences in the `dnas` table contained a match. As with the previous server-side programs, the form is appended to the end of the results to allow the user to quickly execute another search (without having to press the browser's *Back* button). Note that the textarea is initialised with the previous search value, which is done as a convenience to the user. Use these commands to prepare the web server to execute the `db_match_emblCGI` server-side program:

```
su
cp  mersearchmulti.html  /var/www/html
cp  db_match_emblCGI     /var/www/cgi-bin
```

322  *Web Technologies*

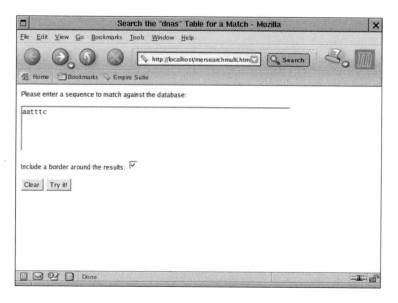

**Figure 15.7** Searching all the entries in the dnas table.

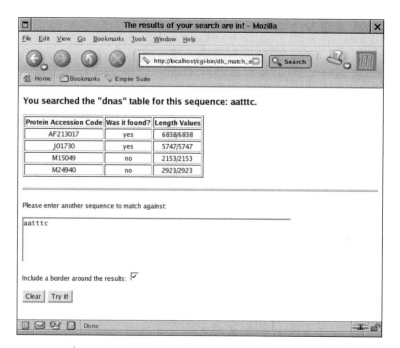

**Figure 15.8** The "results" of the multiple search on the dnas table.

```
chmod  +x  /var/www/cgi-bin/db_match_emblCGI
cp /home/barryp/DbUtilsMER.pm  /var/www/cgi-bin
<Ctrl-D>
```

Note the requirement to copy the `DbUtilsMER.pm` database module to the `cgi-bin` directory. Recall that this module lives in the `bbp` directory. When the `db_match_embl` program executes, it searches this directory location for the module and uses it. However, when the `db_match_emblCGI` server-side program executes, it runs under the user-id of the web server and, as such, cannot access the `bbp` directory and consequently cannot find the module. By copying the module to the `cgi-bin` directory, a server-side program can access its functionality without difficulty. Let's take a look at the entire source code to the `db_match_emblCGI` program, before describing what it does in detail:

```perl
#! /usr/bin/perl -wT

# The 'db_match_emblCGI' program - check a sequence against each EMBL
#                                  database entry stored in the dnas
#                                  table within the MER database.  This
#                                  is the CGI version of this program.

use strict;

use CGI qw/ :standard *table /;
use DBI;
use lib ".";
use DbUtilsMER;

print header, start_html( "The results of your search are in!" );

my $dbh = MERconnectDB
    or die "Connect failed: ", $DBI::errstr, ".\n";

my $sql = qq/ select accession_number, sequence_data,
              sequence_length from dnas /;

my $sth = $dbh->prepare( $sql );

my $to_check = param( "shortsequence" );
my $print_border = param( "printborder" );

$to_check = lc $to_check;

$sth->execute;

print h3( "You searched the \"dnas\" table for this sequence: $to_check." );

if ( $print_border )
{
    print start_table(  { -border => "1" } );
}
else
{
    print start_table(  { -border => "0" } );
```

```perl
    }

    print Tr( { -align => "CENTER" } ), th( "Protein Accession Code" ),
             th( "Was it found?" ), th( "Length Values" );

    while ( my ( $ac, $sequence, $sequence_length ) = $sth->fetchrow_array )
    {
        $sequence =~ s/\s*//g;

        print Tr( { -align => "CENTER" } ), td( "$ac" );

        if ( $sequence =~ /$to_check/ )
        {
            print td( "yes" );
        }
        else
        {
            print td( "no" );
        }

        my $calced_length = length $sequence;

        print td( "$calced_length/$sequence_length" );
    }

    print end_table;

    print p, hr, p;

    print <<MERFORM;

Please enter another sequence to match against:<p>
<FORM ACTION="/cgi-bin/db_match_emblCGI">
  <p>
    <textarea name="shortsequence" rows="4" cols="60">$to_check</textarea>
  </p>
  <p>
    Include a border around the results:
    <input type="checkbox" name="printborder" value="on"
       checked="checked" />
  </p>
  <p>
    <input type="reset"  value="Clear">
    <input type="submit" value="Try it!">
  </p>
</FORM>
MERFORM

    print end_html;

    $sth->finish;

    $dbh->disconnect;
```

Obviously, the majority of db_match_emblCGI resembles that of db_match_embl. Consequently, this description concentrates on the differences between the two programs. The first important point to make is that the magic first line turns taint mode on (always a good idea when it comes to server-side programs). The CGI module is used, importing the :standard set of subroutines as well as those subroutines that allow for the automatic generation of HTML tables. The DBI module is also used, together with the DbUtilsMER module developed earlier in *Bioinformatics, Biocomputing and Perl*. Note the use lib statement that specifies that the current working directory (".") is to be searched for any custom modules. This allows the db_match_emblCGI server-side program to find the DbUtilsMER module in the cgi-bin directory:

```
use CGI qw/ :standard *table /;
use DBI;
use lib ".";
use DbUtilsMER;
```

After starting to create the HTML page and making a connection to the database, the appropriate SQL query is assigned to the $sql scalar:

```
my $sql = qq/ select accession_number, sequence_data,
              sequence_length from dnas /;
```

A straightforward SELECT query extracts three named columns of data from the dnas table. After preparing the statement, the program determines the values associated with the textarea and the checkbox interface elements from the HTML form within the mersearchmulti.html web page. The param subroutine from the CGI module handles this for us:

```
my $to_check = param( "shortsequence" );
my $print_border = param( "printborder" );
```

The value associated with the printborder checkbox is then used to determine whether the table includes a border around each table entry. If printborder is checked, the table is created by invoking the start_table subroutine (included with CGI) with a referenced parameter that sets the -border attribute to one. If the checkbox is *not* checked, the -border attribute is set to zero:

```
if ( $print_border )
{
    print start_table( { -border => "1" } );
}
else
{
    print start_table( { -border => "0" } );
}
```

With the table started, the next thing to do is create a table row for the column headings. Again, the CGI module provides subroutines to help with this: Tr creates a new table row, and th creates a new table heading. Note the table row has its alignment attribute set to centred:

```
print Tr( { -align => "CENTER" } ), th( "Protein Accession Code" ),
      th( "Was it found?" ), th( "Length Values" );
```

A while loop iterates over each row of the results, assigning the three column values returned from the database system to the $ac, $sequence and $sequence_length scalars. Within the loop's body, these scalar values are used to determine the content of each of the rows of the HTML table. The $sequence scalar has any space character removed from its value. The table row is then started, with the protein accession number (contained in $ac) positioned within the first cell of the table (thanks to the td subroutine from CGI). If the value contained in $to_check is found within the $sequence scalar, the next table cell is assigned the value "yes", otherwise it is assigned "no". After calculating the length of the $sequence scalar, the value is used together with the $sequence_length scalar to populate the final table cell of the table row:

```
$sequence =~ s/\s*//g;

print Tr( { -align => "CENTER" } ), td( "$ac" );

if ( $sequence =~ /$to_check/ )
{
    print td( "yes" );
}
else
{
    print td( "no" );
}

my $calced_length = length $sequence;

print td( "$calced_length/$sequence_length" );
```

When the loop ends, that is, when there are no more rows of data to process from the database system, the db_match_emblCGI program ends the table and then displays the HTML form from the mersearchmulti.html web page. Note the use of the $to_check scalar *within* the HERE document to set the value of the textarea within the form[7]. The program ends by concluding the HTML page, then finishing the SQL query and terminating the database connection.

---

[7] We did tell you that any variable can be used within a HERE document, didn't we? No? Well – shame on us – we should have. It's a very useful feature of HERE documents.

## Where to from Here

As stated at the start of this chapter, it is not difficult to produce a static HTML page, either manually or with a visual HTML editor. The ability to install server-side programs onto a web server, thereby extending its services in any number of ways, is also extremely useful, albeit more difficult.

The examples in this chapter can be extended in any number of ways. Start with the exercises at the end of this chapter and take the time to work through the extensive documentation included with the CGI module.

The next chapter extends the last example program, db_match_emblCGI, by automating the user's interaction with the HTML form.

## The Maxims Repeated

Here's a list of the maxims introduced in this chapter.

- *Take the time to learn HTML.*
- *Create static web pages either manually or visually.*
- *Always enable "taint mode" for server-side programs.*
- *Test your web-site on localhost prior to deployment on the Internet.*

## Exercises

1. Take any program that you have written and move it to the web. What advantages does this new version of your program have over the previous one?

2. Rewrite the db_match_emblCGI program to use the *object-oriented style* of programming supported by the CGI module. [Hint: see the CGI documentation for all the details.]

# 16
# Web Automation

*Using Perl to automate web surfing.*

## 16.1 Why Automate Surfing?

Good question, especially as surfing the web for most is so much fun and already very convenient. Recall the db_match_emblCGI program from the end of the last chapter. When combined with the mersearchmulti.html web page, this program allows the user of a web browser to quickly check if a short sequence of DNA matches any of the EMBL entries in the dnas table within the MER database. The user can be anywhere on the Internet, as all that is needed is access to a web browser and the name of the web page to type into the browser's location/address bar.

Now, imagine that a user has 100 sequences to check. Given an average of 45 seconds[1] to enter the sequence into the textarea, click on the *Try it!* button and review the results (noting those of interest), checking the sequences takes a whopping 75 minutes! Yikes! Surely this user's precious time can be better spent? Of course it can, but work is work and the sequences have to be checked. Let's turn to Perl for help and take the grind out of this type of activity. Say "hello" to the WWW::Mechanize module.

---

[1] 45 seconds? Surely we are joking? Well, consider that the HTML form has to load into a graphical browser, the sequence has to be typed into the textarea and the *Try it!* button pressed. Once the results appear, any matches need to be noted (or, perhaps, printed). We think this will take *at least* 45 seconds.

*Bioinformatics, Biocomputing and Perl.* Michael Moorhouse and Paul Barry
© 2004 John Wiley & Sons, Ltd   ISBN 0-470-85331-X

## 16.2 Automated Surfing with Perl

Written by *Andy Lester*, the `WWW::Mechanize` module is a wonderful example of Perl at its most potent. In brief, the `WWW::Mechanize` module allows the Perl programmer to automate interactions with any web-site. It draws a lot of its power and functionality from the `libwww-perl` library that is available on CPAN and written by *Gisle Aas* (and often referred to as *LWP*). The `libwww-perl` library provides a collection of web-programming modules to the Perl programmer. The `WWW::Mechanize` module leverages the facilities of the library to provide a mechanism to turn Perl into an *automated web browser*.

Both the `libwww-perl` library and the `WWW::Mechanize` module install into Perl from CPAN, following the standard install procedure described earlier.

It is not difficult to solve the problem of the fictitious user sitting in front of a web browser. Rather than have this user type 100 sequences into a form on a web page, let's put the sequences into a disk-file, one line per sequence. By way of example, the `sequences.txt` disk-file contains just 15 sequences and looks like this:

```
attccgattagggcgta
aattc
aatgggc
aaattt
acgatccgcaagtagcaacc
gggcccaaa
atcgatcg
tcatgcacctgatgaacgtgcaaaaccacag
agtcgttaaatgttgtaaa
tggtcccgctact
agtactattcccta
cagcaagaaaa
aaattcccgagc
agtagcaacc
ccaaat
```

The sequences in this disk-file are processed one line at a time by a Perl program. Each time the sequence is determined, it is placed by Perl (with some help from the `WWW::Mechanize` module) into the textarea field within the form on the `mersearchmulti.html` web page. Once placed there, the *Try it!* button is clicked, the request goes off to the web server where it is processed and any results are returned to the Perl program. These are scanned by the Perl program and any matched results are extracted.

Sounds complicated, doesn't it? Well, here's a 26-line program, called `automatch`, that takes any number of sequences within a named disk-file and attempts to match each of them against the EMBL entries stored in the `dnas` table within the MER database, using the `mersearchmulti.html` web-page interface:

```perl
#! /usr/bin/perl -w

# The 'automatch' program - check a collection of sequences against
#                          the 'mersearchmulti.html' web page.

use strict;

use constant URL => "http://pblinux.itcarlow.ie/mersearchmulti.html";

use WWW::Mechanize;

my $browser = WWW::Mechanize->new;

while ( my $seq = <> )
{
    chomp( $seq );

    print "Now processing: '$seq'.\n";

    $browser->get( URL );
    $browser->form( 1 );
    $browser->field( "shortsequence", $seq );
    $browser->submit;

    if ( $browser->success )
    {
        my $content = $browser->content;

        while ( $content =~
                 m[<tr align="CENTER" /><td>(\w+?)</td><td>yes</td>]g )
        {
            print "\tAccession code: $1 matched '$seq'.\n";
        }
    }
    else
    {
        print "Something went wrong: HTTP status code: ",
               $browser->status, "\n";
    }
}
```

Let's go through the automatch program in detail. After the standard first line and the switching on of *strictness*, a constant called URL is defined. This constant value is a web address, specifying the mersearchmulti.html web page on the pblinux.itcarlow.ie web server[2]. The WWW::Mechanize module is then used.

A scalar value called $browser is assigned a value as a result of calling the new subroutine associated with the WWW::Mechanize module. In programming terms, this creates a WWW::Mechanize object and assigns it to $browser. Do not

---
[2] Obviously, this constant value needs to change on the basis of the web server on which the web page is installed.

worry about the terminology: when the word *object* is used, think *thingy*. So, this statement creates a WWW::Mechanize *thingy* and assigns it to $browser:

```
my $browser = WWW::Mechanize->new;
```

A while loop reads one line at a time from STDIN and assigns it to the $seq scalar. Within the loop body, the value within $seq is *chomped*, and a message is displayed on screen to provide some feedback to the user. The $browser object is used in the next four statements:

```
$browser->get( URL );
$browser->form( 1 );
$browser->field( "shortsequence", $seq );
$browser->submit;
```

The nice thing about objects is that they have *functionality associated with them* in the form of subroutines that can be invoked against the object. The get subroutine (which is part of the WWW::Mechanize module) takes a web address and retrieves the web page returned from the web server. The web page returned is associated with the $browser object, which highlights another nice thing about objects: they can have *data associated with them*.

> **Technical Commentary:** Objects are useful, and clever programmers use them to *encapsulate* a thingy's data and behaviour (in the form of subroutines). It is beyond the scope of this book to cover the object-creating techniques available to Perl programmers. However, it is not necessary to know how to create objects in order to be able to exploit and use them, as witnessed by the automatch program.

With the returned web page contained in the $browser object, the form subroutine selects the first form contained within the returned web page. Unlike almost everything else in Perl, the form subroutine starts counting from one, *not* zero. It is possible to have more than one form on a HTML page, but the mersearchmulti.html web page has only one, so automatch selects it. The field subroutine provides a mechanism to set a specific field on the form to a value. The invocation within automatch sets the value of the shortsequence textarea to the value contained within the $seq scalar. With this done, the submit subroutine clicks the forms main button, which is the *Try it!* button from the mersearchmulti.html web page.

At this point, the automatch program sends a request to the web server. This results in the CGI on the web server executing, using the $seq value as a parameter. The CGI executes and produces the results page. This is then returned to the automatch program and is assigned, by the WWW::Mechanize module, to the $browser object. An if statement checks to see if the request was successful by calling the success subroutine associated with the $browser object. If it is *not*, a message is displayed on STDOUT. Note that the message contains a status code from the $browser object:

```
            print "Something went wrong: HTTP status code: ",
                      $browser->status, "\n";
```

If the request is successful, the results returned are assigned to a scalar called $content:

```
            my $content = $browser->content;
```

Note that the $content scalar has as its value the HTML results page returned from the web server. A rather scary-looking regular expression is used to look for the word "yes" within the HTML table included within the results. Specifically, the word "yes" has to appear within a table cell (between the <td> and </td> tags). In the table cell immediately before the one that contains "yes", there is the protein accession code associated with the EMBL entry, again surrounded by <td> and </td> tags. Ahead of this is the <tr> tag, which specifies the characteristics of table row. The following regular expression is used to extract the protein accession number from the matched results returned from the web server:

```
<tr align="CENTER" /><td>(\w+?)</td><td>yes</td>
```

Note the use of the grouping parentheses around the \w+? part. This is what allows Perl to extract the protein accession number and remember it in the $1 scalar. Note, too, the use of the non-greedy operator, the ? character. This ensures that the protein accession code and nothing else is extracted[3]. When integrated into the automatch program, the above regular expression is delimited by the [ and ] characters. This avoids the need to escape all the forward-leaning slash characters. Assuming a match, a print statement produces a message indicating success:

```
            while ( $content =~
                      m[<tr align="CENTER" /><td>(\w+?)</td><td>yes</td>]g )
            {
                print "\tAccession code: $1 matched '$seq'.\n";
            }
```

The regular expression is contained within the condition part of a while loop, and is qualified by the use of the g quantifier, which ensures that the regular expression is applied *globally*. This has the effect of continuing to find matches while there are matches to be found within the $content scalar. So, a message is produced for each successful match contained in the HTML results.

The automatch program can be made executable and run against the sequences in the sequences.txt disk-file with these commands:

```
    chmod +x automatch
    ./automatch sequences.txt
```

---

[3] Try removing the ? from the regular expression and see what happens.

which produces the following output:

```
Now processing: 'attccgattagggcgta'.
Now processing: 'aattc'.
    Accession code: AF213017 matched 'aattc'.
    Accession code: J01730 matched 'aattc'.
    Accession code: M24940 matched 'aattc'.
Now processing: 'aatgggc'.
Now processing: 'aaattt'.
    Accession code: AF213017 matched 'aaattt'.
    Accession code: J01730 matched 'aaattt'.
    Accession code: M24940 matched 'aaattt'.
Now processing: 'acgatccgcaagtagcaacc'.
    Accession code: M15049 matched 'acgatccgcaagtagcaacc'.
Now processing: 'gggcccaaa'.
Now processing: 'atcgatcg'.
Now processing: 'tcatgcacctgatgaacgtgcaaaaccacag'.
    Accession code: AF213017 matched 'tcatgcacctgatgaacgtgcaaaaccacag'.
Now processing: 'agtcgttaaatgttgtaaa'.
Now processing: 'tggtcccgctact'.
    Accession code: M15049 matched 'tggtcccgctact'.
Now processing: 'agtactattcccta'.
Now processing: 'cagcaagaaaa'.
    Accession code: AF213017 matched 'cagcaagaaaa'.
Now processing: 'aaattcccgagc'.
Now processing: 'agtagcaacc'.
    Accession code: M15049 matched 'agtagcaacc'.
Now processing: 'ccaaat'.
    Accession code: AF213017 matched 'ccaaat'.
    Accession code: J01730 matched 'ccaaat'.
    Accession code: M24940 matched 'ccaaat'.
```

This took less than 15 seconds on Paul's computer to execute. Remember the 75 minutes from the start of this chapter? Well, by employing Perl and the `WWW::Mechanize` module in the `automatch` program, each sequence check now takes about 1 second. Now, how long would you rather spend completing this task: 75 minutes or 100 *seconds*?[4]

Another question to ask is this: if it is straightforward to mechanise interaction with the `mersearchmulti.html` web page, why not any other? Why not, indeed. Typically, all that's required is an understanding of HTML and an ability to view a web page in *source form*, that is, as HTML. To do this, simply use the *Page Source* option from the browser's *View* menu to display any web page as HTML (as opposed to graphically within a browser window). Figure 16.1 on page 335 shows the *Mozilla* browser's *page source window*. Once viewed in this way, it is a straightforward matter to read the displayed HTML and determine the names of the interface elements in the form (such as `shortsequence`). Of course, if it is possible to view the source to the `mersearchmulti.html` web page, then it is possible to view the source of *any* web page that is viewed in a graphical browser, regardless of the author.

---

[4] For those readers having trouble answering this question, please complete the following simple exercise: *find the nearest brick wall and continue banging your head off of it!*

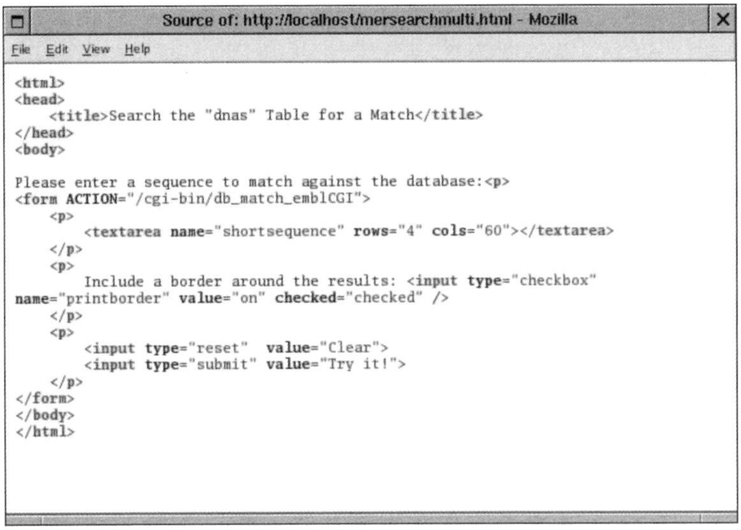

```
<html>
<head>
    <title>Search the "dnas" Table for a Match</title>
</head>
<body>

Please enter a sequence to match against the database:<p>
<form ACTION="/cgi-bin/db_match_emblCGI">
    <p>
        <textarea name="shortsequence" rows="4" cols="60"></textarea>
    </p>
    <p>
        Include a border around the results: <input type="checkbox" name="printborder" value="on" checked="checked" />
    </p>
    <p>
        <input type="reset"  value="Clear">
        <input type="submit" value="Try it!">
    </p>
</form>
</body>
</html>
```

**Figure 16.1** Viewing the source of the `mersearchmulti.html` web page.

The strategy to follow when automating interaction with any web page follows this simple recipe:

- Load the web page of interest into a graphical browser.

- View the HTML used to display the web page by selecting the *Page Source* option from the browser's *View* menu.

- Read the HTML and make a note of the names of the interface elements and form buttons that are of interest.

- Write a Perl program that uses `WWW::Mechanize` to interact with the web page (based on `automatch`, if need be).

- Use an appropriate regular expression to extract the "interesting bits" from the results returned from the web server.

And, of course, find something productive to do with those saved 75 minutes! As long as the web page interacted with remains unchanged, the automation program continues to work and save time. Even if a web page does change, the effort required to amend the automation program should be minimal.

## Where to from Here

This chapter showed just how easy it is to automate interaction with web pages on the WWW using Perl and the `WWW::Mechanize` module. Employing a web automation strategy when repeatedly working with data and web pages on the WWW can be a huge time-saver. The time spent creating a custom program to automate surfing is quickly recovered when there are many manual interactions to perform.

> **Maxim 16.1** *Automate repetitive WWW interactions whenever possible.*

And with that maxim, this part of *Bioinformatics, Biocomputing and Perl* concludes. The final part of this book presents a collection of ready-to-use Bioinformatics programs, many of which are available for download from the Internet.

## The Maxims Repeated

Here's the maxim introduced in this chapter.

- *Automate repetitive WWW interactions whenever possible.*

## Exercises

1. Search CPAN and read about the HTML::Parser and HTML::TokeParser modules. Is it worthwhile using either of these modules to rewrite the automatch program?
2. Pick your favourite Bioinformatics web-site and, using WWW::Mechanize, automate an interaction with it.
3. Move the automatch program to the WWW. Among other things, you will need to provide a mechanism that allows users of your web page to specify multiple sequences *without* having to type them into a HTML form. [Hint: explore the HTML *file-upload* feature.]
4. Automate interaction with the SRS web-site from the *SRS: The Sequence Retrieval System* chapter.

# Part IV

# Working with Applications

# 17
# Tools and Datasets

*Exploring the tools of the trade.*

## 17.1 Introduction

This chapter introduces a collection of Bioinformatics databases and associated tools that are used in the next chapter to characterise a piece of "mystery" DNA. The descriptions here concentrate on installing and running the tools, rather than describing the algorithms they use. This chapter is designed to augment any "installation and usage notes" included with the tools by giving example input data, showing how to run the tools and describing the results expected. For more details on a particular algorithm, consult the original research papers, many of which are available on the Internet[1]. It can also be helpful to talk to other researchers in the field who have used these tools, as there is no substitute for a first-hand account from an expert!

Examples of extracting sequence data are provided elsewhere in this book, and are not repeated here. Although a useful technique (of which every Bioinformatician needs to be aware of), "manually" trawling through a sequence disk-file is at best archaic, and more likely wasteful for everything other than the simplest of searches. The modern Bioinformatician loads the data into a relational database (as described in the *Databases* chapter) and then employs SQL to extract data in a nice, easy-to-use form. Many researchers use technologies such as EMBOSS, Ensembl or Bioperl to do the database populating for them. Such systems tend to scale better, are more flexible and more reliable, while still being generally faster and more robust. That said, the creation of custom programs to process

---
[1] Refer to the appendix entitled *Suggestions for Further Reading*.

*Bioinformatics, Biocomputing and Perl.* Michael Moorhouse and Paul Barry
© 2004 John Wiley & Sons, Ltd   ISBN 0-470-85331-X

Protein Databank (PDB) disk-files is still needed, as the information relating to protein structures within it does not store particularly well in relational databases without the structure (of the database!) becoming overly complicated. Also the `mmCIF` format/database and its associated tools, although often easier to work with, are not widely accepted and used, so expect to trawl through PDB disk-files for some time to come. The remainder of this chapter is split into three broad areas of study, as follows:

1. **Sequence databases** – EMBL, TrEMBL and SWISS-PROT. This section describes what these databases are, what they contain, how to access them and what they are used for.

2. **General concepts and methods** – The concepts relevant to prediction and subsequent validation of those predictions. This section describes the concepts of true and false predictions, using the tRNA gene-finding package *tRNAScan-SE* as an extended example. The use of the two-stage approach is also covered: a fast but inaccurate first stage, coupled to a slow but careful second stage which attains speed with precision.

3. **Bioinformatics tools** – ClustalW and BLAST. ClustalW aligns multiple sequences to identify phylogenetic relationships between them, while BLAST, the Basic Local Alignment Search Tool, is one of the standard tools used to do sequence-similarity searches of databases.

## 17.2 Sequence Databases

There used to be three very distinct types of Bioinformatics databases: Primary, Secondary and Tertiary. The three types still exist, but the boundaries between them are becoming blurred, especially as they become increasingly cross-linked to each other. Add to this the fact that many references between databases are presented as "clickable web-links" and that databases are combined into *metadatabases*. Meta-search systems are also popular, for instance, Sequence Retrieval System (SRS), and to a lesser extent, Bioperl and EMBOSS. Consequently, it is often hard to know from which particular database the information being accessed is originally coming from. The answer is often, at some level, all of them!

Even when systems such as SRS are not used, the cross references built into many of the more popular sequence databases make it easy for a system to build links on the fly, this being one of the "informatics" parts of Bioinformatics. The "bio" part is knowing which links make biological sense. For many years, Bioinformaticians grappled with the informatics part. Nowadays, the availability of scalable relational databases, coupled with the desire and technical competence to use them, has greatly reduced the complexity of many of the raw informatics problems. This allows researchers to concentrate on biology, which is, after all, the whole point of investing time in studying the Bioinformatics tools of the trade.

Given the excellent user guides included with the larger and more popular databases, a detailed discussion of their structures and contents is beyond the scope of this book. Consequently, the description of the EMBL, TrEMBL and SWISS-PROT databases is confined to quoting a few examples of the data found in them, describing why the data are useful and exploring how to use (the data) to do more interesting activities.

There are three main sequence databases:

1. **EMBL/GenBank** – The primary sequence databases for nucleotide sequences, DNA and RNA. If a sequence is published, then it is in one of these databases. The DDBJ nucleotide database in Japan exchanges data with these sequence databases on a nightly basis. While the accession numbers between the three databases are quite different, the contents are very similar because of the constant exchange of information.

2. **TrEMBL** – Short for "Translated EMBL", this database was originally designed to bridge the gap between the nucleotide sequences stored in EMBL and the manually curated SWISS-PROT entries. The TrEMBL database relies on more automatic methods to generate annotation than does SWISS-PROT (and PIR), allowing new sequences to be added quickly. Note that this speed of entry is at the expense of the completeness (and, in some cases, accuracy) of the annotation.

3. **SWISS-PROT/PIR** – Originally the work of researchers working in the field of protein sequence analysis, *Amos Barioch*, along with a team of experts, has lovingly curated the SWISS-PROT database for many years. *Margrate Dayhoff* created the first amino acid substitution matrices, the *Atlas of Protein Structure*[2], which eventually became the PIR (Protein Information Resource) database. Both SWISS-PROT and PIR have richer and more accurate annotation than their TrEMBL counterparts. However, this accuracy comes at a price: there can be a considerable delay between the DNA entry for a protein sequence being published in EMBL/GenBank and its appearance in SWISS-PROT/PIR. As humans take time to gather and evaluate evidence, SWISS-PROT users need patience.

The TrEMBL, SWISS-PROT and PIR databases overlap in some respects and are complementary in others. All are essential to modern Bioinformatics research and, obviously, have to be funded. SWISS-PROT highlighted this point some years ago as a serious "funding crisis". In 2002, the *UniProt Consortium* was granted over US$15 Million towards the task of merging all three of these databases into one.

So what of the Primary, Secondary and Tertiary categories? With all the cross-linking, mergers and general increase in complexity of Bioinformatics data, these terms seem a little obsolete. However, they still have some meaning:

---

[2] Originally published on paper when there were less than 1000 known protein sequences.

### Tools and Datasets

- **Primary** – These are the ground-level databases that contain actual experimental results, such as DNA sequences, protein structures and micro array expression data. Examples are the EMBL/TrEMBL and ArrayExpress, the micro array database and its access software.

- **Secondary** – These contain "derived data". This is information that has been extracted from the primary sequence databases, even though doing so may duplicate material from the primary databases. Examples include SWISS-PROT, PROSITE and HSSP (a compendium of the alignments of similar sequences to known protein structures).

- **Tertiary** – These contain highly abstracted data, such as molecular functions. Examples are GO (Gene Ontology) classification, KEGG (Pathway information) and literature references such as PubMed.

Reproduced below are selected sections of the EMBL data-file for the *Tn501 Transposon*[3] and one of the SWISS-PROT entries cross-linked to it. Accessing these through a database search engine has its disadvantages, as is often the case in the modern Bioinformatics world. One of the biggest is that to formulate search terms, it is necessary to know what to look for before commencing the search. This may be difficult if you are unsure of what you are looking for and there is always the risk of missing something important because the search (inadvertently) excludes the very items you were actually seeking! In such cases, reading the actual data-files themselves is helpful because this presents all the data at once. Such an activity is slow and tedious though, and your author's advice is to do this only as a last resort! That said, this method can often yield surprising insights, allowing for the formulation of hypotheses and the creation of more useful and specific search terms.

1. **Header/annotation** – This is *parsable text* that stores curation and bibliographic information about the sequence, which is often a general description of the sequence including an indication of its function, cross references to other databases and interesting features (such as open reading frames or sequence motifs) found in the sequence. This section can be split up, stored and searched easily using normal informatics methods such as pattern matching, regular expressions and relational databases searches using SQL. It is often quite useful to read the annotation to get an idea as to what the sequence is and how it relates to other database entries. For instance (and as will be shown), the DE and DR lines in EMBL, TrEMBL and SWISS-PROT are used to store brief descriptions of the sequence and cross references to other databases.

---

[3] This is used as an example in the next chapter.

2. **DNA/protein sequence** – This records the original experimental data as well as acting as a record of the sequence to meet publication or legal requirements. This is the "I found it first and it is now *of the Art*" section.

> **Technical Commentary:** If you publish a sequence in an open access database, this creates "Prior Art", which cannot then be protected by a patent. The advantage of a patent is that you can charge for the use of your "invention". Publication and the creation of "Prior Art" can also be used as a blocking technique to prevent someone else taking out patents on sequences. This was one of the motivations behind the reporting of new sequences from the "public" Human Genome Project every night in a publicly accessible form on the WWW. If you have a novel sequence that you think is useful, for example, as a therapeutic target for curing a disease, you may wish to keep quiet about it until you consult a patent lawyer. After the consultation, decide whether publication or patenting (or both) is appropriate.

While the sequence can be searched by common algorithms as used in the *Header section*, such general tools perform poorly because they are designed for *exact* pattern matching. While a DNA or protein sequence can be regarded as a digital bit string (and therefore searched using conventional informatics algorithms), it is a very abstract representation of a real molecule that exists in the world of solutions, solutes concentration and thermodynamic vibrations. In this (literally) chaotic environment, many molecules are "nearly the same" in many different ways at some time or another. While cells are tolerant of this, exact matching algorithms tend not to be. Special sequence comparison and alignment algorithms have been developed to locate *similar* rather than *exact* matches (these are discussed later in this chapter).

## 7.2.1 Understanding EMBL entries

The on-line entry in this section is accessible through the EBI SRS system. Here are selected parts and extracts from EMBL entry ID ISTN501 and Accession Number Z00027. Many sections in the entry's disk-file have XX spacer lines between them. To conserve space, many of these markers, together with some of the sections describing the coding regions, have been removed. It would be helpful while reading this section to have the full entry available, as this will help navigate between the sections highlighted in the discussion here.

The entry starts by indicating the identification and accession number. These are the targets for automatic referencing from some outside source, such as another database or a publication:

```
ID   ISTN501    standard; genomic DNA; PRO; 8355 BP.
```

As well as indicating the accession number, the AC line lists previous entries that have been incorporated into this entry:

```
AC   Z00027; K00031; K01725; X01297; X03406;
```

The SV line specifies the "current" name, the *official reference* to the entry. Note, too, the create and last updated date:

```
SV   Z00027.1
XX
DT   02-JUL-1986 (Rel. 09, Created)
DT   07-JUL-2002 (Rel. 72, Last updated, Version 14)
```

The descriptions in the DE lines and the keywords in the KW lines are often good to search against. The R lines describe the *official* published references (only the first of six are shown here):

```
DE   Transposon Tn501 from Pseudomonas aeruginosa plasmid pVS1 encoding mercuric
DE   ion resistance determinant. Genes: merR (regulation), merT and merP
DE   (transport), merA (reductase), merD (not known); two open reading frames of
DE   unknown function (possibly one is merE); res site, tnpR and tnpA
DE   (transposition).
KW   mercury resistance; plasmid; reductase; transposon;
KW   unidentified reading frame.
OS   Pseudomonas aeruginosa
OC   Bacteria; Proteobacteria; Gammaproteobacteria; Pseudomonadales;
OC   Pseudomonadaceae; Pseudomonas.
OG   Plasmid pVS1
XX
RN   [1]
RP   1302-3048
RX   MEDLINE; 84000429.
RX   PUBMED; 6311258.
RA   Brown N.L., Ford S.J., Pridmore R.D., Fritzinger D.C.;
RT   "Nucleotide sequence of a gene from the Pseudomonas transposon Tn501
RT   encoding mercuric reductase";
RL   Biochemistry 22(17):4089-4095(1983).
         .
         .
         .
```

The DR lines contain the *database cross references* that link to entries in other sequences databases:

```
DR   GOA; P00392.
DR   GOA; P04131.
DR   GOA; P04140.
DR   GOA; P06688.
DR   GOA; P06689.
DR   GOA; P06690.
DR   GOA; P06691.
DR   GOA; P06695.
DR   GOA; Q48362.
DR   SPTREMBL; Q48362; Q48362.
DR   SWISS-PROT; P00392; MERA_PSEAE.
DR   SWISS-PROT; P04131; MERP_PSEAE.
DR   SWISS-PROT; P04140; MERT_PSEAE.
```

```
DR   SWISS-PROT; P06688; MERR_PSEAE.
DR   SWISS-PROT; P06689; MERD_PSEAE.
DR   SWISS-PROT; P06690; MERE_PSEAE.
DR   SWISS-PROT; P06691; TNR5_PSEAE.
DR   SWISS-PROT; P06695; TNP5_PSEAE.
```

The FT lines describe features found in the sequence (those after the first CDS have been omitted):

```
FH   Key             Location/Qualifiers
FH
FT   source          1..8355
FT                   /db_xref="taxon:287"
FT                   /mol_type="genomic DNA"
FT                   /organism="Pseudomonas aeruginosa"
FT                   /plasmid="pVS1"
FT   repeat_region   1..5355
FT                   /transposon="Tn501"
FT   repeat_unit     1..38
FT                   /note="terminal repeat"
FT   misc_signal     505..520
FT                   /note="potential binding site for inducer"
FT   repeat_unit     505..510
FT                   /note="inverted repeat a"
FT   repeat_unit     515..520
FT                   /note="inverted repeat a'"
FT   CDS             complement(114..548)
```

Here is the first of the features, which is a coding sequence:

```
FT                   /db_xref="GOA:P06688"
FT                   /db_xref="SWISS-PROT:P06688"
FT                   /transl_table=11
FT                   /product="merR protein (repressor/inducer)"
FT                   /gene="merR"
FT                   /protein_id="CAA77320.1"
FT                   /translation="MENNLENLTIGVFAKAAGVNVETIRFYQRKGLLLEPDKPYGSIRR
FT                   YGEADVTRVRFVKSAQRLGFSLDEIAELLRLEDGTHCEEASSLAEHKLKDVREKMADLA
FT                   RMEAVLSELVCACHARRGNVSCPLIASLQGGASLAGSAMP"
```

The rest of the features (omitted here) follow, and in this entry are mostly coding sequences. The sequence data then starts, with an SQ line indicating the start:

```
XX
SQ   Sequence 8355 BP; 1560 A; 2709 C; 2650 G; 1436 T; 0 other;
     gggggaaccg cagaattcgg aaaaaatcgt acgctaagct aacggtgttc tcgtgacagc    60
              .
              .
              .
     ttctgcgagc ccccc                                                   8355
//
```

and ends with the "//" end-of-entry marker. This acts as a delimiter between separate entries.

## 17.2.2 Understanding SWISS-PROT entries

Here are selected parts and extracts from the MERP_PSAE/P04131 entry from SWISS-PROT. The overall format is very similar to that of the EMBL, but it tends to be shorter because it describes a *single protein*. The SWISS-PROT line types have almost identical meaning to the EMBL line types:

```
ID   MERP_PSEAE     STANDARD;     PRT;    91 AA.
AC   P04131;
DT   01-NOV-1986 (Rel. 03, Created)
DT   01-NOV-1986 (Rel. 03, Last sequence update)
DT   15-JUN-2002 (Rel. 41, Last annotation update)
DE   Mercuric transport protein periplasmic component precursor
DE   (Periplasmic mercury ion binding protein) (Mercury scavenger protein).
GN   MERP.
OS   Pseudomonas aeruginosa.
OG   Plasmid pVS1.
OC   Bacteria; Proteobacteria; Gammaproteobacteria; Pseudomonadales;
OC   Pseudomonadaceae; Pseudomonas.
OX   NCBI_TaxID=287;
RN   [1]
RP   SEQUENCE FROM N.A.
RC   TRANSPOSON=TN501;
RX   MEDLINE=85014891; PubMed=6091128;
RA   Misra T.K., Brown N.L., Fritzinger D.C., Pridmore R.D., Barnes W.M.,
RA   Haberstroh L., Silver S.;
RT   "Mercuric ion-resistance operons of plasmid R100 and transposon Tn501:
RT   the beginning of the operon including the regulatory region and the
RT   first two structural genes.";
RL   Proc. Natl. Acad. Sci. U.S.A. 81:5975-5979(1984).
```

The CC lines contain human-readable text that gives a fuller description of the protein than do the DE lines:

```
CC   -!- FUNCTION: MERCURY SCAVENGER THAT SPECIFICALLY BINDS TO ONE HG(2+)
CC       ION AND WHICH PASSES IT TO THE MERCURIC REDUCTASE (MERA) VIA THE
CC       MERT PROTEIN.
CC   -!- SUBUNIT: MONOMER.
CC   -!- SUBCELLULAR LOCATION: PERIPLASMIC (PROBABLE).
CC   -!- SIMILARITY: CONTAINS 1 HMA DOMAIN.
CC   -----------------------------------------------------------------------
CC   This SWISS-PROT entry is copyright. It is produced through a collaboration
CC   between the Swiss Institute of Bioinformatics and the EMBL outstation -
CC   the European Bioinformatics Institute.  There are no restrictions on its
CC   use by non-profit  institutions as long as its content  is  in  no   way
CC   modified and this statement is not removed. Usage by and for commercial
CC   entities requires a license agreement (See http://www.isb-sib.ch/announce/
CC   or send an email to license@isb-sib.ch).
CC   -----------------------------------------------------------------------
```

The first DR points to the EMBL entry Z00027:

```
DR   EMBL; Z00027; CAA77322.1; -.
DR   EMBL; K02503; AAA27434.1; -.
DR   PIR; A03557; RGPSHA.
DR   HSSP; P04129; 1AFJ.
DR   InterPro; IPR001802; HG_scavenger.
DR   InterPro; IPR001934; HeavyMe_transpt.
DR   Pfam; PF00403; HMA; 1.
DR   PRINTS; PR00944; CUEXPORT.
DR   PRINTS; PR00946; HGSCAVENGER.
DR   PROSITE; PS01047; HMA_1; 1.
DR   PROSITE; PS50846; HMA_2; 1.
KW   Transport; Mercuric resistance; Periplasmic; Metal-binding; Signal;
KW   Transposable element; Plasmid.
FT   SIGNAL        1     19       POTENTIAL.
FT   CHAIN        20     91       MERCURIC TRANSPORT PROTEIN PERIPLASMIC
FT                                COMPONENT.
FT   DOMAIN       23     89       HMA.
FT   METAL        33     33       HG(2+) (POTENTIAL).
FT   METAL        36     36       HG(2+) (POTENTIAL).
```

The actual protein sequence is contained in the last section, along with a CRC code, which is a computer-generated *verification key*:

```
SQ   SEQUENCE   91 AA;  9491 MW;  6D6DB86B5FCA20CE CRC64;
     MKKLFASLAL AAVVAPVWAA TQTVTLSVPG MTCSACPITV KKAISEVEGV SKVDVTFETR
     QAVVTFDDAK TSVQKLTKAT ADAGYPSSVK Q
//
```

Again, the end-of-entry marker, "//", terminates the entry.

### 17.2.3 Summarising sequences databases

Databases (and datasets) are the foundations upon which Bioinformatics is built. They create a "record of history": what has been found, when and by whom, what it did or did not do. Such information is always useful to individual researchers concerned with specific problems. It also allows for the extraction of information that can help in the compilation of new databases, especially those containing knowledge.

Because of the increasing cross referencing of modern databases, the boundaries between the individual databases are becoming blurred, particularly when multiple databases are loaded into a *metadatabase* such as *InterPro*, or a *Meta Database System* such as SRS. More on these tools later.

## 17.3 General Concepts and Methods

In this section, the *tRNAScan-SE* prediction system is used to help explain and provide real-world examples of important general philosophical concepts. *tRNAScanSE* is an analysis system that finds transfer RNA (tRNA) genes in Genomic

DNA and it is both accurate *and* fast. The actual details of how *tRNA-ScanSE* works are described below in the subsection titled *tRNA-ScanSE, a case study*. First, though, a few general concepts that are useful to know.

## 17.3.1 Predictions and validation

*"...on the Nature of Prediction and Validation of Nature..."*

Bioinformatics often involves making accurate predictions from highly abstract data. In "information space", there are no physical limits to the universe, apart from our ability to represent and handle the amount of data (both of which are important practical limitations). A philosophically fundamental limitation is that any isolated system that can be demonstrated to be *self-consistent* cannot also be *self-validating*. To validate a system (or *model*), external information is needed to act as a reference against which predictions can be tested. In Bioinformatics, the preferred way to do this is to take the predictions generated *in silico* (in a computer) and test them *in vivo* (in life) or *in vitro* (in isolation). Practically, for example, this means purifying the predicted enzyme and testing for catalytic activity in the laboratory (*in vitro* testing) or cloning a *homolog gene* from one organism into another and finding out if it works in the same way (*in vivo* testing).

Whether the problems of prediction are more acute in Bioinformatics than for mainstream biology is debatable. Certainly, these are things that every competent Bioinformatician needs to think about. Ask the following question. *Are the predictions relevant to the context in which they are made?* All scientists have the same problem[4].

Validating a Bioinformatical model with a biological one is, in effect, a merging of the two. This means that the combined model cannot itself be self-validating, just *self-consistent*. In many cases, this is enough, so long as the combined model is *useful*.

> **Maxim 17.1** *Recognise the difference between the validation of a model and the testing of it for self-consistency.*

One thing is certain: such discussion rapidly becomes too deep to be useful. So, here are some examples to help clarify things.

## 17.3.2 True/False/Negative/Positive

The terms "False Positive" and "False Negative" are often found in Bioinformatics discussions, particularly those connected with the analysis of large datasets. Both

---
[4] Those who believe the grass is greener on the other side are in for a shock: it never is!

relate to predictions made and their subsequent testing. Logically, there are four possible outcomes for a comparison between a prediction and the outcome of an event. These are referred to as: *True Positive* (TP), *True Negative* (TN), *False Positive* (FP) and *False Negative* (FN).

As an example, consider predicting tRNA sequences in a piece of DNA. There are two possible options:

1. To predict a tRNA gene at a particular location, which is called a *positive* result.
2. Not to predict a tRNA gene at the same location, which is a *negative* result[5].

The important points are that the results are just predictions, and that they may or may not be correct. Ideally, the results should be tested in some way. Traditionally, this is performed by either going into the laboratory and doing work with the real DNA and organisms or by using test datasets from previous, reliable investigations. In the development of the algorithms (there are actually two coordinated by *tRNAScan-SE*, as discussed below), both validation methods were used. The extra information brought in from outside allowed each individual *positive/negative* classification to be characterised on the basis of *True* (the prediction was correct) or *False* (the prediction was wrong). Consequently, each result can be assigned one of four possible outcomes:

1. **True Positive** – This means that a tRNA gene was predicted at a specific location and one did exist and is functional there. Certain non-functional tRNA genes can be detected but *tRNAScan-SE* places them into the next category.
2. **True Negative** – No tRNA was reported and, indeed, none was present. As the frequency of tRNAs in DNA is low, *False Negatives* are not reported as they would just clutter the output.
3. **False Positive** – The algorithm predicted a tRNA gene, but a corresponding one was not found. A very similar term you may come across is the "False Discovery Rate". From the context and the definition your authors have seen this used in, this term seems to be analogous to what would be called the *False Positive Rate* in the discussion here: the rate at which the algorithm makes *False Positive* errors.
4. **False Negative** – In this case, the algorithm failed to report a tRNA gene when one was present. In everyday language, it was "missed".

While it depends on the particular circumstances, *False Positives* are, generally, the most troublesome type of prediction. The algorithm and, as a consequence,

---

[5] It is usual for programs to print only the positives in their results, as the negatives are so frequent that they would obscure the positives, not to mention that they take up too much space.

statements made by researches using it is claiming a tRNA gene to be present, when *it is not*. Imagine the potential for wasted research effort, not to mention the professional embarrassment, of chasing these particular "ghosts".

While any error is undesirable, *False Negatives* are generally more acceptable, although it depends on the circumstances. With *False Negatives* (or *False Positives*) you can argue that a more *sensitive* system is needed. When an improved algorithm is developed, the same dataset can be analysed again and false negative reports reduced. Further investigation may show that the algorithm is correct and a false report is due to deficiencies in the validation. It is never as reliable as assumed in this discussion. Of course, both the algorithm and the data used for validation may be wrong. This is the way science works.

> **Maxim 17.2** *Generally, False Negative predictions are considered more acceptable than False Positives.*

In some situations, the high *False Positive* rate can be reduced by a second processing step, either Bioinformatics- or wet-lab-based, but the initial phase is needed to maintain a high *True Negative* rate[6].

One point to be aware of when predictions are validated is that the data used for this, often referred to as a *Standard of Truth* or the *Gold Standard*, must be truly independent of the data used to build the model. This is a particular problem if the dataset of experimentally validated items (sequences, structures or functions) is small because this limits both the training and testing datasets. If the same data is used to build the model as well as test it, then the model can be *over-fitted*. This means that the model describes the features of the dataset well and is self-consistent, but is not generally applicable in a wider context. Two commonly used procedures that help avoid this problem are as follows:

1. The available data can be partitioned into a "training set", from which the model is built, and a "test set", against which it is tested.

2. If the amount of data is too small to exclude a proportion (possibly up to half) as a "test set" during the training procedure, special techniques such as "Jackknifing" or "leave N out" can be used. Here, the available data is partitioned into "training" and "test" sets in multiple ways. Different partitions are used so that all the data is used in both sets, but never at the same time. This is held to be considerably more reliable than a simple self-consistency test, but not as good as a fully independent test.

Figure 17.1 on page 351 summarises the assessment/validation procedure, along with the possible outcomes.

---

[6] See the subsection entitled *The use of multiple algorithms to improve performance*, later in this chapter.

# General Concepts and Methods

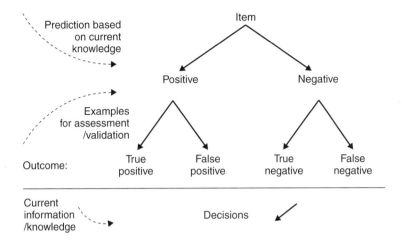

**Figure 17.1** Assessment/validation procedure and possible outcomes.

## 7.3.3 Balancing the errors

Predictions *may* contain errors. Unfortunately, this goes with the territory of making predictions. Often, there is a need to strike a balance between *False Positive* and *False Negative* results, and this can be accomplished by "tweaking" the parameters of the prediction algorithm. In the case of *tRNAScan-SE*, there are at least three *Threshold Scores* that determine whether a particular piece of DNA will be reported as a gene. Let's consider the two extreme cases of "over-tweaking":

- A zero *False Positive* rate can be achieved by simply reporting "No" (Negative) for any particular sequence. The trouble is that the true positive rate is zero as well!

- A zero *False Negative* rate can also be achieved by answering "Yes" for any particular sequence. The problem here is an extremely high *False Positive* rate.

It follows that some optimisation needs to be done to set the reporting threshold, which determines the ratio of *False Positives* to *False Negatives*. Publications tend to include statements or sections that paraphrase to "*... the threshold(s) was set empirically to allow for a conservative number of False Positives, while maintain the True Negative/True Positive rate...*". Often, the threshold is described as "conservative". Usually, this means accepting some *False Negatives* to prevent too many *False Positives*. Deciding on the balance between the two is often more a matter of professional judgement than hard statistics. Here's a maxim to be kept in mind when trying to balance the *False Positive/False Negative* ratio.

> **Maxim 17.3** *With False Negatives, we could come back next year and find the ones we missed, and these are preferred to False Positives, where we can waste time studying them this year, only to find out that the time was wasted. It all depends on the circumstances.*

And here's another maxim to consider (albeit somewhat tongue in cheek).

> **Maxim 17.4** *Sometimes, all those False Positives are maybe, just maybe, trying to tell you something. So, if you aspire to a Nobel prize....*

### 17.3.4 Using multiple algorithms to improve performance

The performance of an algorithm can be measured in different ways. One way is to examine the prediction accuracy as described in the previous subsections: the True/False and Positive/Negative rates. Another very important way is to examine the algorithm's speed. While computational power doubles on average every 18 months or so, the rate of increase of biological databases can match and in many cases exceed this. In essence, new, modern Bioinformatics analyses must run faster simply to stand still. To compound this problem further, as the subject data becomes more complex, so too do the analyses researchers want to perform. There are three alternatives:

1. Do not perform the analysis at all. This is really the *zeroth option*, but in some ways it's the ultimate optimisation. *If you cannot win, do not play the game.* This is less than ideal for many reasons. It is included here because sometimes, just sometimes, it is the best (only) solution.

2. Invent better algorithms or improve the implementation of existing ones. This is very much easier to say than do! Yet there have been cases where optimisation of code has really made an improvement. An excellent example is *tRNA-Scan*, see the quotation from the *tRNAScan-SE* manual later in this chapter.

3. Use multiple algorithms with complementary performance characteristics. A fast screening stage is used to filter the data to prevent a slower, but more accurate, later stage from processing large amounts of data.

The third option has been used most successfully and to great effect in sequence-search algorithms (that align sequences to each other and rank the results in order). Good rigorous algorithms, for example, "dynamic programming algorithms", exist for doing this. When searching large databases, such rigour is to all intents and purposes, *useless* for the vast majority of paired sequences. If they are dissimilar, they will never match well and will, inevitably, have low scores. Running a computationally expensive algorithm with such sequences is very wasteful. Some method of discontinuing the analysis as soon as the discrepancy is clear is often used.

> **Maxim 17.5** *Use a fast, if inaccurate, algorithm to prevent your slow, accurate second-stage algorithm from being overwhelmed with the testing of wholly unsuitable candidate regions/items.*

### 7.3.5 tRNA-ScanSE, a case study

The *tRNA-ScanSE* application searches for transfer RNA (tRNA) genes in Genomic DNA. In essence, tRNAs molecules are adaptor molecules that carry individual, specific, amino acids on their "acceptor stems" for incorporation into new proteins under the direction of Ribosomes, during the process called *Translation*. A set of three nucleotide bases (called the *anticodon*) in the *anticodon arm* denotes specifically which amino acid is attached on the acceptor stem.

During protein synthesis, these anticodons base-pair with the codons in mRNA (messenger RNA) that encode the order of the amino acids in the protein. The

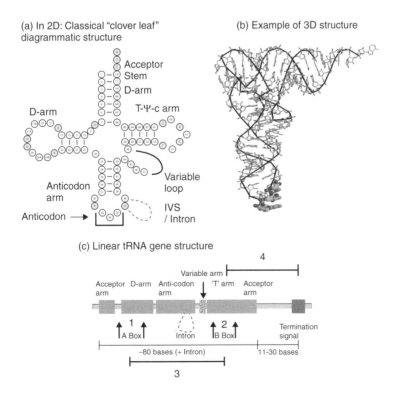

**Figure 17.2** An overview of tRNA: 2D, 3D and gene structure.

ribosomes administer this process and catalyse the formation of peptide bonds. An overview of both the physical and gene structures of tRNAs is shown in Figure 17.2 on page 353.

With reference to Figure 17.2, part (a) is the classic "Clover leaf" two-dimensional diagram of a tRNA. Shaded circles are semi-invariant nucleotide bases. Note how the anticodon is marked at the bottom of the "Anticodon Arm". Part (b) is a three-dimensional structure of PDB entry 1EHZ. The phosphate backbone is traced in black, while the bases forming the anticodon are in heavier "wireframe". Note how they face into solution ready for binding to the codons in the mRNA. Part (c) is the tRNA gene structure. The large numbers indicate the regions targeted by the original tRNAscan program.

An excellent description of how *tRNA-ScanSE* achieves its increase in performance is given in the first paragraph of its manual, which is distributed along with the program code. Rather than attempt to paraphrase the description, here it is verbatim:

```
-------------------------------------------------------------
tRNAscan-SE: a program for improved detection of transfer RNA
genes in genomic sequence

Todd Lowe (1) & Sean Eddy (2)

(1) Dept. of Genetics, Stanford University, Palo Alto, CA
(2) Dept. of Genetics, Washington U. School of Medicine, St. Louis
-------------------------------------------------------------
Current release: 1.21 (October 2000)

Note: An HTML version of this manual can be found on the
web at http://genome.wustl.edu/lowe/tRNAscan-SE-Manual/Manual.html


1. Introduction

   A. Brief Description

tRNAscan-SE identifies transfer RNA genes in genomic DNA or RNA
sequences.  It combines the specificity of the Cove probabilistic RNA
prediction package (Eddy & Durbin, 1994) with the speed and
sensitivity of tRNAscan 1.3 (Fichant & Burks, 1991) plus an
implementation of an algorithm described by Pavesi and colleagues
(1994) that searches for eukaryotic pol III tRNA promoters (our
implementation referred to as EufindtRNA).  tRNAscan and EufindtRNA
are used as first-pass prefilters to identify "candidate" tRNA regions
of the sequence.  These sub-sequences are then passed to Cove for
further analysis, and output if Cove confirms the initial tRNA
prediction.  In this way, tRNAscan-SE attains the best of both worlds:
(1) a false positive rate of less than one per 15 billion nucleotides
of random sequence, (2) the combined sensitivities of tRNAscan and
EufindtRNA (detection of 99% of true tRNAs), and (3) a search speed
1000 to 3000 times faster than Cove analysis and 30 to 90 times
faster than the original tRNAscan 1.3 (tRNAscan-SE uses both a
code-optimised version of tRNAscan 1.3, which gives a 650-fold increase
```

in speed, and a fast C implementation of the Pavesi et al. algorithm).
This program and results of its analysis of a number of genomes have
been published in Nucleic Acids Research (4).

The numbered references, just mentioned, are included at the end of the MANUAL disk-file, and are repeated here:

    1. Fichant, G.A. and Burks, C. (1991) "Identifying potential tRNA
    genes in genomic DNA sequences", J. Mol. Biol., 220, 659-671.

    2. Eddy, S.R. and Durbin, R. (1994) "RNA sequence analysis using
    covariance models", Nucl. Acids Res., 22, 2079-2088.

    3. Pavesi, A., Conterio, F., Bolchi, A., Dieci, G., Ottonello, S. (1994)
    "Identification of new eukaryotic tRNA genes in genomic DNA databases
    by a multistep weight matrix analysis of transcriptional control
    regions", Nucl. Acids Res., 22, 1247-1256.

    4. Lowe, T.M. & Eddy, S.R. (1997) "tRNAscan-SE: A program for
    improved detection of transfer RNA genes in genomic sequence",
    Nucl. Acids Res., 25, 955-964.

In the *tRNAscan-SE* search system, *tRNAscan-SE*, the *fast-steps* are the *Eufindt* and *tRNAscan* algorithms. The *slow-step* is then the *Cove* algorithm which screens the results more thoroughly.

This *fast-step, slow-step* system is also used in other sequence identification tools such as FASTA and BLAST (see the next section). The FASTA search algorithm also uses *k-tuple* words to identify sequence pairs whose alignment can be improved to a "useful level" by the application of a dynamic programming algorithm.

An example of the type of output obtained from *tRNAScan-SE* is shown below. This output was produced as a result of testing contig c1676 from Chromosome 3 of the yeast S. Pombe. This a 1,369,435 bp section of sequence that has been created by the combination of other shorter cosmid sequences[7]. To check the results, use the annotation in the original EMBL file. Here are the results produced:

    tRNAscan-SE v.1.23 (April 2002) - scan sequences for transfer RNAs

        Please cite:
            Lowe, T.M. & Eddy, S.R. (1997) "tRNAscan-SE: A program for
            improved detection of transfer RNA genes in genomic sequence"
            Nucl. Acids Res. 25: 955-964.

        This program uses a modified, optimized version of tRNAscan v1.3
        (Fichant & Burks, J. Mol. Biol. 1991, 220: 659-671),
        a new implementation of a multistep weight matrix algorithm
        for identification of eukaryotic tRNA promoter regions
        (Pavesi et al., Nucl. Acids Res. 1994, 22: 1247-1256),

---

[7] Download the sequence from http://www.sanger.ac.uk/Projects/S_pombe/DNA_download.shtml, then convert it to FASTA format for use with *tRNAScan-SE*.

as well as the RNA covariance analysis package Cove v.2.4.2
(Eddy & Durbin, Nucl. Acids Res. 1994, 22: 2079-2088).

```
------------------------------------------------------------
Sequence file(s) to search:  ../c1676.contig
Search Mode:                 Eukaryotic
Results written to:          ../c1676.op
Output format:               Tabular
Searching with:              tRNAscan + EufindtRNA -> Cove
Covariance model:            TRNA2-euk.cm
tRNAscan parameters:         Strict
EufindtRNA parameters:       Relaxed (Int Cutoff= -32.1)
------------------------------------------------------------
```

| Sequence Name | tRNA # | tRNA Begin | Bounds End | tRNA Type | Anti Codon | Intron Begin | Bounds End | Cove Score |
|---|---|---|---|---|---|---|---|---|
| c1676 | 1  | 8989    | 9059    | Asp | GTC | 0      | 0      | 70.49 |
| c1676 | 2  | 9486    | 9568    | Val | AAC | 9524   | 9532   | 64.30 |
| c1676 | 3  | 19361   | 19461   | Leu | CAA | 19399  | 19417  | 56.07 |
| c1676 | 4  | 22392   | 22463   | Thr | AGT | 0      | 0      | 75.01 |
| c1676 | 5  | 22879   | 22951   | Arg | ACG | 0      | 0      | 75.46 |
| c1676 | 6  | 169745  | 169839  | Ser | GCT | 169784 | 169794 | 74.09 |
| c1676 | 7  | 391586  | 391668  | Lys | CTT | 391625 | 391632 | 74.69 |
| c1676 | 8  | 606265  | 606361  | Ser | TGA | 606302 | 606316 | 78.22 |
| c1676 | 9  | 606369  | 606440  | Met | CAT | 0      | 0      | 69.52 |
| c1676 | 10 | 618451  | 618524  | Asn | GTT | 0      | 0      | 86.64 |
| c1676 | 11 | 705501  | 705572  | Pro | AGG | 0      | 0      | 64.72 |
| c1676 | 12 | 761191  | 761262  | His | GTG | 0      | 0      | 69.87 |
| c1676 | 13 | 779683  | 779754  | Gln | TTG | 0      | 0      | 69.22 |
| c1676 | 14 | 953904  | 953974  | Gly | TCC | 0      | 0      | 64.15 |
| c1676 | 15 | 988632  | 988726  | Ser | GCT | 988671 | 988681 | 74.09 |
| c1676 | 16 | 1056030 | 1055960 | Asp | GTC | 0      | 0      | 70.49 |
| c1676 | 17 | 991614  | 991543  | Gln | TTG | 0      | 0      | 69.22 |
| c1676 | 18 | 692625  | 692544  | Ser | AGA | 0      | 0      | 80.47 |
| c1676 | 19 | 277713  | 277642  | Thr | TGT | 0      | 0      | 79.00 |
| c1676 | 20 | 123371  | 123299  | Val | TAC | 0      | 0      | 76.36 |
| c1676 | 21 | 58816   | 58744   | Phe | GAA | 0      | 0      | 69.56 |
| c1676 | 22 | 56170   | 56088   | Lys | CTT | 56131  | 56124  | 74.69 |
| c1676 | 23 | 23109   | 23039   | Asp | GTC | 0      | 0      | 70.49 |
| c1676 | 24 | 22612   | 22530   | Val | AAC | 22574  | 22566  | 64.30 |
| c1676 | 25 | 14122   | 14051   | Glu | CTC | 0      | 0      | 74.15 |
| c1676 | 26 | 12737   | 12637   | Leu | CAA | 12699  | 12681  | 56.07 |
| c1676 | 27 | 9706    | 9635    | Thr | AGT | 0      | 0      | 75.01 |
| c1676 | 28 | 9219    | 9147    | Arg | ACG | 0      | 0      | 75.46 |

The column headers explain their contents and purpose. The two most interesting points to note are:

1. The tRNA gene structure is consistent enough for *tRNAScan-SE* to be able to predict the amino acid the tRNA will carry, see the "tRNA Type" and "Anti Codon" columns.

2. The "tRNA Begin" marker locations go up and then down again as the system scans along one strand, then back along the complement strand.

## 17.4 Introducing Bioinformatics Tools

This section describes ClustalW and BLAST, two of the most-used Bioinformatics sequence analysis tools. Both are used extensively in the later chapters of *Bioinformatics, Biocomputing and Perl*. The descriptions of how these tools work and how to install them are provided here to avoid impeding the *flow* of the examples using them. The most important concepts of the underlying algorithms and assumptions are described for both tools. These descriptions are brief and intended as primers. In the case of BLAST, the NCBI provides an excellent tutorial that can be accessed from the following web-site:

    http://www.ncbi.nlm.nih.gov/Education/

For both ClustalW and BLAST, good, well-maintained services are available on public web-sites, including those from EBI and NCBI. These types of services typically have "*copy the sequence(s) into the text box and click Run*" type functionality. These are fine for occasional use where the complexity of setting up, running and maintaining a *local service* is difficult to justify. For instance, in the case of BLAST, keeping the databases up-to-date from publicly available sources can be a major chore. Fortunately, many auto-update systems exist. Even so, *these* have to be set up and maintained.

Despite the convenient web-based services, it often arises that both applications need to be executed locally. This is typically the case when dealing with confidential or commercially sensitive data. In certain cases, speed is another motivating factor: there are times when waiting in an on-line queue is acceptable, and there are times when it is not. When prototyping an analysis or doing a small number of searches, it is OK to wait 30 seconds or so for the EBI ClustalW service to align (for example) 12 HMA sequences. However, when performing a large number of such alignments, a local installation is very advantageous. By way of example, running the same ClustalW searches on a 2-GHz Laptop (running Linux) took less than a second!

Another factor is *flexibility*: if an on-line service does not do what's required, then, as a user, there's very little that can be done about it. In such circumstances, running the analysis locally can be advantageous. This is especially true with BLAST when working with a custom database or a subset of a public one. A few commercial Bioinformatics web-based services allow for the upload of custom datasets, but this facility can have a high price: many such providers appear and disappear quite quickly. So, be careful.

> **Technical Commentary:** Some researchers have concerns about the secure transmission of data over the Internet. Using modern informatics and cryptographic systems, it is possible to secure the link. The problem tends to surface at the other end of the connection. With network-based services, data is handed to a third party. This may create legal difficulties with, say, patient confidentiality or regulatory requirements. Additionally, very strict management procedures must be used

to demonstrate consistency in the analysis for drug-acceptance testing. There are credible stories where data has "leaked" from supposedly secure on-line services. One problem is hostile (or otherwise) takeovers of the Bioinformatics outsourcing firms offering such on-line services. Small biotech companies are notorious for going bankrupt and getting consumed by other companies. Your authors are aware of an example where "technology was transferred" between two large commercial rivals by this mechanism. The result was the loss of a patent application, with no chance of legal redress by the company that had performed all the work. "Ouch!", as some would say!

The sections that follow cover how to use these applications[8].

### 17.4.1 ClustalW

ClustalW is a sequence alignment tool that can align two (or more) DNA or protein sequences. Source code and "compile-it-yourself" versions are available for Linux/Unix-type operating systems, as are pre-compiled binary packages for the Microsoft Windows platform. Both ClustalW and its forerunner, ClustalV, are exclusively command-line orientated tools, though they can run in either automated or "interactive" mode. Another version, ClustalX, which uses the same algorithm, has an excellent graphical user interface and postscript output. It also comes pre-packaged with some elementary tree-drawing software, but be aware that specialist packages, such as *TreeView*[9], are vastly superior for producing publication-quality graphics. To draw "serious" trees, the ClustalW documentation suggests *PHYLIP* (written by *Joe Felsenstein*, Department of Genome Sciences, University of Washington). *PHYLIP* has a long history in phylogenetics. An alternative, semi-commercial Phylogenetics package is *PAUP package*, which is a licensed academic work. The ClustalW/ClustalX packages can be downloaded from here:

    http://www-igbmc.u-strasbg.fr/BioInfo/

or from the EBI:

    ftp://ftp.ebi.ac.uk/pub/software

Figure 17.3 on page 359 shows the application running under the *Windows XP* graphical user interface. The sequences are the `MerP/MerA` HMA fragments used in the phylogenetics demonstration in the next chapter. On screen, the amino acids are coloured according to type. ClustalX is a preferred method for the creation of interactive alignments using the ClustalW algorithms.

---

[8] Note that a pre-built, web-based NCBI-BLAST package is available, should a requirement exist to locally provide a BLAST web-site.

[9] For more details, visit the `http://taxonomy.zoology.gla.ac.uk/rod/treeview.html` web-site.

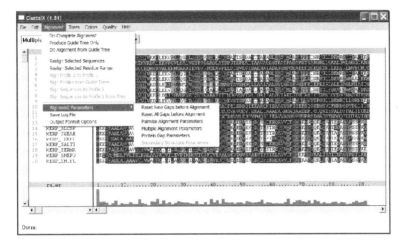

**Figure 17.3** ClustalX operating under Windows XP.

In addition to the references included in the *Suggestions for Further Reading* appendix, further references and pointers to helpful hints and papers are included in the documentation that accompanies these packages.

## 7.4.2 Algorithms and methods

In practice, the alignment of multiple sequences reduces to the comparison of pairs of sequences in order to make the analysis computationally possible. Full rigorous alignment using dynamic programming algorithms takes too long for all but a small number of short sequences. Consequently, a series of "tricks" are used:

- Initially, pairs of sequences are scored in an approximate manner using a fast algorithm (on the basis of "k-tuple words") to determine their interrelationships. For "n" sequences, there are $(n(n-1))/2$ possible alignments, which serves to highlight the main bottleneck in the process. For 20 MerA/MerP HMA sequences, for example, this means 190 separate sequence alignments. This explains why each test needs to be fast!

- From the scores between pairs of sequences (calculated in the previous step), a rough phylogenetic *guide tree* is constructed by what is in essence a standard *average linkage hierarchical clustering* method. The *guide tree* is used to direct the slow, but accurate, alignment of sequences in the later stages.

- As a priming step, the two most similar sequences are aligned together, and the result is converted into a *sequence profile*.

- Using the profile makes it easy to align new sequences to the existing multiple alignment. The order in which the sequences are added is determined by the guide tree/clustering step, and it continues until all the sequences have been merged into the alignment.

- Should it be required, the "Neighbourhood Joining" method is used to produce a phylogenetic tree from the alignment.

ClustalW is very good and it has become the *de facto* standard tool when performing multiple sequence alignment.

There are more recent extensions to the general Clustal algorithm. One example is COFFEE (Consistency-based Objective Function For alignment Evaluation), which is now being developed as T-COFFEE (Tree-based-COFFEE). These tools claim to be better at aligning distantly related sequences. This performance increase is due, in part, to better gap placement and the use of a position-specific scoring matrix during the alignment of sequences. The *Suggestions for Further Reading* appendix provides pointers to available public references.

### 17.4.3 Installation and use

The installation of ClustalW follows the classic UNIX install process. The source code should compile under most distributions of Linux[10]. Use these commands to install ClustalW:

```
gzip -d clustalw1.83.UNIX.tar.gz
tar -xvf clustalw1.83.UNIX.tar
cd clustalw1.83
make
```

ClustalW can be run in interactive mode, but the interface system is entirely textual. Simply execute the program without any options to see this:

```
./clustalw
```

ClustalX accesses the same underlying alignment but has a vastly superior interface specifically designed for human interaction. Many may prefer this to the text-driven menus of ClustalW. That said, for executing ClustalW automatically, there is a series of command-line switches available. These can be used to set such options as input and output files, "k-tuple" word length, phylogenetic tree output format, and so on. All of the command-line options are documented in the supplied manual. A brief summary can be obtained by running ClustalW with the following flag:

```
./clustalw -h
```

---

[10] Don't worry if you receive a few warnings, as these tend to be non-fatal.

Most of the option names are mnemonic, for example:

**-INFILE** – Sets the input disk-file that contains the sequences to be aligned.

**-OUTFILE** – Sets the disk-file to which the alignments are to be directed (saved).

For example, to align the MerP/MerA HMA domains locally, use a command such as this:

    ./clustalw  -INFILE=../MerAHMAs_MerP.swp  -OUTFILE=../Mer.aln

The alignment is placed in the ../Mer.aln disk-file. Note, with this example, that both the input and output disk-files are created in the directory *above* the current one. All of the available options are fully described in the included documentation.

## 7.4.4 Substitution/scoring matrices

Substitution matrices are an important component of pairwise sequence alignment and it is worth understanding the different forms. The default matrices currently used by many pairwise alignment systems are the *BLOSUM* series, which is produced from the alignment of a series of protein *blocks*. The original substitution scoring matrices, the *PAM* series, were produced from the alignments of more closely related sequences than those used for *BLOSUM*. For general database searches, the *BLOSUM* matrices are often more sensitive.

Amino acid substitution matrices have been the subject of many publications resulting in some tens of examples in the scientific literature. All encode the chance of a particular amino acid being found or substituted at the same location in different proteins. The matrices differ because each uses different assumptions and/or is compiled from differing datasets. Generally, the matrices are triangular, also called *half matrices*, because in most substitution models, it is impossible to tell which amino acid was substituted for which: A for C, or C for A. Overall, the amino acid is more likely to be "conserved" or "substituted for itself", rather than mutate so the diagonal values are the highest value.

In the *BLOSUM62* mutation matrix, shown below, this conservation is denoted by the high positive numbers along the diagonal. For example, C (Cysteine) is found to have a high frequency of remaining unchanged, that is, it is likely to be the same, with the "C to C" cell having a score of 9 compared to "C to G" (Glycine) of −3, showing that change is observed less frequently.

Both these examples have been highlighted in **bold** in the reproduced matrix. This is the now-classical *BLOSUM62* matrix as distributed in the ./data/ directory of the NCBI-BLAST search tool:

|   | A | R | N | D | C | Q | E | G | H | I | L | K | M | F | P | S | T | W | Y | V |
|---|---|---|---|---|---|---|---|---|---|---|---|---|---|---|---|---|---|---|---|---|
| A | 4 | | | | | | | | | | | | | | | | | | | |
| R | -1 | 5 | | | | | | | | | | | | | | | | | | |
| N | -2 | 0 | 6 | | | | | | | | | | | | | | | | | |
| D | -2 | -2 | 1 | 6 | | | | | | | | | | | | | | | | |
| C | 0 | -3 | -3 | -3 | 9 | | | | | | | | | | | | | | | |
| Q | -1 | 1 | 0 | 0 | -3 | 5 | | | | | | | | | | | | | | |
| E | -1 | 0 | 0 | 2 | -4 | 2 | 5 | | | | | | | | | | | | | |
| G | 0 | -2 | 0 | -1 | -3 | -2 | -2 | 6 | | | | | | | | | | | | |
| H | -2 | 0 | 1 | -1 | -3 | 0 | 0 | -2 | 8 | | | | | | | | | | | |
| I | -1 | -3 | -3 | -3 | -1 | -3 | -3 | -4 | -3 | 4 | | | | | | | | | | |
| L | -1 | -2 | -3 | -4 | -1 | -2 | -3 | -4 | -3 | 2 | 4 | | | | | | | | | |
| K | -1 | 2 | 0 | -1 | -3 | 1 | 1 | -2 | -1 | -3 | -2 | 5 | | | | | | | | |
| M | -1 | -1 | -2 | -3 | -1 | 0 | -2 | -3 | -2 | 1 | 2 | -1 | 5 | | | | | | | |
| F | -2 | -3 | -3 | -3 | -2 | -3 | -3 | -3 | -1 | 0 | 0 | -3 | 0 | 6 | | | | | | |
| P | -1 | -2 | -2 | -1 | -3 | -1 | -1 | -2 | -2 | -3 | -3 | -1 | -2 | -4 | 7 | | | | | |
| S | 1 | -1 | 1 | 0 | -1 | 0 | 0 | 0 | -1 | -2 | -2 | 0 | -1 | -2 | -1 | 4 | | | | |
| T | 0 | -1 | 0 | -1 | -1 | -1 | -1 | -2 | -2 | -1 | -1 | -1 | -1 | -2 | -1 | 1 | 5 | | | |
| W | -3 | -3 | -4 | -4 | -2 | -2 | -3 | -2 | -2 | -3 | -2 | -3 | -1 | 1 | -4 | -3 | -2 | 11 | | |
| Y | -2 | -2 | -2 | -3 | -2 | -1 | -2 | -3 | 2 | -1 | -1 | -2 | -1 | 3 | -3 | -2 | -2 | 2 | 7 | |
| V | 0 | -3 | -3 | -3 | -1 | -2 | -2 | -3 | -3 | 3 | 1 | -2 | 1 | -1 | -2 | -2 | 0 | -3 | -1 | 4 |

The 62 in *BLOSUM62* refers to the number of identical amino acids between two sequences in the alignments used to produce the matrix.

The *PAM250* matrices are included with the ClustalW and ClustalX packages. As the *PAM* matrices were originally produced from the alignment of closely related sequences, they must be extrapolated to score more distantly related sequences. The number 250 in *PAM250* shows that this matrix is more suitable for general database searching between distantly related sequences. More formally, the sequence is *250 Point Accepted Mutations (substitutions) per 100 amino acids*. It is possible for amino acids to mutate more than once, as there can be multiple substitutions at the same location. In general, the lower the *PAM* number, the closer the evolutionary distance the matrix is optimised for. Unless a special need exists, such as making results comparable with a historical analysis, the advice is to always use the *BLOSUM* matrices.

The need to select an optimal matrix is avoided by these programs, as they adapt the matrices used during execution. ClustalW switches between different matrices in the same series (either *BLOSUM* or *PAM*). See the original ClustalW publication distributed in the ClustalW package as the `clustalw.ms` disk-file for more details.

## 17.5 BLAST

BLAST, the Basic Local Alignment Search Tool, is probably the most widely used tool/algorithm in Bioinformatics. Like ClustalW, BLAST is a sequence alignment algorithm. However, whereas ClustalW is designed for the alignment of multiple sequences, BLAST is designed to test one *query sequence* against a database of sequences in an attempt to find *common regions*[11]. ClustalW is a *global alignment algorithm* that attempts to align every amino acid or nucleotide base to every other one, whereas BLAST is a *local alignment algorithm* that attempts to match the most similar subsections.

Many variants of the original BLAST algorithm, released in 1990, now exist. All use the same general approach:

- Initially, regions of similarity between the query sequence being tested and the sequences from the database are identified. This is accomplished by finding *High Scoring Segment Pairs* (HSSP) referred to as "words". These are analogous to the *k-tuple* pre-screening step of FASTA and the initial *fast alignment step* of ClustalW. Each set of HSSPs can be assigned a score by summing the similarity values from an appropriate substitution matrix. Substitution matrices are only useful with protein (amino acid) alignments. For DNA sequences, a simple *identity matrix* is used, in which each aligned nucleotide scores the same value (usually 1), and unaligned nucleotides score zero. The reason for the different approaches is that there is no adequate model to capture the modulations of the four DNA nucleotide bases, whereas suitable models have been found for the more variable amino acid substitutions.

- If the score of two sequences (based on the HSSPs they contain) is high enough, then an *alignment extension step* is triggered. Otherwise, the algorithm assumes that the two sequences are too dissimilar and moves on to the next sequence in the database.initi

- In the BLAST algorithms, alignment extension is the most computationally expensive step, hence the use of the HSSP scoring step to filter out alignments that will always be too poor to be useful. The extension procedure is to start at the HSSP location and test the surrounding amino acids to determine if they improve the overall alignment. Dynamic programming algorithms, which can make allowance for gaps in the sequence, are sometimes used to accomplish this.

One of the most popular variants of the original BLAST algorithm is NCBI-BLAST. This is maintained by the *National Center for Biotechnology Information (NCBI)*, based in Bethesda, MD, USA. NCBI-BLAST is available for free download and is the version of BLAST described and used within *Bioinformatics, Biocomputing and Perl*.

Another popular version is WU-BLAST, in which the "WU" refers to *Washington University* (in St Louis, Missouri, not Washington, DC). Despite the use of the common name BLAST, only the outline principles used by the NCBI version

---

[11] It is possible to use modern variations of the BLAST algorithm to perform multiple alignments, as well as use ClustalW to search for a single sequence in a database.

and any other version of BLAST are shared. A popular, modern version is PSI-BLAST, the *Pattern Specific Iterative* BLAST. With this tool, an initial BLAST search using a standard matrix (typically *BLOSUM62*) is used to find similar sequences that are then used to compile a position-specific matrix of the frequencies of the amino acids found at particular locations in the alignment. A series of position-specific matrix production/database scans are then produced. Such "personalised" frequency matrices with one *column* per amino acid improve the quality of the alignment and consequently, the performance of the overall search. Choosing which BLAST tool to use often depends on the individual circumstances of the researcher.

> **Maxim 17.6** *Exactly which BLAST is best depends on the circumstances.*

## 17.5.1 Installing NCBI-BLAST

The installation of NCBI-BLAST is designed to be as simple as possible and there are pre-compiled binary packages available for most modern operating systems. The standard Linux/UNIX installation procedure (with a twist) is used to perform an installation. Because of the packaging of the downloaded disk-file, some extra bundled disk-files are included. Use these commands to create a directory for NCBI-BLAST, then copy the downloaded disk-file into the directory and extract the bundled contents:

```
cd
mkdir blast
cp blast-2.2.6-ia32-linux.tar.gz blast
cd blast
gzip -d blast-2.2.6-ia32-linux.tar.gz
tar -xvf blast-2.2.6-ia32-linux.tar
```

The included documentation suggests the creation of a configuration disk-file, called .ncbirc[12]. Typically, configuration disk-files (such as .ncbirc) are found within a user's home directory. On Linux, a user called michael typically has a home directory of /home/michael. The configuration disk-file for NCBI-BLAST would then reside in /home/michael/.ncbirc. The contents of the default configuration disk-file may look something like this:

```
[NCBI]
Data="/home/michael/blast/data"
```

Note that the documentation "suggests" the creation of the disk-file, though when Michael didn't his BLAST installation still worked. So we guess this step is optional to some extent! But we suggest you do as the manual says, just to be safe.

---

[12] The leading dot "hides" the disk-file within a directory. When the standard ls utility is used to list the disk-files in any directory, hidden disk-files are not listed. Use ls -a to force the ls utility to display hidden disk-files.

## 7.5.2 Preparation of database files for faster searching

The BLAST algorithm requires the database of protein or nucleotide sequences be indexed as this significantly improves the speed and efficiency of the searches. The index is created using the included `formatdb` program, which stores the generated indexes in the same directory location from where it is executed. Conveniently, most of the larger sequence databases are available for download in a pre-indexed form (from the usual FTP sites, including those maintained by NCBI, EBI and others). However, a need often exists to create a subset of sequences, in order to prevent wasteful searches or to include sequences not in the public databases.

For demonstration purposes, a database subset containing all of the 55 *Mer Operon* protein sequences present in the SWISS-PROT database has been created. This subset is used to illustrate the automation of the BLAST indexing, search and result-reporting systems.

The sequences were extracted using the SRS service at the EBI, searching for the string "Mer" within the ID field of SWISS-PROT. The relevant entries (those starting with "MerA/B/C/D/E/F/G/P/R/T") were selected manually and saved as FastaSeqs to a disk-file called All_Mer_Proteins.fsa[13].

To avoid cluttering the BLAST directory with index files, create a subdirectory, called `databases`, to house them:

```
mkdir   databases
cd  databases
mv  ../All_Mer_Proteins.fsa  .
```

Then execute the `formatdb` program to create the index:

```
../formatdb  -i  All_Mer_Proteins.fsa  -p  T  -o  T  -n  Merproteins
```

The options used have the following meaning:

**-i** – Specifies the input disk-file (database).

**-p** – Specifies that the input disk-file contains protein sequences (where "T" means true).

**-o** – Specifies that `SeqIdParse` and the creation of extra indexes should be set to true.

**-n** – Creates a database with the provided name, which is `Merprotein` in this example.

---

[13] We could have used *Bioperl* to do this. However, to do so, we would have needed to know each ID in full or its corresponding accession code in order to provide the list of them to *Bioperl*. This is one example in which the interactive search facilities are more convenient than their automated, programmed cousins. More on *Bioperl* in our final chapter.

Execution of the above command-line creates a series of five index disk-files in the `databases` directory, in addition to the `formatdb.log` disk-file. The contents of the `databases` directory, when listed with the "`ls -l`" command, should look something like this:

```
-rw-r--r--   1 michael  users    17044 2003-10-04 22:12 All_Mer_Proteins.fsa
-rw-r--r--   1 michael  users     4744 2003-10-05 12:19 formatdb.log
-rw-r--r--   1 michael  users    12179 2003-10-05 12:19 Merproteins.psq
-rw-r--r--   1 michael  users       99 2003-10-05 12:19 Merproteins.psi
-rw-r--r--   1 michael  users     2162 2003-10-05 12:19 Merproteins.psd
-rw-r--r--   1 michael  users      520 2003-10-05 12:19 Merproteins.pin
-rw-r--r--   1 michael  users     6340 2003-10-05 12:19 Merproteins.phr
```

The `formatdb.log` disk-file records the progress and any errors related to the indexing process. Here is an example log:

```
========================[ Oct 5, 2003 12:44 PM  ========================
Version 2.2.6 [Apr-09-2003]
Started database file "All_Mer_Proteins.fsa"
NOTE: CoreLib [002.003] FileOpen(".formatdbrc","r") failed
NOTE: CoreLib [002.003] FileOpen("/home/michael/.formatdbrc","r") failed
NOTE: [000.000] No number of link bits used found in config  file. Ignoring
NOTE: [000.000] No number of membership bits used found in config file. Ignoring
NOTE: ncbiapi [000.000] SeqIdParse Failure at sw|Q52109|MERA_ACICA
NOTE: ncbiapi [000.000] SeqIdParse Failure at sw|P94188|MERA_ALCSP
NOTE: ncbiapi [000.000] SeqIdParse Failure at sw|P16171|MERA_BACCE
    .
    .
    .
NOTE: ncbiapi [000.000] SeqIdParse Failure at sw|P30345|MERT_STRLI
Formatted 55 sequences in volume 0
```

Of interest is the fact that there are several "failures" reported. Despite this, the output disk-files were still created. In fact, the failure reports relate to a trivial problem with parsing the description: the sequence similarity-searching still works.

As an example, use the MERA_PSEFL protein as a test sequence by placing it in a disk-file called `test_seq.fsa` and do a BLAST search with the command-line:

```
blastall   -p   blastp   -d   databases/Merproteins   -i   test_seq.fsa
```

where the command-line options are:

**-p blastp** – identifies the BLAST type, which is a protein sequence search against a protein database.

**-d databases/Merprotein** – identifies the indexed database disk-file.

**-i test_seq.fsa** – identifies the "query sequence" to test against the database.

When executed, the results were:

```
Reference: Altschul, Stephen F., Thomas L. Madden, Alejandro A. Schaffer,
Jinghui Zhang, Zheng Zhang, Webb Miller, and David J. Lipman (1997),
"Gapped BLAST and PSI-BLAST: a new generation of protein database search
programs", Nucleic Acids Res. 25:3389-3402.

Query= Test_Seq (MERA_PSEFL)
         (548 letters)

Database: All_Mer_Proteins.fsa
           55 sequences; 12,123 total letters

Searching..done
                                                                Score    E
Sequences producing significant alignments:                     (bits) Value

6_Merproteins Mercuric reductase (EC 1.16.1.1) (Hg(II) reductase).  1001   0.0
5_Merproteins Mercuric reductase (EC 1.16.1.1) (Hg(II) reductase).   863   0.0
1_Merproteins Mercuric reductase (EC 1.16.1.1) (Hg(II) reductase).   854   0.0
      .
      .
      .
```

The descriptions are poorly handled by the indexing process, as indicated in the `formatdb.log` disk-file. When properly formatted, the output should look more like this:

```
                                                                Score    E
Sequences producing significant alignments:                     (bits) Value
sp|Q51772|MERA_PSEFL Mercuric reductase (EC 1.16.1.1) (Hg(II) re...  1001   0.0
sp|P00392|MERA_PSEAE Mercuric reductase (EC 1.16.1.1) (Hg(II) re...
 863   0.0
sp|Q52109|MERA_ACICA Mercuric reductase (EC 1.16.1.1) (Hg(II) re...   854   0.0
      .
      .
      .
```

This is more descriptive than before and is nearly in standard format. The standardisation is particularly important if the intention is to automate parsing, for example, when using the EBI BLAST web-based service to create hyperlinks to the SRS database. This particular problem is easily solved, and was caused because of the fact that the `formatdb` program expected the description line in the FASTA disk-file to be in a standard format. The table below, reproduced from the `README.formatdb` disk-file supplied in the NCBI-BLAST package, specifies the description line:

```
Database Name                       Identifier Syntax

GenBank                             gb|accession|locus
EMBL Data Library                   emb|accession|locus
DDBJ, DNA Database of Japan         bj|accession|locus
NBRF PIR                            pir||entry
Protein Research Foundation         prf||name
```

| | |
|---|---|
| SWISS-PROT | sp\|accession\|entry name |
| Brookhaven Protein Data Bank | pdb\|entry\|chain |
| Patents | pat\|country\|number |
| GenInfo Backbone Id | bbs\|number |
| General database identifier | gnl\|database\|identifier |
| NCBI Reference Sequence | ref\|accession\|locus |
| Local Sequence identifier | lcl\|identifier |

The FASTA sequence format generated by the EBI SRS server is:

```
sw|Q52109|MERA_ACICA Mercuric reductase (EC 1.16.1.1) (Hg(II) reductase).
MTTLKITGMTCDSCAAHVKEALEKVPGVQSALVSYPKGTAQLAIEAGTSSDALTTAVAGL ...
```

Apart from using `sw` instead of `sp` to denote the database type, this format is almost the same as the format used by the MERA_ACICA report from earlier. Unfortunately, it is different enough to be wrong. Sadly, there are no defined standards for FASTA-formatted descriptions. It is possible to use a custom Perl program to convert the format of the database downloaded from EBI `FastaSeqs` to a format that the NCBI BLAST `formatdb` program can process. However, a better tool (in this particular instance) is the `sed` utility. The `sed` utility is a "serial editor", and it uses the same regular expression syntax as Perl. Execute this command in the `databases` directory to convert the FASTA-formatted file (that performs the necessary substitution on every matching line in the disk-file):

```
sed 's/sw|/sp|/' All_Mer_Proteins.fsa > Mer_db.prot
```

The newly created MER_db.prot disk-file can now be indexed, overwriting the previous database index disk-file:

```
../formatdb -i Mer_db.prot -p T -o T -n Merproteins
```

This command execution results in the following text being appended to the `formatdb.log` disk-file:

```
========================[ Oct 6, 2003  1:23 PM ]========================
Version 2.2.6 [Apr-09-2003]
Started database file "Mer_db.prot"
NOTE: CoreLib [002.003] FileOpen(".formatdbrc","r") failed
NOTE: CoreLib [002.003] FileOpen("/home/michael/.formatdbrc","r") failed
NOTE: [000.000] No number of link bits used found in config  file. Ignoring
NOTE: [000.000] No number of membership bits used found in config file. Ignoring
Formatted 55 sequences in volume 0
```

which is, thankfully, a lot cleaner than before! There are still some "failed" comments, but these relate to the absence of the disk-files that would modify the default database and are not serious. Once the database has been created, the original FASTA disk-file is no longer needed. It is safe to delete it or move it elsewhere for storage. Re-running the `blastp` search gives the expected output:

```
Reference: Altschul, Stephen F., Thomas L. Madden, Alejandro A. Schaffer,
Jinghui Zhang, Zheng Zhang, Webb Miller, and David J. Lipman (1997),
"Gapped BLAST and PSI-BLAST: a new generation of protein database search
programs",  Nucleic Acids Res. 25:3389-3402.

Query= Test_Seq (MERA_PSEFL)
       (548 letters)

Database: Mer_db.prot
          55 sequences; 12,123 total letters

Searching..done
                                                                Score    E
Sequences producing significant alignments:                     (bits)  Value

sp|Q51772|MERA_PSEFL Mercuric reductase (EC 1.16.1.1) (Hg(II) re...  1001   0.0
sp|P00392|MERA_PSEAE Mercuric reductase (EC 1.16.1.1) (Hg(II) re...   863   0.0
sp|Q52109|MERA_ACICA Mercuric reductase (EC 1.16.1.1) (Hg(II) re...   854   0.0
sp|P94188|MERA_ALCSP Mercuric reductase (EC 1.16.1.1) (Hg(II) re...   852   0.0
      .
      .
```

## 7.5.3 The different types of BLAST search

Using the principle of "three nucleotides to codon for one amino acid", there are different types of translations/reverse translations that can be performed internally by the generic blastall program. The type of BLAST search is specified with a switch. Earlier, "-p (blastp)" was used to search protein databases with protein query sequences. The full list of switches is as follows:

**blastp** - Protein (amino acid) query sequence used to search a protein database.

**blastn** - Nucleotide (DNA) query sequence used to search a nucleotide database.

**blastx** - Nucleotide query sequence that is translated in all reading frames by BLAST into an amino acid query sequence that is then used to search a protein sequence database.

**tblastn** - Protein query sequence reverse translated in all six reading frames by BLAST into a nucleotide sequence that is used to search a nucleotide database.

**tblastx** - Nucleotide sequence translated in all reading frames by BLAST into a protein sequence and used to search a nucleotide database that is also being translated in all reading frames by BLAST.

Of course, not all of these make sense in all situations. For instance, running blastx against a protein query sequence is meaningless. Note that there are also considerable differences in execution time. As one would expect, the more the algorithm has to do, the longer it takes: blastp and blastn are fastest, next fastest is blastx and blastn, with the slowest being tblastx.

The parsing of BLAST outputs and the interpretation of the results is covered later in this book, when *Bioperl* is introduced. For now, a working introduction to these topics is presented, rather than a detailed discussion. For output parsing, BLAST is probably as good a parser as will ever be needed. Other useful tools, supplied with the NCBI-BLAST package, include `fastacmd` and `blastclust`.

### The `fastacmd` tool

This tool can extract individual sequences from databases indexed by `formatdb`. The `fastacmd` tool can also summarise the contents of the database. By way of example:

```
fastacmd  -d  databases/Merproteins  -I
```

produces the following output:

```
Database: Mer_db.prot
        55 sequences; 12,123 total letters

File name:
databases/Merproteins
        Date: Oct 5, 2003  3:50 PM    Version: 4    Longest sequence: 631 res
```

The `fastacmd` tool can also extract a single sequence from the database. This is a useful feature when storing the original FASTA disk-file from which the indexed version was created, either on compressed disk or on a "slower" access medium such as Tape, CD or DVD. This example command-line:

```
fastacmd  -d  databases/Merproteins  -s  MERA_SHIFL
```

produces the following output:

```
sp|P08332|MERA_SHIFL Mercuric reductase (EC 1.16.1.1) (Hg(II) reductase)
MSTLKITGMTCDSCAVHVKDALEKVPGVQSADVSYAKGSAKLAIEVGTSPDALTAAVAGLGYRATLADAPSVSTPGGLLD
KMRDLLGRNDKTGSSGALHIAVIGSGGAAMAAALKAVEQGARVTLIERGTIGGTCVNVGCVPSKIMIRAAHIAHLRRESP
FDGGIAATTPTIQRTALLAQQQARVDELRHAKYEGILEGNPAITVLHGSARFKDNRNLIVQLNDGGERVVAFDRCLIATG
ASPAVPPIPGLKDTPYWTSTEALVSETIPKRLAVIGSSVVALELAQAFARLGAKVTILARSTLFFREDPAIGEAVTAAFR
MEGIEVREHTQASQVAYINGVRDGEFVLTTAHGELRADKLLVATGRAPNTRKLALDATGVTLTPQGAIVIDPGMRTSVEH
IYAAGDCTDQPQFVYVAAAAGTRAAINMTGGDAALNLTAMPAVVFTDPQVATVGYSEAEAHHDGIKTDSRTLTLDNVPRA
LANFDTRGFIKLVVEEGSGRLIGVQAVAPEAGELIQTAALAIRNRMTVQELADQLFPYLTMVEGLKLAAQTFNKDVKQLS
CCAG
```

### The `blastclust` tool

This tool automatically clusters database sequences using their similarity scores from a single linkage clustering method. For example:

```
blastclust -d databases/Merproteins | head
```

produces the following output:

```
Oct 5, 2003  4:21 PM  Start clustering of 55 queries
P94700 P04140 P04336 P13112 P94185 Q51769 Q52106
P04129 P04131 P13113 P94186 Q51770
P00392 P94702 Q52109 P94188
P08664 P77072 Q91UN2
P06688 P07044 P13111
P08653 Q8CU52
P06689 P08654
P94703 Q52110
P94701 Q52107
  .
  .
  .
```

### 7.5.4 Final words on BLAST

BLAST is a useful algorithm, whether it is available on-line (attached to well-curated databases), or in use locally (for accessing bespoke or custom sequence databases). The NCBI-BLAST package is easy to download, configure and use. Despite this, others, such as WU-BLAST, are worthy of consideration. Pre-built "drop-in" web-based server packages that expose a BLAST server to an internal (or external) network via the WWW are also available. Be aware that interpretation of the results can be tricky, as is demonstrated later in this book.

## Where to from Here

This chapter provides an introduction to some well-known databases, concepts and tools commonly used in Bioinformatics sequence analysis. The installation of two of the tools, ClustalW and BLAST (used extensively in the next chapter) were described. Some extra features of these tools were also examined. The brief examples presented in this chapter are expanded upon in the next chapter.

## The Maxims Repeated

Here's a list of the maxims introduced in this chapter.

- *Recognise the difference between the validation of a model and the testing of it for self-consistency.*

- *Generally, False Negative predictions are considered more acceptable than False Positives.*

- *With False Negatives, we could come back next year and find the ones we missed, and these are preferred to False Positives, where we can waste time* studying them this year, only to find out that the time was wasted. It all depends on the circumstances.

- *Sometimes, all those False Positives are maybe, just maybe, trying to tell you something. So, if you aspire to a Nobel prize....*
- *Use a fast, if inaccurate, algorithm to protect your slow, accurate second-stage algorithm.*
- *Exactly which BLAST is best depends on the circumstances.*

# 18
# Applications

*Using standard Bioinformatics applications.*

## 18.1 Introduction

In this chapter, a common set of Bioinformatics applications are used to analyse an example piece of DNA. The example used here has already been well characterised experimentally by laboratory work, and it was selected for this very reason. In the real world, the option to "look up the answer in the book" does not typically exist: all you get are a set of predictions. For now though, sit back and relax. Just bear in mind that life won't always be this easy! Figure 18.1 on page 374 summarises the activity of this chapter.

Bear in mind that as this chapter's material is worked through, the results produced may differ slightly from those presented here. This has to do with the fact that the underlying databases (and the data contained therein) are continuously being updated. For instance, the "missing genes" identified later in this chapter may well be incorporated into the annotation after the publication of this book.

---

*Bioinformatics, Biocomputing and Perl.*  Michael Moorhouse and Paul Barry
© 2004 John Wiley & Sons, Ltd   ISBN 0-470-85331-X

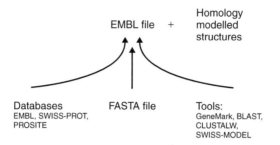

**Figure 18.1** The overall plan for the regeneration of high-quality annotation information as contained in the EMBL disk-file ISTN501.

## 18.2 Scientific Background to Mer Operon

This section describes the *Mer Operon*, its function, structure, regulation and mobility.

### 18.2.1 Function

The Mer Operon is a system by which bacteria can detoxify reactive $Hg^{2+}$ (mercury) ions into metallic mercury ($Hg^0$) using the compound NADPH, nicotinamide adenine dinucleotide phosphate. The system has been well characterised in terms of functionality tests in laboratory experiments, mutagenesis experiments and the features of the DNA/protein sequences for many different organisms.

The proteins of the Mer Operon are one of the major detoxification mechanisms in mercury-contaminated soils, which often occurs as the result of human activities, especially gold mining. It is also the principal resistance mechanism bacteria possess against the mercurial compounds that were used in some of the first antibiotics, even though it is now somewhat dubious how effective many of these treatments were in the first place.

### 18.2.2 Genetic structure and regulation

An operon is a group of genes that are either expressed or not, depending on environmental conditions. An operon contains an *operator* to which a series of control proteins termed *repressors* or *activators* binds. In the normal *off* state, the control proteins are expressed at low levels and the *Structural Genes* are not transcribed[1].

When the functionality of the structural genes is needed, the repressors detach from the second control region and transcription starts, that is, the genes are

---

[1] These are called "Structural" for historic reasons: the first operons were discovered in viruses coding for the structural parts of the virus coat/head. In this sense, they were the "structural proteins", rather than an operator concerned with DNA replication.

*expressed.* In some systems, other activator proteins may also bind to increase the level of expression: this is a role postulated for the MerC. Alternatively, another protein might be required if expression of the operon is to occur at all. For instance, the *CAP* (Catabolite Activator Protein) in the Lactose Metabolism System coded for by the *Lac Operon*.

## 8.2.3 Mobility of the Mer Operon

The Mer Operon is found in a very similar form in many bacteria because it exists in the form of a plasmid (a short section of circular DNA) on transposons (a mobile piece of DNA) and in cellular DNA, the genome of an organism. Plasmids, in particular, are frequently transferred between different species of bacteria as well as within the same species. This is one of the mechanisms by which antibiotic resistance spreads.

A *transposon* is a short piece of DNA that has the ability to move around and between the genomes of cells. A typical feature of transposons is the *inverted terminal repeats*, which are complementary sections of sequence found at both ends of the sequence. These repeats assist with the transpsonon's integration and excision from a cell's genome, and they have a set of enzymes called *transponases* that catalyse these processes.

The principal proteins and their functions are summarised in Figure 18.2 on page 376[2].

Cysteine amino acids are highlighted as these are what the $Hg^{2+}$ ions bind to in a *reversible way*, so each one is potentially of great functional significance. The gene structure (near the bottom of Figure 18.2) is taken from the EMBL database, entry ISTN501/AC Z00027.

The known genes, as parsed and plotted by the Embl_plot.pl program (excluding the unknown/probable ORFs), are:

```
[Feat. No] [Gene] [Spans nucleotides]

Feature# 0   merR  (  548 to   114)
Feature# 1   merT  (  620 to   970)
Feature# 2   merP  (  983 to  1258)
Feature# 3   merA  ( 1330 to  3015)
Feature# 4   merD  ( 3033 to  3398)
Feature# 5   tnpR  ( 4792 to  5352)
Feature# 6   tnpA  ( 5356 to  8322)
```

---

[2] This is based on *Figure 1*, Hobman, J. L., Brown, N. L., in Metal Ions in Biological Systems Volume 34 *Mercury and its effects on environment and biology* (Sigel, A., Sigel, H., editors), Marcel Dekker Inc, 1997, p. 527–569, and in *Operon Mer: Bacterial Resistance to mercury and potential for bioremediation of contaminated environments*, Nascinmento, AMA & E. Chartone-Souza, Genetics and Molecular Research 2 (1) 92-101, 2003 and Summers, A. O., 1986, Ann. Rev. Microbiol, 40, p. 607.

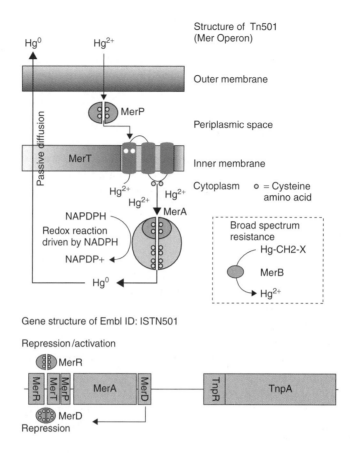

**Figure 18.2** The principal proteins, their functions and the arrangement of the genes in the Mer Operon.

The Mer Operon proteins and their function are summarised as follows:

| Protein | Function/Comment |
|---|---|
| MerA | Mercury Reductase: Reduces Hg2+ to Hg0 |
| MerP | Sequestration Protein: Complexes with Hg2+ |
| MerT | Transport Membrane Protein: increases passive diffusion inner membrane |
| MerR & MerD | Regulatory Proteins |
| MerB | Organo Mercurial Lyase: Cleaves Hg2+ from Organic compounds (not found in all bacteria) |

```
     ----------          ------------------------------------
     MerC & MerE         Currently unknown (or uncertain function)
                             MerC: Regulation?
                             MerE: Broad spectrum resistance?
     ----------          ------------------------------------
```

## 18.3 Downloading the Raw DNA Sequence

The starting point for this chapter's investigation is the full DNA sequence of the Tn501 transposon from *Pseudomonas aeruginosa*, EMBL ID ISTN501. Obtain a copy of this sequence by any of the following methods:

1. Using the *Bioperl* modules, as discussed in Chapter 20 (starting on 441).

2. Searching one of the on-line retrieval systems such as the SRS at the EBI, as discussed in Chapter 14 (starting on page 297).

3. Retrieving directly from a "local copy" of the EMBL database[3].

For a single sequence, it is often convenient to use SRS, as doing so also provides access to "clickable" database cross references, which are especially useful for accessing information about any particular protein found. Use any web browser to surf to the SRS homepage, start a new session and enter the ISTN501 into the *Quick Text Search* box, then click the *Search* button. The requested sequence contains the full Tn501 transposon with all the Mer Operon genes, in addition to a few extra transposon genes, with a total length of close to 8300 bases. To download the sequence, click on the ID, then follow the *Save* link. Depending on how the browser is configured, the entry may be saved directly to a disk-file or displayed within the browser window. In either case, be sure to save the entry as Text (not HTML) that is typically selectable in the browser's download dialogue box. Download ISTN501 in two different formats[4]:

1. **Complete entries** – This format is the EMBL flat file and contains the full annotation. The disk-file can be named ISTN501.embl. The length of the saved disk-file indicates that there is a lot known about this particular sequence. Note that this format is used within this chapter to check predictions.

2. **FASTA sequence only** – This format contains the name and the DNA sequence data, which closely mimics the original data (less all the

---

[3] A subject that is beyond the scope of this book. However, a good starting point is the *EMBOSS* package, available from http://www.emboss.org.

[4] The SRS help pages are very helpful and describe how to save the different formats.

378    Applications

annotations) as obtained from a sequencing project. Name this disk-file ISTN501.fsa.

The goal for the remainder of this chapter is to obtain the same annotation (or better!) as in the EMBL-formatted version of the disk-file (ISTN501.embl) from the raw DNA sequence in the FASTA-formatted disk-file (ISTN501.fsa).

## 18.4 Initial BLAST Sequence Similarity Search

The first step is to compare the raw DNA sequence in the FASTA disk-file against the SWISS-PROT database, giving some clues as to the type of proteins that are coded for by the DNA sequence in the disk-file. Although far from perfect (discussed in detail later), this analysis is quick and easy, and is usually a good place to start.

Use the NCBI-BLAST server at EBI to search the SWISS-PROT database using the BLASTx version of the algorithm. Be sure to do a 6-frame translation. This ensures that the search covers all the possible ORFs (see section 18.5 for what an 'ORF' is).

Note that the valid options on the BLAST web page can change depending on the selections made, and this can be a little disconcerting at first. For example, it does not make sense to use the blastn program variant to compare a protein sequence with the SWISS-PROT database, use blastp instead. Note that exactly what drives the contents of which box does take a little getting used to but, remember, the BLAST web page does not know whether the sequence is DNA or protein, until after the *Run* button has been clicked.

For a sequence of this size (8300 bases), searched against SWISS-PROT database, results should be returned within a few of minutes (peak times) or in under a minute (non-peak times, usually weekends). Here is an extract of the results returned:

```
                                                              Score    E
Sequences producing significant alignments:                   (bits)  Value
SW:TNP5_PSEAE  P06695  Transposase for transposon Tn501.       1957   0.0
SW:TNP9_ECOLI  P51565  Transposase for transposon Tn1721.      1923   0.0
SW:TNP7_ECOLI  P13694  Transposase for transposon Tn3926.      1446   0.0
SW:TNP2_ECOLI  P06694  Transposase for transposon Tn21.        1445   0.0
SW:MERA_PSEAE  P00392  Mercuric reductase (EC 1.16.1.1) (Hg(II) re... 1092  0.0
SW:TNP5_ECOLI  P08504  Transposase for transposon Tn2501.      1040   0.0
SW:MERA_ENTAG  P94702  Mercuric reductase (EC 1.16.1.1) (Hg(II) re... 1008  0.0
SW:MERA_ACICA  Q52109  Mercuric reductase (EC 1.16.1.1) (Hg(II) re... 1003  0.0
SW:MERA_ALCSP  P94188  Mercuric reductase (EC 1.16.1.1) (Hg(II) re...  960  0.0
SW:MERA_SHIFL  P08332  Mercuric reductase (EC 1.16.1.1) (Hg(II) re...  931  0.0
SW:MERA_PSEFL  Q51772  Mercuric reductase (EC 1.16.1.1) (Hg(II) re...  915  0.0
SW:MERA_THIFE  P17239  Mercuric reductase (EC 1.16.1.1) (Hg(II) re...  843  0.0
SW:TNP6_ENTFC  Q06238  Transposase for transposon Tn1546.       751   0.0
SW:TNPA_BACTU  P10021  Transposase for transposon Tn4430.       699   0.0
SW:MERA_SHEPU  Q54465  Mercuric reductase (EC 1.16.1.1) (Hg(II) re...  596  e-169
```

```
SW:MERA_SERMA P08662 Mercuric reductase (EC 1.16.1.1) (Hg(II) re...   593   e-168
SW:TNPA_ECOLI Q00037 Transposase for transposon gamma-delta (Tra...   462   e-129
   .
   .
   .
```

The results show that the top "hits" are against the *transposase* of various transposons, and Tn501 is the highest scoring hit from the correct organism, *Pseudomonas aeruginosa*. In fact, the first few hits are so high, the *E-value* (expected value) is reported as 0, which indicates that they are *off the scale*. Remember: for BLAST scores, "smaller, that is, closer to zero, is better". This is somewhat counter-intuitive. However, the score is linked to the probability that the alignment exists by chance.

> **Maxim 18.1** *With BLAST scores, up is down and smaller is better and more significant.*

Remember that E-values are raised to a negative power: the larger the exponent, the smaller the number.

The *Mercuric Reductase* hits are next, with the MERA_PSEAE entry achieving the highest score (as expected), which is the correct organism. These entries are then followed by other organisms intermixed with the other transposon related proteins.

If the number of reported hits is increased to the maximum of 250, the search finds more of the lower-scoring hits corresponding to proteins such as MerP, MerC and MerD. However, this larger list is full of hits against *Mercury Reductase*, *Glutathione Reductase* and *Dihydrolipoamide Dehydrogenase*, all of which are perfectly valid having low scores and good alignments, despite the fact that all these extra hits do not add significantly to our knowledge. It is clear that this sequence refers to a particular protein. Further, when different proteins are reported, there are a lot of them and they are all mixed up among each other.

The first problem, involving multiple hits, is easily solved. There are simply too many examples of similar proteins in the SWISS-PROT database. For instance, there are 12 MerA protein sequences. As demonstrated later, this can be very useful, but for a simple identification, one example of each protein is all that's needed. The solution to this problem is to use a non-redundant dataset, as discussed elsewhere in this book.

The second problem has to do with the way BLAST scores an alignment. The longer, higher quality alignments have higher scores. When, as above, there is a range of protein sizes, many of the shorter but high-scoring alignments are pushed down to the bottom of the list by lower quality but longer alignments. It turns out that the BLAST program is doing exactly what it is meant to. The problem has to do with the way BLAST is being applied.

## 18.5 GeneMark

It is helpful to pre-cut the DNA into regions that code for individual proteins before using BLAST (or any other tool) to identify what the particular protein is. This is a task for *gene-prediction* programs, of which the GeneMark algorithm is one example. GeneMark is available through the web-based EBI interface. The GeneMark homepage is located here:

    http://opal.biology.gatech.edu/GeneMark/

In this chapter, a bacterial system is used to demonstrate these tools, as the accurate prediction of genes is often difficult when not using bacteria. This is especially true when working with "higher organisms", such as human or yeast.

Bacterial proteins are generally coded for in one continuous long stretch of amino acids, and has a 1:3 mapping to the corresponding DNA sequence. Additionally, the *Start* and *Stop* codons, as well as the clear initiation/termination signals[5] mark the general region of the gene that codes for the translated protein. This is the definition of an ORF, *open reading frame*, we use here. Although appearing nice and simple, complexities abound in that start/stop codons and the diagnostic binding site sequences vary slightly between different organisms. If these are wrong, the predictions are weak unless the algorithm is robust. Typically, assuming a simple "look between the start and stop codons" strategy is not sufficient.

It is even more problematic in higher eukaryotic organisms, such as yeast, mice and men, as the entire gene expression system is more complex. The protein-coding regions can be discontinuous and for anything more complex than yeast, they usually are. This means that only a small number of regions of a gene is transcribed into RNA and hence, into protein. Add to this the fact that different *splice variants* exist, and the process becomes more complicated. Gene expression in Eukaryotes in general is a topic beyond the scope of this book. However, here we will use a "simple" bacterial system for demonstration.

> **Technical Commentary:** Before moving on, let's mention *GenScan*, one of the best programs for gene prediction in Eukaryotes. While some versions of GeneMark have been optimised for finding genes in Eukaryotic organisms, GenScan does not operate on bacterial sequences, as it is optimised for human or vertebrate searching. For bacteria use GenMark; for higher organisms use GenScan.

For illustration purposes, let's use the GeneMark web-based interface running at EBI (refer to Figure 18.3 on page 381). To begin, select an organism to illustrate the search. Unfortunately, the *Pseudomonas Aeruginosa* is not listed. As an alternative, let's use *Escherichia coli*, as this has been known to carry Mer Operon. Additionally, the plasmids/transposons appear to be quite interchangeable between different organisms.

---

[5] For example, polymerase binding site sequences.

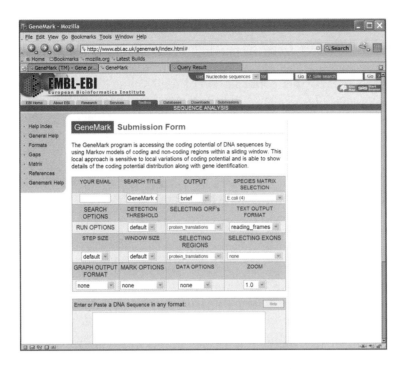

**Figure 18.3** The web-based interface to GeneMark as running at EBI.

Set the TEXT OUTPUT FORMAT to "reading_frames" and set SELECTING ORFs to "protein_translations". These options ask GeneMark to report the sequences of the predicted proteins in an appropriate format, and suppress some output that is less relevant. Set the other options as shown in Figure 18.3, then paste the ISTN501 sequence fragment into the text box. When the search concludes, GeneMark returns a series of seven ORFs, which it places in a disk-file called ISTN501_ecoli(4).orf.txt. The start of the disk-file contains that which is of interest:

```
;                          GENEMARK OUTPUT

; Protein sequences translated from ORFs:

> orf_1 (embl|Z00027|ISTN501 Transposon Tn501 from Pseudomonas aeruginosa \\
plasmid pVS1  encoding mercuric ion resistance determinant. Gene, 538 -    \\
164) translated
VETIRFYQRKGLLLEPDKPYGSIRRYGEADVTRVRFVKSAQRLGFSLDEIAELLRLEDGTHCEEASSLAE
HKLKDVREKMADLARMEAVLSELVCACHARRGNVSCPLIASLQGGASLAGSAMP*


> orf_2 (embl|Z00027|ISTN501 Transponson Tn501 ...
```

## 18.5.1 Using BLAST to identify specific sequences

The results of the search from the previous section indicate that GeneMark thinks seven amino acid sequences are proteins. Let's refer to these as "hypothetical proteins" until there is further confirmation of their validity. One such confirmation technique is to use BLAST against a well-curated database such as SWISS-PROT.

Once again, the web-based interface at the EBI can be used. On this occasion, select the NCBI `blastp` version of the BLAST algorithm, and the SWISSPROT database. For each of predicted proteins/ORFs in turn, paste the sequence into the text box and click 'Run' to see if a similar protein has already been found. Assuming the analysis system is operating correctly, the expectation is to receive several hits. Inspect the top hits and check that:

- they all have high scores and
- they are all broadly consistent with each other.

More generally, an unknown ORF/predicted protein may be found to have similarity to many seemingly disparate proteins, even though they have similar high scores. This is reasonable: many proteins have regions that have specific functions that are not unique to that particular protein. These regions may be a few amino acids or an entire subsection of a protein, known as a *domain*. The discussion that follows highlights an example of this in relation to the `MerA` protein.

Here is a summary of the BLAST outputs. The more interesting results are then discussed in detail.

| GeneMark Prediction | | BLAST Results | | SWISS-PROT Searches | | |
|---|---|---|---|---|---|---|
| Code | Sequence Length | Probable Protein | Gaps in Alignment of Top Hit? | Probable ID (Actually Correct?) | Probable Protein Length | Good agreement between prediction and reality? |
| ORF1 | 124 | MerR | No | P06688 (Y) | 144 | N (see text) |
| ORF2 | 91 | MerP | No | P04131 (Y) | 91 | Y |
| ORF3 | 561 | MerA | No | P00392 (Y) | 561 | Y |
| ORF4 | 128 | MerD | No | P06689 (Y) | 121 | Y |
| ORF5 | 388 | TNPM_ECOLI (Transposon Modulator)? | Yes | P04162 (?) | 116 | N (see text) |
| ORF6 | 186 | TNR5_PSEAE (Transponoson Resolvase) | No | P06691 (Y) | 186 | Y |
| ORF7 | 988 | TNP5_PSEAE | No | P06695 (Y) | 988 | Y |

In general, GeneMark does a good job of finding those proteins that it was expected to find. That is, the results provided enough hints to enable an educated guess to be made as to what the proteins were. Before discussing what's missing, let's examine the predictions for ORFs 1 and 5.

## The ORF1 (MerR) prediction

Examining the results of the BLAST search, it appears that this protein is the MerR (regulator) protein, in that ORF1 and MERR_PSEAE align almost perfectly. A subsequent ClustalW alignment of ORF1 against the full SWISS-PROT MerR entry uncovers that GeneMark incorrectly identified the protein *start point*, as shown here:

```
CLUSTAL W (1.82) multiple sequence alignment


MERR_PSEAE         MENNLENLTIGVFAKAAGVNVETIRFYQRKGLLLEPDKPYGSIRRYGEADVTRVRFVKSA 60
ORF1               -------------------VETIRFYQRKGLLLEPDKPYGSIRRYGEADVTRVRFVKSA 40
                                      *****************************************
```

The reasons for this incorrect identification are difficult to determine, but the sub-optimal settings of the E. coli(4) model may be to blame. This issue is returned to later in this chapter.

## The ORF5 prediction

The selected test system was chosen because it is a good demonstration of what can happen during analysis. ORF5 is typical of the type of results that can often be confusing. Except for ORF5, all the other ORFs have similar, well-characterised sequences within SWISS-PROT. However, in the case of ORF5, high BLAST scores against many "hypothetical proteins" indicate that similar proteins/ORFs have been found before, but that they are poorly characterised. The top hits from the BLAST output, with the two most interesting hits highlighted in bold, are:

```
Sequences producing significant alignments:                        (bits)  Value

SW:YNTC_AZOCA Q04855 Hypothetical 80.5 kDa protein in ntrC 5'reg...  192   1e-48
SW:YD54_MYCTU Q11024 Hypothetical protein Rv1354c/MT1397/Mb1389c.    183   5e-46
SW:TNPM_ECOLI P04162 Transposon Tn21 modulator protein.              183   6e-46
SW:PHY2_SYNY3 Q55434 Phytochrome-like protein cph2 Bacteriophyt...   182   1e-45
SW:YI95_SYNY3 P74101 Hypothetical protein sll1895.                   181   2e-45
SW:Y4LL_RHISN P55552 Hypothetical 91.8 kDa protein Y4LL.             163   7e-40
SW:YDDU_ECOLI P76129 Hypothetical protein yddU.                      162   1e-39
SW:YD57_MYCTU Q11027 Hypothetical protein Rv1357c/MT1400/Mb1392c.    160   6e-39
```

Reproducing the highlighted BLAST alignments gives this:

```
SW:TNPM_ECOLI P04162 Transposon Tn21 modulator protein.
Length = 116

Score =  183 bits (464), Expect = 6e-46
Identities = 88/116 (75%), Positives = 96/116 (81%), Gaps = 1/116 (0%)
```

```
Query: 272  MDVVAEGVETSASLDLLRQADCDTGQGFLFAKPMPAAAFAVFVSQWRGATMNASDSTTTS  331
            M+VVAEGVET   L LRQA CDT QGFLFA+PMPAAAF FV+QWR TMNA++ +T S
Sbjct: 1    MEVVAEGVETPDCLAWLRQAGCDTVQGFLFARPMPAAAFVGFVNQWRNTTMNANEPST-S  59

Query: 332  CCVCCKEIPLDAAFTPEGAEYVEHFCGLECYQRFEARAKTGNETDADPNACDSLPS     387
            CCVCCKEIPLDAAFTPEGAEYVEHFCGLECYQRF+ARA T  ET   P+ACDS PS
Sbjct: 60   CCVCCKEIPLDAAFTPEGAEYVEHFCGLECYQRFQARASTATETSVKPDACDSPPS     115
```

and this:

```
SW:PHY2_SYNY3 Q55434 Phytochrome-like protein cph2 (Bacteriophytochrome cph2).
Length = 1276

Score = 182 bits (462), Expect = 1e-45
Identities = 103/247 (41%), Positives = 145/247 (58%), Gaps = 6/247 (2%)

Query: 72   ELAQAVERGQLELHYQPVVDLRSGGIVGAEALLRWRHPTLGLLPPGQFLPVVESSGLMPE  131
            +L QA+    + L++QP V L +G ++G EAL+RW+HP LG + P  F+P+ E  GL+
Sbjct: 613  DLRQALTNQEFVLYFQPQVALDTGKLLGVEALVRWQHPRLGQVAPDVFIPLAEELGLINH  672

Query: 132  IGAWVLGEACRQMRDWRMLAWRPFRLAVNASASQVGPDFDGWVKGVLADAE---LPAEYL  188
            +G WVL  AC   + +    R R+AVN SA Q    +   W+ VL    +P E L
Sbjct: 673  LGQWVLETACATHQHFFRETGRRLRMAVNISARQFQDE--KWLNSVLECLKRTGMPPEDL  730

Query: 189  EIELTESVAFGD-PAIFPALDALRQIGVRFAADDFGTGYSCLQHLKCCPISTLKIDQSFV  247
            E+E+TES+    D     L LR+ GV+ A DDFGTGYS L  LK  PI  LKID+SFV
Sbjct: 731  ELEITESLMMEDIKGTVVLLHRLREEGVQVAIDDFGTGYSSLSILKQLPIHRLKIDKSFV  790

Query: 248  AGLANDRRDQTIVHTVIQLAHGLGMDVVAEGVETSASLDLLRQADCDTGQGFLFAKPMPA  307
              L N+ D  I+  VI LA+GL ++ VAEG+E+ A L L++    C  GQG+    +P+PA
Sbjct: 791  NDLLNEGADTAIIQYVIDLANGLNLETVAEGIESEAQLQRLQKMGCHLGQGYFLTRPLPA  850

Query: 308  AAFAVFV                                                       314
            A   ++
Sbjct: 851  EAMMTYL                                                       857
```

A good question to ask is which of these is "the best"? According to the BLAST scores, they are both very similar with E-values of **6e-46** and **1e-45**. The first alignment is shorter than the second, and has more identical or similar amino acids, which normally implies a better alignment. However, this is not always the case.

The other proteins in the list are all hypothetical and have little or no functional annotation attached to them. Consequently, little insight can be gained from these. The two high-scoring hits have such radically different functions that choosing between them is not easy.

The first hit is against protein TNPM_ECOLI/P04162, which is described in SWISS-PROT as "transposon Tn21 modulator protein". Given that the other proteins listed identified around it[6], it is highly likely that the predicted ORF really is this protein.

---

[6] Referred to as the "biological context".

The second hit is against PHY2_SYNY3/Q55434, which is described in SWISS-PROT as "Phytochrome-like protein cph2 (Bacteriophytochrome cph2)". Its *function field* contains the following:

```
PHOTORECEPTOR WHICH EXISTS IN TWO FORMS THAT ARE REVERSIBLY INTERCONVERTIBLE
BY LIGHT: THE R FORM THAT ABSORBS MAXIMALLY IN THE RED REGION OF THE SPECTRUM
AND THE FR FORM THAT ABSORBS MAXIMALLY IN THE FAR-RED REGION
```

Could this realistically be what the ORF predicted? It seems unlikely given the biological context of a transposon. What is a phytochrome doing in a transposon? Although it is possible that the second hit is valid, the first hit seems far more plausible.

Referring back to the original EMBL disk-file that contained the full annotation of the ISTN501 test sequence, there is an ORF-2 listed that *seems* to be very similar to that which GeneMark predicted and called ORF5. Sadly, the link to SPTREMBL describing ORF-2/ORF5 is uninformative:

```
FT   CDS         3628..4617
FT               /db_xref="GOA:Q48362"
FT               /db_xref="SPTREMBL:Q48362"
FT               /note="orf-2"
FT               /transl_table=11
FT               /protein_id="CAA77326.1"
FT               /translation="MSAFRPDGWTTPELAQAVERGQLELHYQPVVDLRSGGIVGAEALL
FT               RWRHPTLGLLPPGQFLPVVESSGLMPEIGAWVLGEACRQMRDWRMLAWRPFRLAVNASA
FT               SQVGPDFDGWVKGVLADAELPAEYLEIELTESVAFGDPAIFPALDALRQIGVRFAADDF
FT               GTGYSCLQHLKCCPISTLKIDQSFVAGLANDRRDQTIVHTVIQLAHGLGMDVVAEGVET
FT               SASLDLLRQADCDTGQGFLFAKPMPAAAFAVFVSQWRGATMNASDSTTTSCCVCCKEIP
FT               LDAAFTPEGAEYVEHFCGLECYQRFEARAKTGNETDADPNACDSLPSD"
```

To help clarify this situation, use ClustalW to do an alignment between the three protein sequences: the GeneMark predicted ORF5, the protein sequence from SWISS-PROT for TNPM_ECOLI and the translated sequence included in the ISTN501 EMBL entry (also called orf-2). Here is an extract from the ClustalW output:

```
Sequence 1: ORF5         388 aa
Sequence 2: ISTN501      329 aa
Sequence 3: TNPM_ECOLI   116 aa
       .
       .
       .

ORF5         AELPAEYLEIELTESVAFGDPAIFPALDALRQIGVRFAADDFGTGYSCLQHLKCCPISTL 240
ISTN501      AELPAEYLEIELTESVAFGDPAIFPALDALRQIGVRFAADDFGTGYSCLQHLKCCPISTL 181
TNPM_ECOLI   ------------------------------------------------------------

ORF5         KIDQSFVAGLANDRRDQTIVHTVIQLAHGLGMDVVAEGVETSASLDLLRQADCDTGQGFL 300
ISTN501      KIDQSFVAGLANDRRDQTIVHTVIQLAHGLGMDVVAEGVETSASLDLLRQADCDTGQGFL 241
TNPM_ECOLI   -----------------------------MEVVAEGVETPDCLAWLRQAGCDTVQGFL   29
                                          *:********..*  **** *** ****
```

```
ORF5         FAKPMPAAAFAVFVSQWRGATMNASDSTTTSCCVCCKEIPLDAAFTPEGAEYVEHFCGLE 360
ISTN501      FAKPMPAAAFAVFVSQWRGATMNASDSTTTSCCVCCKEIPLDAAFTPEGAEYVEHFCGLE 301
TNPM_ECOLI   FARPMPAAAFVGFVNQWRNTTMNANE-PSTSCCVCCKEIPLDAAFTPEGAEYVEHFCGLE  88
             **:*******  **.***.:****.:  .:*****************************

ORF5         CYQRFEARAKTGNETDADPNACDSLPSD 388
ISTN501      CYQRFEARAKTGNETDADPNACDSLPSD 329
TNPM_ECOLI   CYQRFQARASTATETSVKPDACDSPPSG 116
             *****:***.*..**...*:**** **.
```

What is apparent is that ORF5 is slightly longer than, and has the same sequence as, orf-2. The TNPM_ECOLI protein has a high amount of similarity in relation to part of the other two. Why just this region? Are there multiple domains of the modulator protein? And if so, does our protein here have an extra one? Does TNPM_ECOLI lack this domain? Or is it present but not acknowledged in the SWISS-PROT annotation? It is not possible to tell from this information alone.

What is clear (from the evidence) is that ORF5/orf-2 is most likely the Tn501 version of the Tn21 modulator protein. This would usually be acknowledged as:

```
Tn501 modulator protein - by similarity
```

or some other similar note within a typical annotation scheme.

### 18.5.2 Dealing with false negatives and missing proteins

It is important to consider if any proteins are missing. According to the published literature on Tn501, it is normal to find two other proteins:

1. MerT - which codes for a major part of the Mer Operon functionality, specifically, the transport of Mercury ions through the cell membrane.

2. MerE - which is often found in Tn501, but which has (to date) some unknown role.

Neither of these was identified in the ORFs column of the BLAST results on page 382. Yet in the EMBL-formatted disk-file, the MerT protein is listed and coded for by nucleotides 620 through 970. This helps explain its inclusion in the gene diagram of the Mer Operon presented at the start of this chapter (Figure 18.2 on page 376). Note that GeneMark failed to predict this protein. This is possibly due to the sub-optimal settings specified. It could also indicate that the MerT protein has special features that caused it to be missed. However, given its position in the operon and its biological function, this seems unlikely and it should be considered 'missed' by GeneMark i.e. MerT is false negative.

In the case of MerE, it is necessary to consider whether there is a gene actually present to predict at all. MerE genes have been found in Mer Operons, but only recently, and certainly not in all cases. To date, no specific function has been associated with them; this is in spite of many years of research. Possibly its function (along with the MerG gene) is to broaden the range of mercury compounds that the operon is effective against. Although possible, this is a tentative speculation at best.

It turns out that ISTN501 does have a sequence that might well be a MerE gene, as the EMBL annotation implies:

```
FT   CDS         3395..3631
FT               /db_xref="GOA:P06690"
FT               /db_xref="SWISS-PROT:P06690"
FT               /note="orf-1 (merE protein?)"
FT               /transl_table=11
FT               /protein_id="CAA77325.1"
FT               /translation="MNNPERLPSETHKPITGYLWGGLAVLTCPCHLPILAVVLAGTTAG
FT               AFLGEHWVIAALGLTGLFLLSLSRALRAFRERE"
```

From this annotation, it is clear that the annotators were unsure as to the classification of this protein that is identified and listed as orf-1, with the more speculative (merE protein?) appended as a suggestion of function. It seems likely that ISTN501 does indeed contain a MerE gene now that this sequence has been highlighted, so for the sake of argument, let's assume GeneMark missed predicting this gene, generating a *false negative prediction*.

## 8.5.3 Over-predicted genes and false positives

It is important to consider whether GeneMark incorrectly predicted any regions that were in fact just pieces of DNA with no seemingly specific purpose. There were none, but given the test system, there is not much scope for this type of error. Nearly all the DNA sequence supplied coded for proteins! The identification using BLAST could have been erroneous, but this would not have altered the GeneMark *false positive rate*.

Had the analysis been performed on DNA from higher organisms, then the false positive rate may well have been a problem, as some organisms have long stretches of *non-coding DNA*. Additionally, coding regions are often discontinuous (as already noted), that is, they contain *Introns*, which can be a serious problem.

A crude "trick" is to perform a quick *reality check*, to effectively reduce the false positive rate is to do a BLAST search against the sequence databases for each ORF. If low scores result from this search, implying that nothing similar has been seen before, then the ORFs can be dismissed as *false positive predictions*. Further, it's possible to check the alignment to see if the start and end points are valid.

Unfortunately, this technique has a number of drawbacks, not least of which is that it excludes anything not previously identified, including truly novel sequences that are often the most interesting!

### 18.5.4 Summary of validation of GeneMark prediction

An overview of the results of the GeneMark predictions is as follows:

| Gene Name | Correct Prediction? True Positive | Missed Prediction? False Negative | ORF Name |
|---|---|---|---|
| MerP | Yes | | ORF1 |
| MerP | Yes | | ORF2 |
| MerA | Yes | | ORF3 |
| MerD | Yes | | ORF4 |
| TNPM | Yes | | ORF5 |
| TNR5 | Yes | | ORF6 |
| TNP5 | Yes | | ORF7 |
| MerE | | Yes | |
| MerT | | Yes | |

The test system does not really have sufficient non-coding regions to assess the *False Positive* or *True Negative* rates.

## 18.6 Structural Prediction with SWISS-MODEL

This section describes how to produce full 3D structural models for the predicted proteins. SWISS-MODEL is an automated homology modelling server operating at the ExPASy[7] web-site, located at *The Swiss Institute of Bioinformatics*:

```
http://www.expasy.org/swissmod/
```

SWISS-MODEL uses one of the most accurate forms of structural prediction, using known protein structures as a framework upon which to build the model of the new sequence. This requires that a suitable template be identified in the set of known structures. That is, at least one structure with a sequence that has detectable homology needs to exist with the sequence to be modelled. As a post-processing step, energy minimisation is performed on the resultant model to optimise it.

The *DeepView* program (previously called *Swiss-Pdb Viewer*) can be used to prepare sequence alignments to protein structure templates, then submit these to SWISS-MODEL for modelling. The structural models returned can be viewed and manipulated. Previously saved models can also be loaded from disk-file and modelled. The models may be aligned structurally on the basis of selected amino

---
[7] Yes, that's how it is spelt.

acids or atoms, they may have angles or distances between atoms measured and they can be corrected prior to being rendered to a very high quality using SWISS-MODEL's internal (or some other external) Ray Tracing algorithm[8]. DeepView is available for free download for most desktop operating systems, including Apple Macintosh, Microsoft Windows, GNU/Linux and SGI Irix.

Of note is that SWISS-MODEL uses a BLAST search of the Protein Databank (PDB) structural database to find suitable template structures.

Use the web-based interface to the BLAST service at EBI to mimic the SWISS-MODEL search. This allows the results to be inspected more easily. Normally, this would be unnecessary as SWISS-MODEL provides a similar service. However, when working with a set of unknown proteins, the ability to explore the PDB using the clickable links returned in the EBI search output is useful. It is possible to gain an impression of the structures that could be used as templates to model the sequences. The settings for the BLAST search should be:

```
Database     = PDB
Program      = blastp
Search Title = ORF + (the index of the ORF sequence).
```

Here is a summary of the top BLAST search hits against the PDB database[9]:

| ORF/Protein | \multicolumn{3}{c}{Details of best PDB hits from the NCBI BLAST service} | |
|---|---|---|---|---|
| | PDB ID | E-value | Description (truncated by BLAST) | Can be used to predict structure? |
| ORF1/MerR | 1EXJ | 0.012 | Multidrug-Efflux Transport ... | - |
| ORF2/MerP | 2HQI | 1e-30 | Mercuric Transport Protein | Yes |
| ORF3/MerA | 1EBD | 1e-63 | Diydrolipoamide Dehydroge ... | Yes |
| ORF4/MerD | 1JBG | 1e-05 | Transcription Activator Of ... | Yes |
| ORF5/TNPM | - | - | *** no hits found *** | - |
| ORF6/TNR5 | 1GDT | 2e-21 | Gamma-Delta Resolvase | Yes |
| ORF7/TNP5 | 1COI | 2.7 | D-Amino Acid Oxidase | - |

From these results, it would appear that it is worth submitting SWISS-MODEL requests for ORFs 2, 3, 4 and 6.

Incidentally, it is rare indeed that BLAST reports "no hits found" for ORF5! Normally, BLAST finds *something*, even though it might be with a very poor E-score, for example, above 0.01. It is worth remembering that the PDB is a small, specialised database compared to SWISS-PROT or EMBL, and this would often make the chance of finding something similar quite low.

---

[8] *Ray Tracing* is a general method used in high-quality image generation.

[9] Remember: using the EBI service mimics the search by the SWISS-MODEL homology server, but allows the results to be *inspected*.

390    Applications

> **Maxim 18.2** *The major limitation of "homology modelling" is that homology to a known structure is needed.*

### 18.6.1 Alternatives to homology modelling

In those cases in which BLAST cannot find a similar structure, an alternative structural prediction technique, such as *Ab initio* or *Threading* can be used. Unfortunately, the results produced by these techniques are sometimes very speculative. Most of the other common structural prediction techniques will *eventually* give a structure that is broadly correct. The problem is that the structure may be "buried" in a set of 20 or 30 incorrect structures, all of which have subtle errors that are hard to detect. Recall that the structural prediction algorithms have already done their best, so non-experts (like us!) working through the models manually are unlikely to do better.

If secondary structure prediction is OK, there are good methods available (such as *PhD*), which are now incorporated into the *PredictProtein* system. These assign each amino acid to the same broad classes as the STRIDE algorithm: Helix, Sheet/Extended and Turn/Coil, but it is uncertain what advantage this is. Interesting it may be, but is it useful? Well, only for certain special cases – one of which is as a basis for the prediction of the full 3D structure by another method.

### 18.6.2 Modelling with SWISS-MODEL

As there is no extra information to optimise the alignment, just copy and paste the sequence data into the text box on the *First Approach* mode of the SWISS-MODEL web page. The system prompts for an e-mail address to send results to as a PDB data-file[10]. Enter a meaningful title (such as a tag to identify the ORF) into the *Request Title* field, then click on *Send Request*.

An acknowledgement, in the form of a "Welcome to SwissModel" e-mail, confirms that the request is being processed. The speed with which any further e-mails arrive depends on what structures SWISS-MODEL finds as templates.

When the modelling completes, a summary of the process is sent as an e-mail titled "Tracelog". This e-mail contains two parts:

1. The *AlignMaster* output lists what similar templates were found, how similar they are and how they overlapped with each other in the form of a schematic table. It also shows which part(s) of the protein sequence it will pass forward for modelling. These are often the most interesting statistics as they indicate how complete the model will be. If the sequence shows definite homology to two or more groups of structures (if it is a multi-domain protein, for example), then *AlignMaster* initiates modelling on each section separately.

---

[10] Be careful: each e-mail may be large, often some megabytes, and a lot of them arriving over a short period of time may rapidly fill up your e-mail inbox.

2. The *ProMod* output takes the alignments produced by *AlignMaster* (or an alignment optimised in a package such as *DeepView*) and does the actual modelling. Refer to the SWISS-MODEL manual for more details.

When further e-mails then arrive, each region of the sequence modelled will then arrive in a separate email. These e-mails contain data in PDB format. Here is a summary of the expected trace logs:

```
                          Region(s) modelled      Number of accepted
ORF/Protein   ORF Length  (index of amino acids)  templates
-----------   ----------  ----------------------  ------------------
ORF2/MerP         91      15 to 91                        9
ORF3/MerA        561       1 to 73 (A)                   83
                          95 to 551 (B)                   2
ORF4/MerD        128      -                                0
ORF6/TNR5        186      23 to 186                        7
```

There are several interesting aspects to these results. With ORF2 and ORF6, the prediction does not start at amino acid 1, which may seem strange at first. This is OK, as there is no absolute requirement for it to do so. *AlignMaster* detects homology and it has coincident starts for very similar proteins that have conserved domains, but for less similar proteins, the homology could start anywhere in the structure.

ORF3/MerA is predicted in two different regions with a gap of nearly 20 amino acids in between. By inspecting the *AlignMaster* output, it is possible to determine a particular template matches either one part or the other, but not both. This implies that the protein has two distinct domains as shown by this fragment of the alignment table:

```
Target Sequence:     |=========================================================|
1ebdA.pdb            |          ---------------------------------------
      .
      .
      .
1aw0_.pd             |------
2aw0_.pdb            |------
```

ORF4/MerD was not predicted, but the EBI BLAST search found a suitable homologous structure. SWISS-MODEL imposes stricter criteria than our simple BLAST search, as follows:

```
BLAST search P value : < 0.00001
Global degree of sequence identity (SIM) : > 25 %
Minimal projected model length = 25 aa.
```

While the local BLAST alignment was good enough, with 34 of 109 (31%) of amino acids identical, the alignment to 1JBG was rejected. This happens when all 128

**392** *Applications*

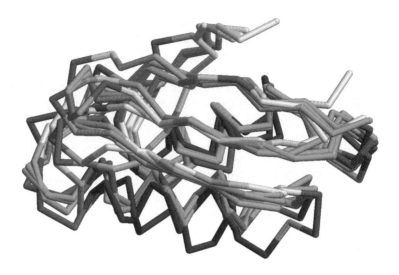

**Figure 18.4** The SWISS-MODEL predicted structure of ORF2/MerP.

amino acids are considered by *AlignMaster* using the *global alignment criterion*, and where too few "identities" were acceptable.

Why the difference? It is important to remember that the NCBI-BLAST and the *AlignMaster* BLAST may be the same or different algorithms. Additionally, they may use different parameters and, consequently, produce different alignments. In a borderline case (as described here), this difference is important. However, do not worry about this. The model produced from a single template with so little homology would *probably* have been quite poor anyway! Still, any model at all is often better than no model, and *DeepView* can always be used to manually prepare an alignment. Just remember that the resultant model may be of dubious quality.

The next step is to extract the structural models from the e-mails and load them into a graphical viewer. A useful tool is *Open Rasmol*, which can be used to get an impression of the overall structure. Consider the `ORF2/MerP` example shown in Figure 18.4 on page 392:

Looking at the model on screen (Figure 18.4), it is often difficult to see any relationships. However, the model does give some impression of the deviation between the different templates and the original structures on which the model was based. This is due to the fact that SWISS-MODEL returns both of these in the same PDB data-file. This is useful, but it is often necessary to view the target structure on its own. A good technique is to cut the PDB data-file at the boundary between the `TARGET` and the first of the original structure models. To help with finding the correct position, load the PDB data-file into a text editor and *cut* as indicated in this extracted fragment:

```
SPDBVa   69   70   71   72   73   74   76   77   78   79
SPDBVa   80   81   82   85    5
```

```
SPDBVn    r  TARGET
SPDBVE

<<<<<<<<<<<<<<<<<<<<<<<<<  Cut needed here!

COMPND    ?
REMARK    File generated by Swiss-PdbViewer  3.70b17
REMARK    http://www.expasy.org/spdbv/
CRYST1    1.000    1.000    1.000  90.00  90.00  90.00 P 1           1
ATOM       1  N   ALA    1      17.219   2.023  -0.443  1.00  0.00
```

The code fragment also excludes any extra lines added by SWISS-MODEL that confuse *DeepView*. *DeepView* expects the other structures to be present, *Open Rasmol* does not care! Here is a fragment of Perl program code to perform the cut:

```
while (<>)
{
    if (/^SPDBVE/)
    {
        print "END\n";
        last;
    }

    if (/^SPDBV/) { next; }

    if (/^SEQALI/) { next; }

    print $_;
}
```

This while loop terminates when the SPDBVE line is encountered, while skipping the SPDBV and SEQALI lines. All other lines (up to the SPDBVE line) are printed.

For a low number of structures such as we have here (only five), it is a viable option to use a text editor to delete all the remaining lines in the file. Be careful not to "over-code" solutions to problems and create custom programs only when you really need to.

To distinguish these single structures from those also containing the templates, use memorable names such as ORF2_Target.pdb and ORF3_A_Target.pdb and so on. A simple *shell script* can help here, as follows:

```
#!/usr/bin/tcsh

./Target_Parse.pl ORF2.pdb > ORF2_Target.pdb
./Target_Parse.pl ORF3_A.pdb > ORF3_A_Target.pdb
./Target_Parse.pl ORF3_B.pdb > ORF3_B_Target.pdb
./Target_Parse.pl ORF6.pdb > ORF6_Target.pdb
```

SWISS-MODEL returns multiple models. The first is the predicted model, and each subsequent model is of the sections of the original proteins on which the model

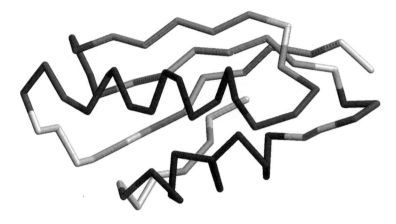

**Figure 18.5** The SWISS-MODEL predicted structure of ORF2/MerP, second version.

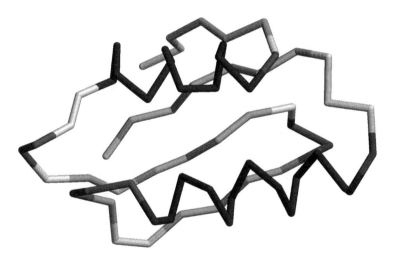

**Figure 18.6** The SWISS-MODEL predicted structure of ORF3/MerA (A).

was based. To produce these views in *Open Rasmol*, the *Display* is set to *Backbone* and *Color* is set to *Structure*[11]. The produced models are shown in Figures 18.5 through 18.8.

Simply by looking at the structures from ORF2/MerP and ORF3/MerA (A), they seem to be very similar. The next section describes how to confirm or refute this suspicion.

---

[11] We have rotated the model for publication purposes.

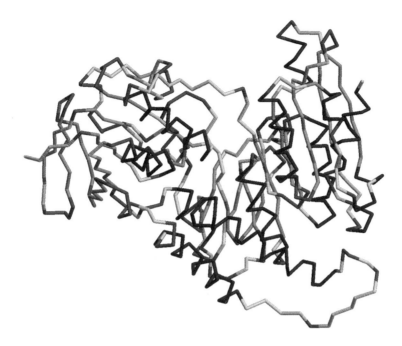

**Figure 18.7** The SWISS-MODEL predicted structure of ORF3/MerAB.

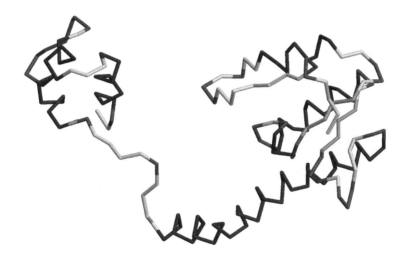

**Figure 18.8** The SWISS-MODEL predicted structure of ORF6/TNR5.

## 18.7 DeepView as a Structural Alignment Tool

In this section, the *DeepView* program is used to extract protein sequences, perform an alignment on the basis of the structure and measure some bond distances. Assuming *DeepView* is installed, load the two data-files (ORF2_Target.pdb and ORF3_A_Target.pdb) into *DeepView*. The resultant display should look like the screen-shot in Figure 18.9 on page 396.

The images in Figure 18.9 have been rendered to show helices, sheets and turns in order to demonstrate the rendering capability of *DeepView*. To generate a similar view, select *Preferences*, *Ribbons*, then check the "Render as Solid Ribbon" option. In the figure, extensive alterations to the defaults were made to help distinguish the different secondary structures in black and white. Originally, helices had a red theme, sheets were yellow and coils were blue. The original image is quite beautiful[12]. To separate the two proteins as shown in the Figure 18.9, be sure to uncheck the "can move" option on the top right of the *Control Panel* for one structure and use a combination of the *Move* and *Rotate* functions. Some experimentation with the various options will be necessary.

To align the two structures on the basis of their 3D spatial coordinates, use the *Iterative Magic Fit* facility. Select *Fit*, then *Iterative Magic Fit* from the menu system. A dialogue box appears, as shown in Figure 18.10 on page 397.

The result is that the two structures align with a root mean squared deviation (RMSD) of 1.59 Angstroms. This is very close, but there are still some differences,

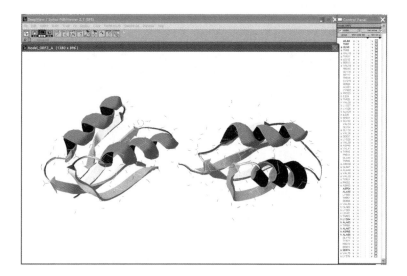

**Figure 18.9** The ORF2 and ORF3_A structures loaded into DeepView prior to structural alignment.

---

[12] It can be seen in all its colourful glory on the *Bioinformatics, Biocomputing and Perl* web-site.

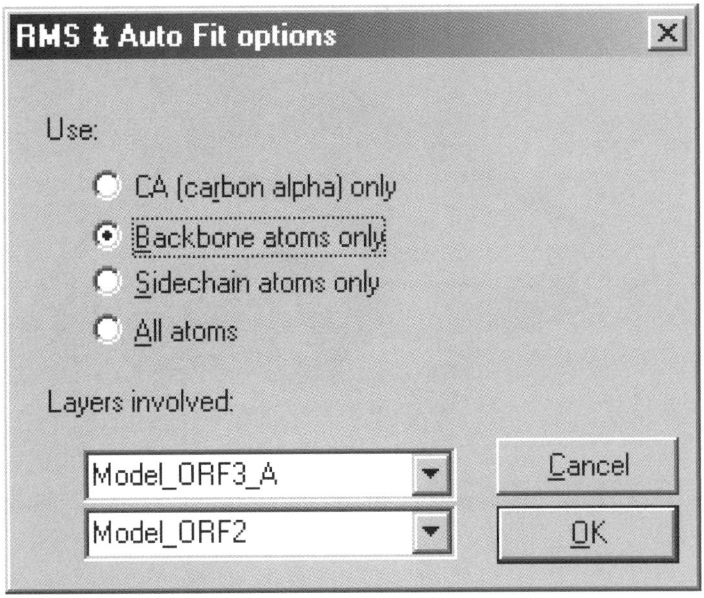

**Figure 18.10** DeepView's *Iterative Magic Fit* dialogue box.

see Figure 18.11 on page 398. Poor fits may be greater than 10 Angstroms or more, although SWISS-MODEL warns when radically different proteins do not fit well. As an example, attempt the alignment of the ORF3_A and ORF3_B structures and see what happens. DeepView issues a warning and aborts, effectively saving us from ourselves!

The two aligned models can be investigated further in *DeepView* (such as by measuring the distances between atoms) or exported in their current orientation. To demonstrate using *DeepView*, click *Select*, then *All*. Then choose *Edit*, then *Create Merged Layer from Selection*[13]. Then select the new copies of the merged proteins, which have been created on a new layer called _merge_, by using the drop down menu in the *Control Panel* window, which is marked by a small black triangle (refer to Figure 18.12 on page 398). When ready, use *File*, then *Save*, then *Layer* to save the structure to a PDB formatted disk-file. This is a less-than-obvious process, so refer to Figure 18.12 as necessary.

*DeepView* also has the ability to export the sequence to a FASTA-formatted disk-file. This allows two sequences to be aligned in ClustalW to see how the two models align in terms of sequence. Export each structure (but *not* the merged layer) in turn using *File*, then *Save*, then *Sequence (FASTA)* to appropriately named disk-files, such as ORF2_DeepView.fsa and ORF3_A_DeepView.fsa. Perform a

---

[13] *DeepView* uses the concept of layers. Each structure is on a separate layer, until assigned otherwise.

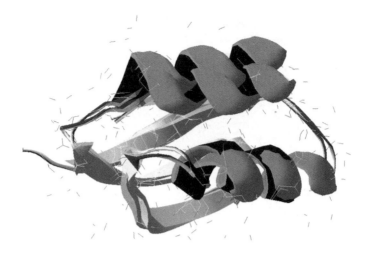

**Figure 18.11** Structural alignment created using the DeepView's *Iterative Magic Fit* facility.

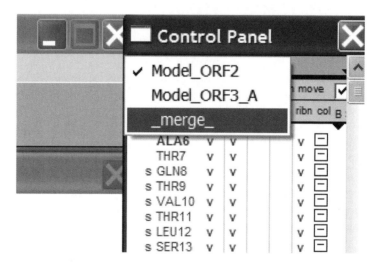

**Figure 18.12** Selecting the current "layer" in DeepView.

sequence-based alignment using ClustalW, producing the following results:

```
CLUSTAL W (1.82) multiple sequence alignment

Model_ORF2      ATQTVTLSVPGMTCSACPITVKKAISEVEGVSKVDVTFETRQAVVTFDDAKTSVQKLTKA 60
Model_ORF3_A    ---MTHLKITGMTCDSCAAHVKEALEKVPGVQSALVSYPKGTAQLAIVPG-TSPDALTAA 56
                .  *.:.****.:*.  **:*:..:* **... *:: .  * :::   . ** :  ** *

Model_ORF2      TADAGYPSSVKQ 72
Model_ORF3_A    VAGLGYKATLAD 68
                .*. ** ::: :
```

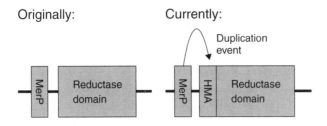

**Figure 18.13** Possible explanation behind MerA/HMA duplication event.

These results indicate why the structural alignment was so good. The sequences are very similar, but not absolutely the same. This latter point is important, as it indicates that neither we nor *GeneMark* got confused along the way and analysed the same thing twice, which is easily done. Now that it is known that the results are not in error, it is worth investing more time in interpreting them. The conclusion, thus far, is that the ORF2/MerP and the ORF3_A/MerA sequences and structures are very similar.

Is this a logical conclusion? Yes, it is because the MerP and MerA domains are physically next to each other on the piece of DNA. So it is reasonable to speculate that duplication occurred at some time in the evolutionary past and became preserved, see Figure 18.13 on page 399. It seems that the function of the MerP protein is to sequester $Hg^{2+}$ ions in the Periplasmic space to prevent them from reacting with anything vital – at least until they can be passed into the cell interior and detoxified by the Mer system.

It is plausible that otherwise MerT would be may simply be dumping these reactive ions into the cytoplasm. It would be advantageous to the cell to again sequester these ions until it detoxifies them ... and where better to do so than close to the MerA protein, thus increasing the local concentration in the vicinity of the detoxification system?

Before moving on, it is worth comparing the results of the homology model to a real Mercury Reductase structure in a *semi-blind* test. Recall that neither the PDB database search nor SWISS-MODEL's found a similar structure. It turns out that one does indeed exist, although it is of medium quality and from Bacillus. On top of this, it is published in the top-quality journal *Nature*. How can this be?

The explanation has to do with *timing*. The *Mercury Reductase* structure was published in *Nature* before the editorial policies changed to those that required submission of the structure to public databases prior to – and as a condition of – publication. In fact, there has never been any attempt by the researchers to suppress this structure and it was gladly e-mailed to your authors upon request[14].

---

[14] If you want to try the structure yourself, consider its acquisition a modern-day, web-based, biological treasure hunt. Please do not ask us to send it to you, as it is not ours to give away.

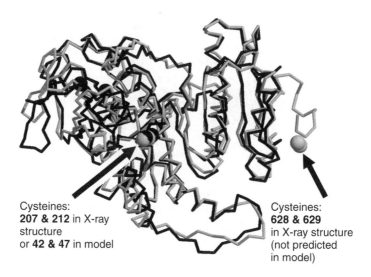

*Figure 18.14* The structural alignment of ORF3_B and the "official" Mercury Reductase X-ray structure.

Figure 18.14 on page 400 shows the ORF3_B/MerA reductase model aligned to the "official" Mercury Reductase X-ray structure with an RMSD of close to 1.50 Angstroms.

On Figure 18.14, the X-ray determined structure is coloured grey and the model structure is coloured black. For clarity, only the backbone traces are shown. The sulphur atoms of the cysteine amino acids, which are important for catalysis, are shown *spacefilled*. In general, the backbone is well predicted and follows the "real" structure closely. However, the important structural details around the two cysteine residues labelled 628 and 629 in the X-ray structure (corresponding to residues 558 and 559 in the ORF3 sequence) are missing. The general shapes are the same, but some regions of the backbone have minor deviations. In fact, this could well be the case. It should not be expected that the two structures from different species are absolutely identical.

The main point is that SWISS-MODEL cannot predict the locations of the two key cysteine amino acids at the end of the structure because this is an apparently unique arrangement and there cannot be anything similar in the PDB database. Hence, the model stops just when everything is getting interesting, as can be seen from this manual alignment with the two key amino acids highlighted in bold:

```
SWISS-MODEL Can Predict    >...GLKLAAQTFN
Actual Sequence            >...GLKLAAQTFNKDVKQLS CC AG
```

Such is life when homology modelling.

> **Maxim 18.3** *Homology modelling can only model protein sequences similar to those that are already known.*

To get accurate positions for these residues with current technology, the best option is to re-determine the structure experimentally. It may be possible to use molecular modelling techniques, but doing the *wet-lab work* may be easier. It is certainly possible, as somebody has already done it! *Good luck with your efforts.*

## 18.8 PROSITE and Sequence Motifs

*PROSITE*, *PRINTS* and *BLOCKS* are sequence motif databases that document short fragments of protein sequences found in proteins of similar function. A *motif* is a short section of protein (or DNA for that matter) that refers to a particular piece of sequence or structure that re-occurs frequently and, as such, is related to a particular function. For instance, a group of six or eight amino acids that are rigorously conserved in or around the active site of an enzyme or binding site. Longer motifs are also found that are more diagnostic of general domains. These are generally more variable and stored as sequence-scoring profiles.

The underlying premise is that to obtain an idea of a sequences function, all that's required is the identification of a few key amino acids within the sequence in a particular order. While the rest of the protein is important for supporting these active residues, from a diagnostic perspective, it is just "fluff".

Simple *PROSITE* patterns are essentially regular expressions that encode positions of amino acids relative to each other in the sequence. For example:

```
PROSITE Pattern:    G-M-T-K-[GM]-x-C
Regex:              /GMTK[GM].C/
```

where "x" denotes any amino acid. Patterns are identified by inspecting sequence alignments of multiple proteins for regions where there is good conservation at the same position throughout all the examples (that is, in the same column for a typical ClustalW output). The aim is to produce the longest *consensus pattern* possible, one that has the least variability and consequently, the highest specificity.

The problem of employing pattern matching using short motifs is the high *False Positive Rate*. On purely statistical grounds, the probability that short patterns occur by random chance is high. Consider as an example a pattern with five amino acids in it (for example, APLIK), that has been identified as a diagnostic feature of a protein under study. Assuming an equal probability for each amino acid of 1 in 20, this gives an expected frequency of the pattern occurring once in every 3,200,000 amino acids by chance alone (as 20 raised to the power of 5 is 3,200,000). When multiplied by three (as three nucleotides are needed to code for a single amino acid), the chance of a pattern occurring is 9,600,000 to one. That is one occurrence in every 9.6 million bases. By starting to translate DNA into protein in all six reading frames, as would be done for a sequence similarity search analogous to using `blastx`, there would be an occurrence of the pattern on average every 1.6 million bases (9.6 divided by 6). This is far too high a value, especially when scanning entire genomes in the giga-base range! Even when considering dissimilar proteins, the amino acids are not uniformly distributed[15] and this increases the probability of random matches occurring.

> **Maxim 18.4** *Searching large datasets with non-specific, short sequence fragments results in many false positives.*

The root of this problem is that each of the amino acids in the pattern either matches or does not. Therefore, the patterns must be kept short, or they fail to match real examples. That is, the patterns need to be short to help maintain a high *True Positive Rate*. When they are less specific, they match other regions in error, therefore increasing the *False Positive Rate*, which is not good.

Although more complex, a better solution is the use of *Sequence-scoring Profiles*[16]. Profiles are superior because they use the stored frequency profile of individual amino acids at a particular location, rather than simply trying to determine whether a particular amino acid is found or not. Determining the frequency profile from an alignment is not difficult: count up the number of times each type of amino acid is found at each location, then divide by the expected frequency of finding these there. A sequence is then scored by comparing its amino acid contents to the frequencies stored in the profile, then adding up the likelihood of finding each amino acid at each location. If the result exceeds a certain value (usually empirically determined), then the sequence is reported as containing that profile.

Tricks such as the use of log-odd scores and dynamic programming algorithms are often used with sequence scoring profiles to make applying them computationally more efficient. It is worth mentioning that sequence-scoring profile methods are closely related to *Markov Models*.

## 18.8.1 Using PROSITE patterns and matrices

Rather than create new PROSITE patterns or scoring matrices (this is quite easy to do if you have a multiple sequence alignment), let's apply some pre-compiled patterns. When it comes to testing sequences, there are two usual options:

1. Use a web-based interface to a hosted service.

2. Download the programs and run them.

---

[15] The sequence space is biased.

[16] This technique is used extensively in the *tRNAScan-SE* search system and set of programs.

In either case, it is a good idea to enable the filter that blocks patterns that occur frequently, unless specifically interested in them. Running the first search configured to report these might be useful, so that in the future it is possible to know what is missing (why there is no need to worry about it). The current *PROSITE* homepage is:

    http://www.expasy.org/prosite/

Using this web-site is straightforward: simply follow the on-screen instructions. Supply a sequence (copy and paste a sequence into the text box), supply a SWISS-PROT accession code or queue a file for upload. It is often prudent to check the "Exclude patterns with a high probability of occurrence" check box prior to starting the search, or your results will be buried in a lot of trivial patterns that are non-specific and occur very frequently.

## 8.8.2 Downloading PROSITE and its search tools

The *PROSITE* database and search tools are available for download, which is free to academics and non-profit organisations. Pre-compiled binary versions are available for different operating systems, and the source code is also available. Download the pre-compiled Linux version from the /tools directory and decompress and unpack it using these commands:

```
gzip  -d  ps_scan_linux_x86_elf.tar.gz
tar   -xvf  ps_scan_linux_x86_elf.tar
```

These commands create a new directory called ps_scan, which contains the main scanning program (pfscan) and its associated, easy-to-use interface program (ps_scan.pl). The disk-file containing the PROSITE database is also required. This is called prosite.dat, be sure to place it in the same directory as the pfscan program.

To start a PROSITE scan, use ps_scan.pl against the SWISS-PROT or FASTA-formatted sequence disk-file of interest. If there are multiple sequences in the same file, then the -e parameter can be used to identify the particular sequence to be scanned.

The ps_scan.pl program is an administrator interface to pfscan. It also parses the patterns from the PROSITE database file, converts them into regular expressions in order to perform the scans and then it calls pfscan, which does the actual matrix searches.

In its default mode, ps_scan.pl prints a list of matches, or prints nothing when there are no matches. The command lines that follow show how to execute the PROSITE scan on the ORF sequences predicted by *GeneMark*. The -s switch suppresses the most frequent and non-specific patterns.

```
./ps_scan.pl  -s  seqs/ORF1.fsa  >  ../ORF1.prosite
./ps_scan.pl  -s  seqs/ORF2.fsa  >  ../ORF2.prosite
```

No PROSITE patterns or sequence profiles were found in ORF1, ORF4 or ORF7, so, although these disk-files exist, they are empty. ORF2, ORF3, ORF5 and ORF6 all produce results, and these are analysed in the subsections that follow.

**ORF2 and ORF3**

ORF2 (MerP) and ORF3 (MerA) both report the same pattern and sequence profile for their *Heavy-metal-associated* domains:

```
orf_2 : PS01047 HMA_1 Heavy-metal-associated domain.
        28 - 57  VpgMtCsACpitVkkaIsevegvskvd.VtF
orf_2 : PS50846 HMA_2 Heavy-metal-associated domain profile.
        23 - 89  TVTLSVPGMTCSACPITVKKAISEVEGVSKVDVTFETRQAVVTFDDAKTSVQKLTKATAD L=0
                 AGYPSSV
```

Notice how with HMA_1, the sequence pattern is shorter (amino acids 28 through 57). Contained inside HMA_2 is the scoring profile (amino acids 23 through 89). This is a classic demonstration of patterns being smaller and targeting subsections of a larger corresponding profile for the reasons discussed earlier.

The results for ORF3 are understandably similar, as would be expected given the structural evidence from SWISS-MODEL discussed in the previous section. Here though, HMA_1 runs from amino acid 6 through 35, and HMA_2 runs from 1 through 66. There's also a PYRIDINE_REDOX_1 pattern found at residues 133 through 143:

```
orf_3 : PS00076 PYRIDINE_REDOX_1 Pyridine nucleotide-disulphide oxidoreductases \\
        class-I active site.
        133 - 143  GGtCVnvGCVP
```

This is consistent with the reductase functionality of MerA and the SWISS-MODEL results.

**ORF5**

The output for ORF5 contains an EAL domain profile, as follows:

```
orf_5 : PS50883 EAL EAL domain profile.
        66 - 318  DGWTTPELAQAVERGQLELHYQPVVDLRSGGIVGAEALLRWRHPTLGLLPPGQFLPVVES L=0
                  SGLMPEIGAWVLGEACRQMRDWRMLaWRPFRLAVNASASQV-GPDFDGWVKGVLADAELP
                  AEYLEIELTESVAF-GDPAIFPALDALRQIGVRFAADDFGTGYSCLQHLKCCPISTLKID
                  QSFVAGLANDRRDQTIVHTVIQLAHGLGMDVVAEGVETSASLDLLRQADCDTGQGFLFAK
                  PMPAAAFAVFVSQWR
```

This output highlights the problem with this method of scanning PROSITE: the descriptions can be quite brief! To be fair, the `prosite.dat` disk-file is designed for super-fast scanning. Consequently, the human-readable descriptions are in the disk-file called `prosite.doc`. The two disk-files can be linked together using the ID code (an example of which is PS50883 from above). Creating a Perl program to create this linkage is left as an exercise for the reader.

In the absence of a program to do this particular "grunt work", a manual text search of `prosite.doc` finds this entry:

```
END
PDOC50883
PS50883; EAL
BEGIN
***********************
* EAL domain profile *
***********************
```

The EAL domain is an around 250-amino acid signalling domain. It is made of four conserved regions and has been named EAL according to a conserved sequence within the second of these regions. The EAL domain is found in a large number of eubacterial multi-domain proteins involved in signal transduction. The EAL domain is found in association with other domains of the prokaryotic two-component signal transduction systems, such as the GGDEF domain (see <PDOC50887>), the **response regulatory domain** (see <PDOC50110>), the PAS repeat and the PAC domain (see <PDOC50112>), the MHYT domain, the HAMP domain (see <PDOC50885>), the GAF domain, the TPR repeat, or the FHA domain (see <PDOC50006>). It has been proposed that the EAL domain might function as a diguanylate phosphodiesterase. Accordingly, it contains several conserved acidic residues that could participate in metal binding and potentially might form a **phosphodiesterase active site** [1,2,3].

Some proteins known to contain an EAL domain are listed below:

 - Acetobacter xylinum diguanylate cyclase (DGC).
 - Acetobacter xylinum phosphodiesterase A (gene pdeA).
 - Bacillus subtilis hypothetical protein ykoW.
 - Bordetella pertussis bvgR protein.
 - Escherichia coli rtn protein. It is involved in resistance to phages N4 and lambda.
 - Escherichia coli hypothetical protein yahA.
 - Escherichia coli hypothetical protein yciR.
 - Escherichia coli hypothetical protein yddU.
 - Escherichia coli hypothetical protein yfeA.
 - Escherichia coli protein yhjK.
 - Escherichia coli hypothetical protein yjcC.
 - Klebsiella pneumoniae fimK protein.
 - Mycobacterium tuberculosis hypothetical protein Rv1354c.
 - Rhizobium sp. strain NGR234 hypothetical protein Y4LL.
 - Synechocystis sp. strain PCC 6803 nitrogen fixation positive activator protein.
 - Synechocystis sp. strain PCC 6803 hypothetical protein slr0359.

The profile we developed covers the entire EAL domain.

-Sequences known to belong to this class detected by the profile: ALL.
-Other sequence(s) detected in Swiss-Prot: NONE.
-Last update: December 2002 / First entry.

[ 1] Tal R., Wong H.C., Calhoon R., Gelfand D., Fear A.L., Volman G.,
    Mayer R., Ross P., Amikam D., Weinhouse H., Cohen A., Sapir S.,
    Ohana P., Benziman M.
    J. Bacteriol. 180:4416-4425(1998).
[ 2] Merkel T.J., Barros C., Stibitz S.
    J. Bacteriol. 180:1682-1690(1998).
[ 3] Galperin M.Y., Nikolskaya A.N., Koonin E.V.
    FEMS Microbiol. Lett. 203:11-21(2001).

## Applications

```
+----------------------------------------------------------------------+
| This PROSITE entry is copyright by the Swiss Institute of Bioinformatics |
| (SIB). There are no restrictions on its use by non-profit institutions as |
| long as its content is in no way modified and this statement is not |
| removed. Usage by and for commercial entities requires a license agreement |
| (See http://www.isb-sib.ch/announce/ or email to license@isb-sib.ch). |
+----------------------------------------------------------------------+
```

Considering how the other evidence points to this protein being involved with Tn501 transposition, in addition to the top BLAST hit from earlier being against an E-coli protein called `TNPM_ECOLI` (Transposon Modulator), this particular result is plausible. Two phrases have been highlighted in bold: `response regulatory domain` or `phosphodiesterase active site`. These phrases could be involved with administering DNA insertion/excision of the transposon.

### ORF6

The `ORF6` PROSITE patterns are definitely indicative of the suspected function, which is Transponoson Resolvase according the BLAST search of SWISS-PROT:

```
orf_6 : PS00397 RECOMBINASES_1 Site-specific recombinases active site.
        8 - 16    YVRVSSfdQ
orf_6 : PS00398 RECOMBINASES_2 Site-specific recombinases signature 2.
        55 - 67   GDtvVvhsMDRLA
```

### A quick word about InterPro

Recently, a combined *meta-database* of sequence pattern and profiles, referred to collectively as "signatures", has been produced and is called *InterPro*. This is an excellent technology, and its home on the Internet is:

http://www.ebi.ac.uk/interpro/

The sequence signatures from all the major protein motif databases (including *PRINTS, Pfam, PROSITE, ProDom, Smart, TIGRFAMs, PIR SuperFamily* and *SUPERFAMILY*) are combined and then scanned using a single top-level program called *InterProScan*.

*InterProScan* is written in Perl, and it administers database searches in much the same manner as `scan_ps.pl`, but does so on a much larger scale. *InterProScan* is *fully parallelised*, allowing each database search to take place in a different computer. In this way, individual search failures can be handled in a way that does not impact any of the other searches that are taking place. Take the time to explore the features of the *InterPro* system.

### 8.8.3 Final word on PROSITE

PROSITE scans are often very useful in clarifying the functions of proteins. In many circumstances, *sequence motif searches* are complementary to BLAST sequence similarity searches. Sequence motif searches identify-specific regions of sequence similarity that have been associated with a particular function, allowing these to be linked to specific easy-to-understand human concepts. Sequence similarity search tools merely identify similar sequences and leave it to the human researcher (or another analysis system) to interpret any results. Which method (manual or automated) is most appropriate depends on the particular task at hand and as shown here, the results of both are often informative.

From a certain perspective, sequence motifs are a digested version of the parts of previous sequence similarity searches that were found to be useful. These were annotated and stored in databases such as PROSITE, PRINTS and BLOCKS with varying levels of human curation. The next level of abstraction is *Gene Ontology*, and more details on the *GO Consortium* can be found here:

    http://www.geneontology.org

Ontologies describe what genes (and hence the proteins they code for) actually *do* in a cell, with the description presented as a set of linked human conceptual terms. This research is still in its early stages, but along with the associated work in the identification of *Pathways*, the results to date are very encouraging. For more details and an example, see the *KEGG Pathway* databases located at:

    http://www.kegg.org

## 18.9 Phylogenetics

No modern Bioinformatics text would be complete without some mention of *phylogenetics*, which is the study and identification of relationships between different forms of life. This section presents an example that investigates the relationships between the HMA domains from the MerA and MerP proteins.

### 8.9.1 A look at the HMA domain of MerA and MerP

The evidence garnered so far from the discussion of the MerA and MerP proteins suggests that the HMA domain of MerA is in fact a duplicated version of the MerP protein. The most significant points are as follows:

- There is high sequence similarity between the two regions at the protein level.

- They lie physically next to each other in the `Tn501` transposon DNA sequence. This implies a gene duplication event, giving rise to two copies of the same gene from a common ancestor.

- Their functions, such as have been characterised experimentally (though this is tentative at best for the `MerA` HMA domain), are very similar. Both complex with $Hg^{2+}$ ions in a reversible way, thus sequestering them until they can be detoxified.

Consequently, it is likely that there is some *close relationship* between the two domains and this is assumed for the purpose of this demonstration. The first step is to extract the HMA domain of the `MerA` proteins from the full `MerA` sequence. There are two ways to do this, namely:

1. Automatically, using Bioperl (discussed in this book's final chapter).
2. Manually, using any text editor.

Whichever method is used, the HMA domain is the first 92 amino acids of the full `MerA` protein sequence (more on this assumption later). So, either all residues after 92 need to be deleted (using a text editor) or just the first 92 need to be included in the new sequence (using Bioperl). With this example, the proteins identified below were used:

```
-------------------------------  -------------------------------
SWISS-PROT IDs of MerP Proteins  SWISS-PROT IDs of MerA Proteins
-------------------------------  -------------------------------
              MERP_ACICA                       MERA_ACICA
              MERP_ALCSP                       MERA_ALCSP
              MERP_PSEAE                       MERA_BACSR
              MERP_PSEFL                       MERA_ENTAG
              MERP_SALTI                       MERA_PSEAE
              MERP_SERMA                       MERA_PSEFL
              MERP_SHEPU                       MERA_SERMA
              MERP_SHIFL                       MERA_SHEPU
                                               MERA_SHIFL
                                               MERA_STAEP
                                               MERA_STRLI
                                               MERA_THIFE
-------------------------------  -------------------------------
```

These proteins were aligned using the EBI ClustalW web-based service and gave the alignment as shown in Figure 18.15 on page 409.

The alignment in Figure 18.15 indicates that the core of the HMA regions, centred on the `GMTCxxC` pattern, contains the two key cysteine amino acids that complex the $Hg^{2+}$ ion and are well conserved between the two domains. The large number of end gaps, denoted by the "-" character, indicate that the ends of the

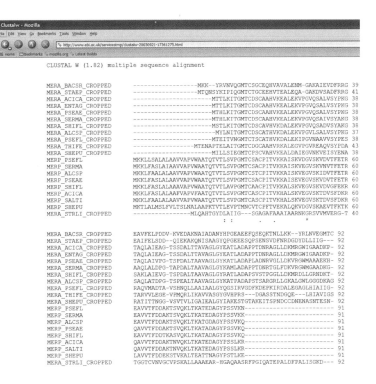

**Figure 18.15** The multiple sequence alignment of the example proteins.

domain are not conserved. Further, the alignment splits into two groups, MerP and MerA, with the MerA sequences being padded with end gaps at the start, and the MerP coming at the end. From this, it is possible to conclude that the two domains are genuinely different from each other.

Included as a standard report by the EBI web-based service is a tree that shows graphically how the sequences relate to each other. This gives a better indication of the phlyogenetic relationships between the sequences than simply "manually" looking at the alignment. ClustalW uses the *Neighbour-Joining* method to produce "true" phlyogenetic trees (not to be confused with its internal *Guide Trees* that direct which pair of sequences to align next). Scrolling down through the results page of the above alignment reveals the tree shown in Figure 18.16 on page 410:

> **Technical Commentary:** The tree shown in Figure 18.16 is drawn by a flexible Java applet, called ClustalTree.class, which is available on the EBI WWW Site. This applet provides various formatting options for changing the view of the tree. Note that the merge points between sequences are denoted by the vertical lines and the lengths of the horizontal lines indicate the difference/similarity between one group (possibly a single sequence) and another.

ClustalW outputs the most similar sequences toward the top. The MerA HMA domains dominate the upper part of the tree, as expected. These sequences are

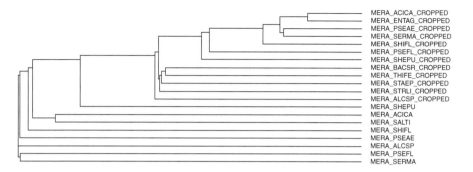

**Figure 18.16** The EBI's tree graphical display.

more similar to each other than they are to the `MerP` sequences, which in turn are more similar to each other than to the `MerA` sequences.

# Where to from Here?

There are many other things that could have been searched for within this system, both in the DNA and protein sequences, as well as in the structures. Two interesting areas not looked into are as follows:

1. The terminal repeats at the ends of the DNA sequences that allow the `Tn501` to integrate into host genomes.
2. Promoter regions at which transcription begins.

Consider the exploration of these areas as extended exercises for the interested reader! It is important to be careful not to generalise too much from the small study presented in this chapter, which is based on a gene-rich region of 8 Kb of nucleotides, in a single (simple) bacterium. This is far from a representative sample of all DNA sequences known, even for bacteria! For a more representative view we would need to study other examples, which leads to this rather tongue-in-cheek maxim:

**Maxim 18.5** *Whenever you make a statement, call for more research (money)!*

Here, we highlight (deliberately, to add a little light-hearted humour) that the end of a study always seems to be accompanied by a plea for more money! The point is that the end of a study, or the money to investigate further, is not the end of the investigation that the scientific community *could* do if only it had more resources. The knowledge gained during one study influences the questions to be addressed in the next.

The above example is deceptively easy. A well-characterised piece of DNA was taken and analysed using two standard (state-of-the-art) tools. The results were

# Phylogenetics

then used to direct and guide standard database queries. Yet as the pages of the resulting discussion demonstrated, this leads to more questions. There are a lot of other techniques and analyses, which (owing to space constraints) have not been covered in this book. The importance of good-quality database annotation cannot be overstated.

> **Maxim 18.6** *Database annotation is hard to do well, so be prepared to update it on a regular basis.*

The system studied is well characterised, so your authors had a good idea as to which predictions would be found. In effect, this provided a *self-validation* of the analysis tools. Generally, these performed well, but care is needed in interpreting the results. Automation would have helped in some aspects. For example, the running of *GeneMark* on the original DNA sequence in addition to a couple of *BLAST* searches on the resultant ORFs. However, the results still need to be scanned/reviewed by a suitable knowledgeable human.

> **Maxim 18.7** *Automation can be very helpful when creating annotation, but to achieve the highest quality, humans are needed to make some value judgements.*

Further, the annotation needs to be updated as new knowledge from experiments or data-mining techniques becomes available. So, the annotation process is an ongoing one. Also, it is important to acknowledge that any conclusions drawn at any particular time relate to that time only. It is prudent to re-examine the predictions by re-analysing the same system at some point in the future, when either the databases relied upon include new knowledge or the analytical tools have improved.

> **Maxim 18.8** *Conclusions are based on the available data, which, in this case, is the database annotation (which may or may not be current).*

The study in this chapter required a considerable amount of interpretation, even though there was a good idea as to what to look for. Now consider the effort and skill required to annotate a full genome. Such a "study" is not at all trivial.

## The Maxims Repeated

Here's a list of the maxims introduced in this chapter.

- *With BLAST scores, up is down and lower is better.*
- *The major limitation of "homology modelling" is that homology to a known structure is needed.*

412    Applications

- *Homology modelling can only model protein sequences similar to those that are already known.*
- *Searching large datasets with non-specific, short sequence fragments results in many false positives.*
- *Whenever you make a statement, call for more research (money)!*
- *Database annotation is hard to do well, so be prepared to update it on a regular basis.*
- *Automation can be very helpful when creating annotation, but to achieve the highest quality, humans are needed to make some value judgements.*
- *Conclusions are based on the available data, which, in this case, is the database annotation (which may or may not be current).*

## Exercises

1. Repeat the analysis used in this chapter on a Tn21 transposon DNA sequence. Follow the steps as outlined in this chapter. Are you surprised by the amount of work this entails?

# 19
# Data Visualisation

*A picture is worth a thousand words.*

## 19.1 Introducing Visualisation

Visualisation techniques are an important part of Bioinformatics. The increasing amounts of data, together with its associated complexity, mean that better *human–data interfaces* are needed. The days of simple, flat disk-files printed to an 80 times 40 character terminal are long gone. In the modern Bioinformatics world, much more effective presentation systems are required and they usually need to provide extra interactive capabilities.

Before describing typical examples, it is helpful to first cover some general concepts. Strictly, "visualisation" means the generation and presentation of pictures that help people understand a particular feature of a dataset, making data *mean something to people*. This definition of visualisation might be a little too specific as pictures (visions or graphics) are only one of a number of ways of representing information – it is just that the most successful forms are, well … *visual*. Visualisation is the most widely used of a general class of *perception technologies*.

There is no conceptual reason why the other human senses – smell, taste, touch or hearing – cannot be used as representations of biological data. However, there are some very practical limitations:

1. The need for an appropriate method of representation. How would you hear, taste or feel representations of biological data? Could you hear DNA

intron splicing sites or smell Microarray clustering results? Maybe people can, but it is the success of the use of visual graphics has discouraged the use of the other senses for representation of data – but if you have a good idea we want to hear it.

2. The "bandwidth" of the other senses has an effect. In this context, this refers to the amount of information that can be communicated per unit time. The problem with using the other senses is that the vision system in humans is so highly developed that it has an extremely large processing capability in comparison to the other senses. Probably the closest is hearing. However, even this is a poor substitute when compared to vision.

Consider, too, the motivation behind presenting data: in many cases, it is to summarise information and provide for the identification of *patterns*. The human visual system excels at pattern recognition as there has been constant pressure, over millions of years, to select for this ability. For humans and their ancestors, a good vision system was essential for survival, both for finding food and for avoiding becoming something else's food. Even in the modern technological world, this is still useful for practical activities such as avoiding cars while out shopping. This very same visual system is also excellent for analysing abstract diagrams derived from biological datasets, hence the popularity of visual representations within the biosciences.

High-end, leading-edge technologies within the visualisation area can be quite spectacular. Walk-in rooms that display data on three walls and in the central region in three dimensions are now common[1]. These allow researchers to have the sensation of walking around within their datasets, very much like the *Holodeck* from the *Star Trek* TV series.

Why is visualisation so important? The answer – as already hinted at – is "because people are". People ultimately drive science, make it what it is, as well as shape what it will become. The early successes in the Artificial Intelligence community encouraged expectations of a utopian future of *intelligent thinking machines*. However, this is still the realm of science fiction[2]. In the real world, it is people alone who are truly creative and it is people who know when they have an idea. And it is people who have the resources for scientific development and people allocate them. Clever algorithms and integrated databases are no more than useful tools for people who understand how to use them (and know what they are doing). Visualisation technologies help researchers gain insight into the world because they present data and information in a way that is *meaningful to humans*. In a similar (and somewhat loose) sense, statistics have the same function in computational numerical analysis.

---

[1] Michael's department at *Erasmus MC* is getting one of these. The one up the road he walked around in at SARA, Amsterdam (see: http://www.sara.nl/projects/projects 07 03 eng.html) was *very* nice. Paul is not jealous!

[2] One interesting example is the "Robots" books written by *Issac Azimov*. Dr Azimov's day job was as Professor of Biochemistry at Boston University, USA.

> **Maxim 19.1** *People are processing tools, too, especially when it comes to processing visual information.*

In overview, good data visualisation is arguably one of the most important challenges facing modern biologists. If it troubles biologists and it concerns computers or databases, then chances are it ends up in the in-tray (or more likely the "Inbox") of the Bioinformatician. Although entire books have been written about data visualisation, it is worthwhile including an introduction to the production of visual representations of biological data. This helps by aiding researchers in identifying patterns that relate to the underlying processes. Three simple but effective techniques, by way of example, are presented in this chapter, and they are:

1. Using HTML tables to list SWISS-PROT IDs.

2. Plotting an EMBL entry to show the arrangement of the *Mer Operon* genes.

3. Using Grace (a graph drawing program) to draw plots.

All three of these techniques can be used on a home computer with minimal processing power and free software. So, no walk-in, 3D room here – sorry!

## 19.2 Displaying Tabular Data Using HTML

In this section, HTML is used to generate meaningful, visual displays. Recall (from Chapter 15) that there are two common situations in which HTML is used on the world wide web:

1. To create static web pages. This is the simplest and most common use of HTML. The browser requests the web page from the web server, which responds by sending an appropriately formatted text disk-file containing the HTML mark-up. This disk-file is interpreted by the browser, producing a more visually pleasing representation of the document than flat text alone typically does.

2. To create on-the-fly dynamic web pages from server-side programs. In this case, HTML web pages are generated for every request by a program executing on the web server. Typically, input is provided to the program as part of the initial web request from the browser (using a HTML form).

A third situation involves producing HTML dynamically based on data parsed and processed from some other data source. Rather than store the static HTML on a web server or produce the dynamic HTML on a web server, with this technique, HTML is saved to a local disk-file for later viewing as an "off-line" static web page. At some later date, the page can be made available through a web server if necessary. However, the use of HTML (in this situation) provides a local visualisation representation of data. A custom program can take the data source(s) and produces an HTML visualisation as output. This allows for some very effective – be they complex or quite simple – visualisations to be easily created.

Figure 19.1 on page 416 and Figure 19.2 on page 417 present two examples of what's possible. The HTML used to present these visualisations was produced by custom Perl programs, which generated HTML disk-files. The disk-file were then viewed in the *Mozilla* web browser[3].

Figure 19.1 uses the background colour of the cell (which is grey-scaled in the figure because of printing restrictions) to identify which of four possible states each amino acid pair is in:

1. Significantly different from an average state.

2. Similar to an average state.

3. Indeterminate.

4. Those pairs with too few examples for an assessment to be made.

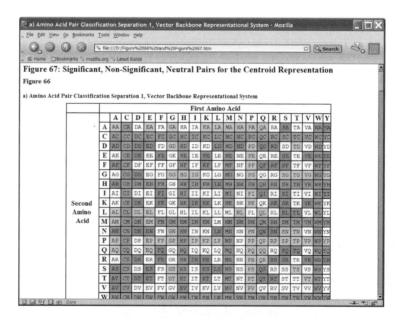

**Figure 19.1** Example HTML visualisation: identifying amino acid states.

---

[3] Both of these visualisations are adapted from Michael's Ph. D. thesis. Colour versions are available on the *Bioinformatics, Biocomputing and Perl* web-site.

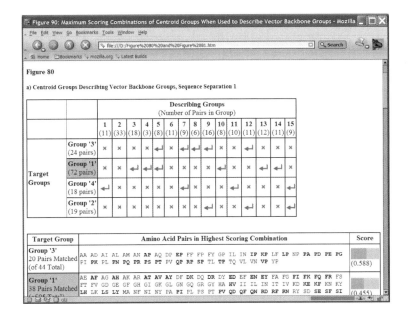

**Figure 19.2**  Example HTML visualisation: grouping amino acids.

The production of this HTML representation is the final high-level summary of a structural analysis pipeline, which itself contains four other stages.

Figure 19.2 describes which groups of amino acids from one analysis can be combined to give a similar group from another analysis. The bar graph display (used to represent the scores) is created from two small image disk-files that are included as many times as necessary using multiple HTML image tags (<IMG>). This is a simple "trick" that works very well.

Despite the perceived complexity of these visual representations, the amount of HTML used in each is quite small: each representation uses no more than ten individual HTML tags repeated over and over.

### 9.2.1 Displaying SWISS-PROT identifiers

In this visualisation, ID codes are sourced from the FASTA protein sequence disk-file containing the 55 *Mer operon* genes found in the SWISS-PROT database. The idea is to extract and format these into a HTML table. The original disk-file is the same one as that used to demonstrate the creation of Basic Local Alignment Search ToolBLAST indexed databases earlier in this book.

This example was chosen for two reasons: it is relatively straightforward and it demonstrates an important point, which is that producing the HTML mark-up is often the easy part. What's harder is having the idea, the acquisition and extraction of the data, and its storage within a custom program, and so on. The

**418**    *Data Visualisation*

custom program requires the FASTA disk-file to be in the format expected by the NCBI-BLAST package or supplied by the EBI SRS web-based service, for example:

```
sw|Q52109|MERA_ACICA Mercuric reduct ...
MTTLKITGMTCDSCAAHVKEALEK ...
```

Further, the program uses a combination of hard-coded HTML tags, as well as generated tags produced by the CGI module. Despite the availability of a `table` sub-routine with CGI, in your author's experience, the use of hard-coded <TABLE> and </TABLE> is more convenient in certain situations. Another useful technique is the inclusion of newline characters as part of the resulting HTML disk-file. By default, web browsers generally ignore newline characters (as HTML provides the <BR> tag). Including them though makes the resultant HTML disk-file more readable by a human. Here's the `Mer_Table.pl` program:

```perl
#! /usr/bin/perl -w

# Mer_Table.pl - produce a HTML table from the 55 Mer Operon proteins.
#
#               Designed to take a FASTA formatted protein sequence
#               file and create and HTML overview table of sequences
#               it contains.  Further demonstrates a mix of CGI.pm
#               subroutine calls and "hard-coded" HTML tags.  Uses a
#               Hash of Hashes data structure: see "perldoc perldsc"
#               for further explanation.

use CGI qw/:standard/;
use strict;

my %Seq_Details;

while ( <> )
{
    unless ( /^>s[w|p]\|/ )
    {
        next;
    }

    my ( $tmp, $Accession, $ID ) = split( /\|/, $_ );

    $ID = $ID =~ m/(^[\w\d]*) /;

    my $Gene = $ID =~ m/(.*)_/;

    $Seq_Details{ $Gene }{ $Accession } =  $ID;
}

print start_html( "Summary of SWISS-PROT 'Mer' Operon Genes" ), "\n";

print h1( "Summary of 'Mer' Genes" );

print "<TABLE WIDTH=100% BORDER = 2>\n";
```

```perl
    print Tr, th( "Gene" ), th( "Accession Codes" ), th( "Gene IDs" ), "\n";

    foreach my $Gene (sort keys %Seq_Details)
    {
        print "<TR>\n", th ($Gene), "\n<TD>";

        foreach my $Accession (sort keys %{$Seq_Details{$Gene}})
        {
            print code ($Accession), ", " ;
        }

        print "</TD>\n<TD>";

        foreach my $Accession (keys %{$Seq_Details{$Gene}})
        {
            print code ($Seq_Details{$Gene}{$Accession}), ", " ;
        }
        print "</TD></TR>\n";
    }
    print "</TABLE>\n";
    print hr;
    print i( system( "date" ));
    print end_html,"\n";
```

Let's work through this program to see what's going on. After the usual first line and an appropriate comment, the CGI module is imported and *strictness* is switched on. A hash, called %Seq_Details, is then declared. This hash stores information on the accession number and gene identifiers extracted from the program's input disk-file:

```perl
    use CGI qw/:standard/;
    use strict;

    my %Seq_Details;
```

A while loop cycles through the disk-file, one line at a time:

```perl
    while ( <> )
    {
```

Each iteration begins by skipping any lines that are not header lines:

```perl
        unless ( /^>s[w|p]\|/ )
        {
            next;
        }
```

Assuming the line is a header line, it is split on the *pipe* symbol, and the results are used to initialise three scalars (note that the $tmp scalar is not actually used in this program):

## 420    Data Visualisation

```perl
my ( $tmp, $Accession, $ID ) = split( /\|/, $_ );
```

Two regular expressions extract the accession code and the gene identifier, assigning them to appropriately named scalar variables:

```perl
$ID = $ID =~ m/(^[\w\d]*) /;

my $Gene = $ID =~ m/(.*)_/;
```

With the accession code and gene identifier known, the next line of code records the details in the $Seq_Details hash, then the loop iteration ends:

```perl
    $Seq_Details{ $Gene }{ $Accession } =    $ID;
}
```

The code that updates the hash demonstrates a new technique. Typically (and by design), hashes contain scalars as the value-part. Here, Perl's ability to store a hash within a hash is exploited[4].

With the entire disk-file processed, the Mer_Table.pl program proceeds to create the HTML visualisation. It starts by beginning the HTML web page, and printing a HTML Level 1 header:

```perl
print start_html( "Summary of SWISS-PROT 'Mer' Operon Genes" ), "\n";

print h1( "Summary of 'Mer' Genes" );
```

A print statement starts the table, then a collection of invocations of the table row and heading producing subroutines from the CGI module occur:

```perl
print "<TABLE WIDTH=100% BORDER = 2>\n";

print Tr, th( "Gene" ), th( "Accession Codes" ), th( "Gene IDs" ), "\n";
```

A foreach statement is used to cycle through the data stored within the %Seq_Details hash. Note how the name-parts of the hash are extracted (using keys), sorted (using sort) and then assigned with each iteration to the $Gene scalar. The body of the foreach loop starts by creating a HTML table row:

```perl
foreach my $Gene ( sort keys %Seq_Details )
{
    print "<TR>\n", th ($Gene), "\n<TD>";
```

An inner foreach loop then processes the accession codes associated with the gene, populating another table cell with the list stored within the hash of hashes that is %Seq_Details:

---

[4] A full discussion of these techniques are, sadly, beyond the scope of *Bioinformatics, Biocomputing and Perl*. Refer to the perldsc manual page for a good introduction.

```
foreach my $Accession ( sort keys %{ $Seq_Details{ $Gene } } )
{
    print code( $Accession ), ", " ;
}
```

Of interest, above, is how the CGI module's code sub-routine is used to display the accession codes in typewriter font.

The end of the table cell is marked, then another begun. Another inner foreach loop processes the gene identifiers associated with the gene:

```
print "</TD>\n<TD>";

foreach my $Accession ( keys %{ $Seq_Details{ $Gene } } )
{
    print code( $Seq_Details{ $Gene }{ $Accession } ), ", " ;
}
```

The outer foreach loop concludes by terminating the HTML table row:

```
    print "</TD></TR>\n";
}
```

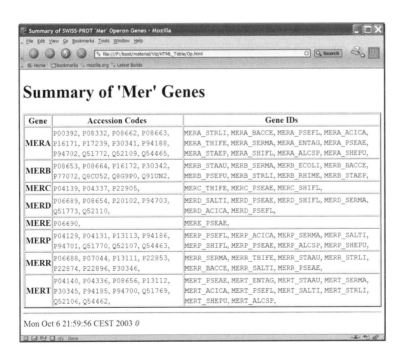

**Figure 19.3** Overview of the Mer Operon proteins in the SWISS-PROT database.

With the entire disk-file processed, all that remains is to terminate the table, insert a horizontal rule, add the date to the web page and then terminate with the end-of-HTML tag:

```
print "</TABLE>\n";
print hr;
print i( system( "date" ));
print end_html,"\n";
```

When executed, the resultant HTML is displayed within the *Mozilla* browser as shown in Figure 19.3 on page 421. Keep in mind that these accession codes and IDs exist. A refinement to the Mer_Table.pl program could include the creation of HTML hyperlinks to any of the on-line databases for each code. By clicking on the hyperlink, users are directed (via their web browser) to the appropriate on-line database.

## 19.3 Creating High-quality Graphics with GD

High-quality graphics are a very useful aid to data visualisation, both as an interactive, on-the-fly service attached to web pages, as well as when producing material for publications.

One of the best interfaces to primitive graphic functions[5] from within Perl programs is the GD module written by *Lincoln D. Stein*. This well-documented module hides a lot of the underlying complexity and links to the gd graphics library written by *Tom Boutell*. Depending on the functionality required, gd invokes a series of other libraries installed along with the operating system. For example, to use gd to create PNG images requires the services of the libpng library (and the zlib library libpng calls). Likewise, using TypeType fonts requires the installation of the **FreeType** library.

For scientific work, libpng (and hence zlib) and *FreeType* are two of the most useful libraries to have installed. The gd library can produce also JPEG images if your system has jpeglib installed. However, as JPEGs tend to 'blur' images, the emphasis in this chapter is on producing PNG images, as they tend to preserve any crisp, sharp lines that exist within the image.

The downside to using the GD module is that installation can be, at best, tricky. At worst, it fails and nothing works! What with the GD module requiring the gdlib library, which in turn requires the libpng library, which in turn requires the services of the zlib compression library, a chain of *installation dependencies* is created that often frustrates the user of the easy-to-use GD.pm Perl module.

Your author's best advice is to start the installation process with the GD module from CPAN if you are unsure as to which libraries are installed on your system.

---

[5] *Graphics primitives* include circles, lines and rectangles, as well as colour definitions and direct pixel access techniques. These can be combined together to make more shapes or effects that are "less primitive" (more complex).

Any missing libraries/functionality should be highlighted during the (attempted) installation of GD. If something is missing, source it on the Internet and install it, before returning to the GD module and continuing the installation. See the Bioinformatics, Biocomputing and Perl WWW site more details if you are still having problems.

Note that the installation of the GD module follows the standard Perl module installation process first introduced in Chapter 5:

```
perl   Makefile.PL
make
make   test
su
make   install
<Ctrl-D>
```

During the `perl Makefile.PL` step, the module asks if JPEG, *FreeType* and XPM support should be built. Be sure to answer "yes" to these questions so as to match the libraries that are installed. Included with the module is a `demo` directory. If the `ttf.pl` program within this directory executes with no errors, the module is very likely successfully installed. If it executes with errors, some additional installation work is still needed. Be sure to execute the `ttf.pl` program with this command-line:

```
ttf.pl | display[6]
```

which should produce the test image shown in Figure 19.4 on page 423.

**Figure 19.4**   The test image produced by the GD module.

---

[6] The `display` program is part of the `ImageMagick` package as used later in the next section.

## 19.3.1 Using the GD module

The GD module works around the concept of *image canvases*. A canvas is a *workspace* upon which an image is manipulated. Images can be created, loaded from an existing disk-file, drawn over, copied from other canvases or written to a disk-file in any of the supported graphic formats. The module's documentation is very extensive and can be viewed on screen using the standard perldoc utility included with Perl[7]:

```
perldoc GD.pm
```

Some examples help explain how the module is used.

The GD module is loaded, and a new image is created using the module's new sub-routine, as shown below. Note the image size is specified as 100 by 100 pixels, and the image accessed by a scalar variable called $image:

```
use GD;

my $image = new GD::Image( 100, 100 );
```

With the image in existence, colours are added to its canvas using a sub-routine called colorAllocate. Colours are added using the standard RGB notation (Red, Green, Blue). Note how extra whitespace keeps this code snippet nice and neat:

```
$white = $image->colorAllocate( 255, 255, 255 );
$black = $image->colorAllocate(   0,   0,   0 );
$red   = $image->colorAllocate( 255,   0,   0 );
$blue  = $image->colorAllocate(   0,   0, 255 );
```

The background is then set to be transparent and interlaced, as follows:

```
$image->transparent( $white );
$image->interlaced( 'true' );
```

A black line (or frame) is drawn around the picture using the rectangle sub-routine. A blue oval is added to the image using the arc sub-routine, which is then filled with the colour red:

```
$image->rectangle( 0, 0, 99, 99, $black );

$image->arc( 50, 50, 95, 75, 0, 360, $blue );

$image->fill( 50, 50, $red );
```

After ensuring that the output produced by this code is binary (as opposed to the more normal text), the image is printed to STDOUT in the format of a PNG image:

---

[7] Michael keeps a copy of the module's documentation close to his computer. It is about 20 A4 pages of text. Constantly referred to, it is invaluable when working with the subroutines provided by GD, as there can be a lot of parameters to specify when invoking the module's functionality.

```
binmode STDOUT;

print $image->png;
```

The resulting output is a fully compliant PNG image of one box inside another, which can be viewed with any number of programs. Note how the image stored in the $image scalar has attributes (colours) and functionality (fill) associated with it. This is an example of a Perl-based *object-oriented* interface, which is a very powerful programming abstraction[8].

Often, the most complex part of drawing images is working out where to put the various drawing components on the canvas, which is always the hardest part of drawing images using the graphics primitives. There are two techniques that can help reduce the problems caused by this complexity:

1. Lots of planning – take the time to plan a complex graphic in a vector graphics program or on paper *before* starting to code. Make a note of some meaningful variable names and annotate the drawing with the values of any constant "off-sets". This can be very helpful!

2. Use multiple canvases – the GD module can create a series of canvases, each of which can be manipulated *separately*. These are then merged together into one *parent canvas* prior to producing the image on STDOUT. This prevents many "off-sets" in each graphics call and moves the problem to one "copy image" statement for each sub-canvas. These calls can still be formidable, but it is one that has to be right only *once*. This is where the image plan described in the previous point is most helpful.

By way of example, Figure 19.5 on page 426 is a simplified version of Michael's plan for a piece of code that plots a correlation coefficient matrix as a "heat map". With reference to Figure 19.5, it is possible to deduce the coordinates of "P", the point at which the *Colour Scale* image would be included in the final, merged image. For example, the x coordinate of "P" is calculated as:

```
Labels_Width + ( Title_Bar_Width - Colour_Scale_Width ) / 2
```

The same planning techniques are useful for both determining the positions of sub-images within a larger canvas and for positioning of elements on an individual canvas.

**Maxim 19.2** *Producing plans avoids problems before problems surface.*

---

[8] The details of it are beyond the scope of this book. However, it is not necessary to know how to program objects in order to be able to use and exploit them, as this example code demonstrates.

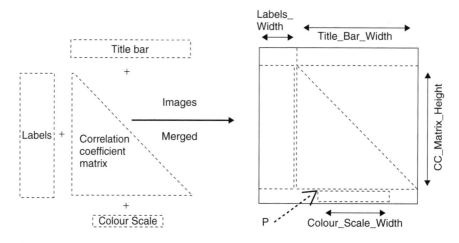

**Figure 19.5** A sample image plan for a "heat map".

An important point to consider is the eventual size of the image. A general rule of thumb suggests the bigger the image, the better. It is straightforward to convert a large image to a smaller size by averaging the information already present than it is to expand an image. The "Cost of Canvas" in the GD module is low, allowing large images to be generated. For instance, the program that produces Figure 19.5 produces an image 6500 pixels wide by 3500 pixels high. This results in an image that prints across four A3 pages, producing a 1-metre wide by 80-cm high visualisation. On first glance, this seems excessive, but each element in the matrix is only 10 pixels wide by 10 pixels high (about 4 mm at the print size mentioned). Even at this size, the labels on the left of the printed output are barely readable. The image is still generated in seconds on even the most modest of personal computers. Sadly, it takes a further 15 minutes to print and manually cut the paper edges away, before taping the resulting image together, producing the final result!

### 19.3.2 Displaying genes in EMBL entries

Two images produced by the GD module appear elsewhere in this book: the contact map on page 196 and the *Mer Operon* gene arrangement on page 376. The program used to produce the latter image is described in detail later in this section. Before presenting the program, some important points to note are as follows:

- The program is designed to generate the base graphic for the image from the data contained in the original EMBL entry *as a guide*. The alternative involves the (tedious) manual inspection of the EMBL entry, most probably on-line, followed by significant box-drawing work in a vector graphics package.

## Creating High-quality Graphics with GD

- The program is designed to demonstrate the GD module's drawing capabilities in its most general form. However, the program is *not* a general solution that can be used with any EMBL entry. Although it may work with many other EMBL entries, the regular expressions may fail in certain circumstances, rendering the resulting image incorrect.

- The code uses the object-orientated interface provided by the GD module. Although strange at first, the object-oriented syntax is comprehensible and easy to relate to and use.

- The generated image is large. The image is over 8000 pixels wide and is designed to be resampled (discussed below) rather than used as-is[9]. The next section describes how the *ImageMagick* tool *mogrify* is used to resize the image to something more sensible.

- The program uses a specific font, identified in the code as `albr85w.ttf` within the `/windows/C/WINDOWS/Fonts` directory.[10] This font may or may not be installed on every computer and an alternative may need to be substituted. Note that the GD module provides a generic font called `Generic.ttf`.

Here is the source code to the program, called `Embl_plot.pl`:

```perl
#! /usr/bin/perl  -w

# Embl_plot.pl <EMBL entry disk-file> - producing a plot of interesting
#                                       genes within an EMBL entry.

use GD;
use strict;

my $start;
my $end;

my $Image_Size_X;
my $Image_Size_Y = 600;

my $line_width = 20;

my @Features;

while ( <> )
{
    chomp;
```

---

[9] This might seem excessive, maybe even lazy, but working at this scale does save the use of many scaling factors when producing the drawing. As the graphic is black, white and grey, PNG compression reduces the disk-file produced to less than 19,000 bytes.

[10] In fact this font was installed by a program in a standard location on the Microsoft Windows section of Michael's computer that his copy of GNU/Linux automounts as `/windows/C`. You may not have this font installed in this specific location in which case: adjust the path is the program to point to a TrueType font you do have. Remember `GD.pm` comes with the 'Generic.ttf' font in it's '/t' directory for testing purposes as shown in Figure 19.4.

```perl
        if ( /^FT   source        / )
        {
            ( $Image_Size_X ) = m/(\d*)$/;
            print "D: Image_Size = '$Image_Size_X'\n";
        }
        if ( /^FT   CDS/ )
        {
            ( $start, $end ) = m/(\d*)\.\.(\d*)/;
        }
        if ( /\/gene=/ )
        {
            ( my $gene ) = m/\"(\w*)\"/;
            print "D: Gene       = '$gene'\n";
            push @Features, [ $start, $end, $gene ];
        }
    }

    my $image = new GD::Image( $Image_Size_X, $Image_Size_Y );

    my $White = $image->colorAllocate( 255, 255, 255 );
    my $Black = $image->colorAllocate(   0,   0,   0 );
    my $Half  = $image->colorAllocate( 128, 128, 128 );

    $image->filledRectangle( 0, $Image_Size_Y/2 - $line_width,
                             $end,
                                $Image_Size_Y/2 + $line_width,
                                    $Black );

    foreach my $C_Feature ( 0 .. $#Features )
    {
        my $start = $Features[ $C_Feature ][ 0 ];
        my $end   = $Features[ $C_Feature ][ 1 ];
        my $name  = $Features[ $C_Feature ][ 2 ];

        printf( "Feature# %1i %5s   (%5i to %5i)\n",
                $C_Feature, $name, $start, $end );

        $image->filledRectangle( $start, 1, $end,
                                 $Image_Size_Y-1, $Black );
        $image->filledRectangle( $start + $line_width, 1 + $line_width,
                                 $end - $line_width,
                                     $Image_Size_Y-1 - $line_width, $Half );
        $image->stringTTF( $Black, "/windows/C/WINDOWS/Fonts/albr85w.ttf",
                           60, 0, $start + 2 * $line_width,
                               $Image_Size_Y /2, $name );
    }

    open OUTPUT_FILE, ">Embl_sequence_graphic.png"
        or die "Cannot open output file; $!.\n";

    print OUTPUT_FILE $image->png;
    close OUTPUT_FILE;
```

which, when executed, produces the image shown in Figure 19.6 on page 429. The image was originally 8355 pixels wide, with one pixel in the horizontal dimension representing one nucleotide base.

**Figure 19.6** A plot of the interesting genes identified in EMBL entry ISTN501.

Take the time to read through the Embl_plot.pl program in conjunction with the GD module's on-line documentation. Note the use of an array or arrays when populating the @Features array[11].

Figure 19.6 provides a good overview of the arrangement of the coding regions in the DNA. It even exposes the gaps where two extra ORFs (open reading frames) are found. These are probably genes, but this needs to be confirmed.

Other uses for the GD module are limited only by the programmer's creativity, imagination and ability. The GD module is very flexible, but be careful not to reinvent the wheel when it comes to preparing visualisations. For some jobs, using programs such as *ImageMagick* or *The Gimp* can be just as effective.

### 9.3.3 Introducing `mogrify`

There are times when a single image is to be produced at a number of different resolutions or when the resolution of an image is too high, as in the EMBL visualisation in Figure 19.6. Typically, different resolutions are required when printing on paper, e-mailing an image or for inclusion on a web page (either as a full graphic or as a thumbnail). *ImageMagick* is maintained and developed by *ImageMagick Studio*, a non-profit organisation "... *dedicated to making software imaging solutions freely available ...* " according to the licence agreement. The interfaces to the *ImageMagick Application Program Interface* (API) is available for use with many programming languages, not just Perl. The API is also embedded within many command-line tools and utilities, such as `display`, `mogrify` and `montage`.

The `mogrify` utility, among other things, can transform images by reducing the number of pixels they use, when invoked with either the -size <geometry> or -sample <geometry> switches. The following command-line reduces the number of pixels in the EMBL image from Figure 19.6 to 1600 pixels wide, adjusting the number of pixels vertically to 115:

```
mogrify  -resize  1600  Embl_sequence_graphic.png
```

whereas this command-line resizes the image to be 100 pixels high and consequently, 1393 pixels wide:

```
mogrify  -resize  x100  Embl_sequence_graphic.png
```

---

[11] Again, you are referred to the `perldsc` for more details on complex data structures.

**Figure 19.7**  The difference between resampling and resizing. This is resized.

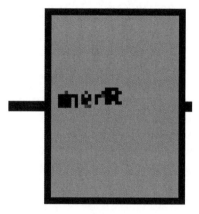

**Figure 19.8**  The difference between resampling and resizing. This is resampled.

For a complete description of all the options and switches, consult the mogrify help pages, accessible via the utility's man page:

```
man mogrify
```

or via the *ImageMagick* web-site:

```
http://www.imagemagick.org/
```

An important point is that the resized image overwrites the original image disk-file. So, if the original, full resolution image is needed, keep a back-up copy as follows:

```
cp  Embl_sequence_graphic.png  Embl_sequence_graphic.original.png
mogrify  -resize  1600  Embl_sequence_graphic.png
```

Another point to keep in mind is that there is some difference between the `-sample` and `-resize` functions.

Referring to Figure 19.8, the image is resampled, whereas the image in Figure 19.7 is resized. The latter image is more blurred, but preserves the text better, whereas the former image is sharper (look at the lines), whereas the text is all but unreadable. As to which resolution to use in which circumstances, the advice is always to consider the output device on which the image will ultimately be displayed. Make the maximum dimension of the image the same as the maximum dimension of output device.

These command-lines drop the resolution of a 4500 by 3000 pixel image, designed for printing, to one that is 1024 pixels wide, suitable for use on a web page or as a reasonably sized e-mail attachment. The original disk-file is approximately 600,000 bytes in size, whereas the sampled image is 60,000 bytes in size:

```
cp   large_sequence_graphic.png   lite_sequence_graphic.png
mogrify   -sample   1024   lite_sequence_graphic.png
```

## 19.4 Plotting Graphs

Plotting data in the form of a pictorial graph is often a very useful way to view numerical data.

> **Technical Commentary:** We use the word "pictorial" here as the word "graph" is also used in the context of graph theory. This is a general mathematical description of interlinked *vertices* (points or nodes) that are connected by *edges* (links). This terminology is used from time to time in Bioinformatics and bioscience literature to represent *networks* of how things are related to each other or linked together in *pathways*. Not knowing the difference between the two can be confusing.

The section describes the Grace program, also known as `xmgrace`. Grace is an excellent interactive scientific, open source program developed from the popular *Xmgr* (ACE/gr) program, originally written by *Paul Turner*. Many of the graphs published in scientific papers, as well as in this book, are produced by one of these packages – the style of the image produced is quite distinctive. More on Grace in just a moment.

Another popular plotting tool is `gnuplot`. This classic scientific graph-plotting tool has been around for two decades, and has a large, loyal user base. Despite its usefulness, `gnuplot` is not discussed further in *Bioinformatics, Biocomputing and Perl*, as it has been (in the opinion of your authors), superseded by modern, GUI-driven alternatives. We cannot help but think that gnuplot is "yesterday's solution".

## 432 Data Visualisation

Quick 'n' dirty graphs can be created within a program using Perl's GD::Graph modules. The use of these modules is especially advantageous when used within a server-side CGI program, when creating dynamic web content.

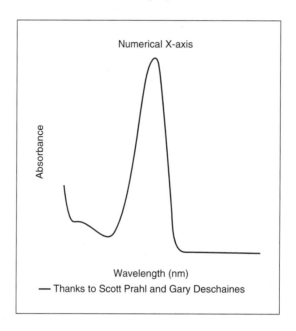

**Figure 19.9** Example line graph from the GD::Graph module.

### 19.4.1 Graph-plotting using the GD::Graph modules

The GD::Graph and GD::Graph3d modules, available on CPAN, interface with the GD library, providing a convenient, programmer-controlled way to produce high-quality graphs. The GD::Graph module produces the standard set of graph types: bars, stacked bars, lines, XY points and pies. The GD::Graph3d module adds extra shading and perspective, simulating a 3D look and feel.

Figure 19.9 on page 432 shows an example line graph included with GD::Graph. Figure 19.10 on page 434 shows a sample pie chart from the GD::Graph module.

The source code used to produce the image in Figure 19.9 is shown below. To execute this code, a fully working version of the gdlib library and the GD module, as well as the GD::Text::Align module (available from CPAN) is required:

```
use GD::Graph::lines;

require 'save.pl';

@data = read_data( "sample54.dat" );

$my_graph = new GD::Graph::lines();
```

```perl
$my_graph->set(
    x_label           => 'Wavelength (nm)',
    y_label           => 'Absorbance',
    title             => 'Numerical X axis',
    y_min_value       => 0,
    y_max_value       => 2,
    y_tick_number     => 8,
    y_label_skip      => 4,
    x_tick_number     => 14,
    x_min_value       => 100,
    x_max_value       => 800,
    x_ticks           => 1,
    x_tick_length     => -4,
    x_long_ticks      => 1,
    x_label_skip      => 2,
    x_tick_offset     => 2,
    no_axes           => 1,
    line_width        => 2,
    x_label_position  => 1/2,
    r_margin          => 15,
    transparent       => 0
);

$my_graph->set_legend( 'Thanks to Scott Prahl and Gary Deschaines' );

my $gd = $my_graph->plot( \@data )
    or die $my_graph->error;

open( IMG, '>file.png' )
    or die $!;

binmode IMG;
print IMG $gd->png;
```

This program extract is taken from the sample56.pl program included with the GD::Graph distribution (and has been reformatted to make it easier to read). The read_data sub-routine is provided in the sample program, but not reproduced here, and its function is to retrieve the data to plot on the graph. The values set in the invocation of the set sub-routine are self-explanatory. Refer to the documentation included with GD::Graph for additional explanations.

### 9.4.2 Graph-plotting using Grace

Grace is designed to be used for scientific graphing, and is especially good at XY *scatter plots*. It is freely available from the following web-site:

http://plasma-gate.weizmann.ac.il/Grace/

The Grace executable is called xmgrace. There are three common approaches to plotting graphs with Grace, as described in this section:

1. Interactive graph generation using a GUI-based application program.
2. Command-line "batch" plotting using the xmgrace command-line utility.

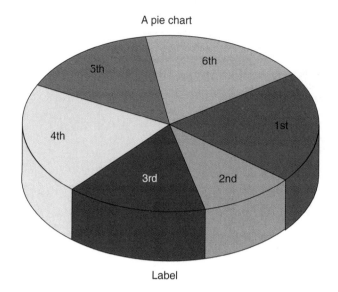

**Figure 19.10** Example pie chart from the GD::Graph module.

3. Programmer-generated graphs using the Chart::Graph::Xmgrace module from CPAN.

## Interactive plotting

Figure 19.11 on page 435 shows the same "Absorbance" test dataset used earlier (with GD::Graph) within the GUI-based Grace program. To recreate this graph, load the required data, as follows:

- From the menu, select *Data*, then *Import*, then *ASCII* to identify the sample54.dat disk-file included with the GD::Graph module.
- Highlight the disk-file name in the *Grace: Read Sets* dialogue box (as shown in the screen shot), then click the *OK* button. The data set loads and the graph appears, autoscaled to fit within the display.
- Double click on the drawn line to open the *Grace: Set Appearance* dialogue box.
- Set the *Line Width*, in the *Line properties* section on the *Main* panel, to 3.
- Sit back and admire your handiwork!

With Grace, all the configuration windows are completely detachable, which is a wonderful feature, avoiding the incessant opening and closing of two windows/dialogues or tab switching, which is an annoying feature of many other

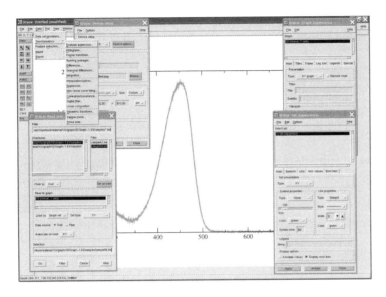

**Figure 19.11** The GUI-based Grace application program.

programs. Grace's interface can result in a fair amount of screen clutter. This can be reduced by dragging windows off to one side of the screen, or onto another screen if your computer supports more than one monitor.

**Selecting: File: Print Setup displays the Grace:** *Device setup* dialog, in which PNG can be selected as the 'Device'. If so, *Print* -ing the graph creates a PNG image at 792 by 812 pixel resolution (the default resolution of Michael's computer), though a better quality image would be created by 1600 by 1200 pixels (use 'Size...Custom' and set the dimensions in the Grace: Device setup dialogue to do this). The created PNG disk-file is named `Untitled.png`, unless specified otherwise. Other image formats are supported, including *Postscript* and *JPEG*.

When Grace is instructed to save an image (using *File*, then *Save* or *Save As* from the menu), the resulting disk-file contains standard ASCII text. The disk-file produced by saving the above image is called `sample56.agr`, and it contains 2146 lines detailing the features of the image. 1842 of the lines are the actual data to plot, the rest are' very useful when automatically generating plot in "batch mode" or when moving the image from one plotting program to another.

### Batch plotting

Grace has a complex internal command language that can be used to produce graphs and plots non-interactively. It is possible to avoid the use of this command language when the plotting requirements are straightforward. For instance, to turn the `sample54.dat` disk-file into a PNG image called `sample54.png`, colour the line plotted, then add some axis labels and a title, use this command-line:

    xmgrace sample54.dat -hardcopy -hdevice PNG -batch batch.grace

where each of the parameters each has the following meaning:

**`sample54.dat`** – the data disk-file to plot (in tab or space delimited format).

**`-hardcopy`** – instructs `xmgrace` to plot to a disk-file. Without this parameter, `xmgrace` plots the graph, then launches Grace in fully interactive mode, allowing for the plot to be inspected or changed, if need be.

**`-hdevice PNG`** – specifies that a PNG image is to be created.

**`-batch batch.grace`** – specifies that the series of commands included in the disk-file `batch.grace` are to be executed.

The `batch.grace` disk-file used to produce the "Absorbance" image from earlier looks like this:

```
s0 line color 3
s0 line linewidth 3.5
title "Mass Spec Data"
xaxis label "M/Z Ratio"
yaxis label "Relative Intensity"
autoscale
page size 1024, 768
```

These commands override the default graph-plotting settings configured into `xmgrace`. The meaning of most of the above commands is fairly obvious. Refer to the Grace and `xmgrace` documentation for details as to their individual meaning or examine the ASCII disk-file representation of a plot saved by the Grace application.

Note that Grace datasets are numbered "s0", "s1", "s2", and so on. Additionally, formatting directives within the saved ASCII disk-file start with an "@" symbol. These should be omitted in `xmgrace` batch command disk-files. For example, this extract from the saved disk-file (`sample56.agr`):

```
@    s0 line type 1
@    s0 line linestyle 1
@    s0 line linewidth 1.0
@    s0 line color 3
@    s0 line pattern 1
```

is specified as follows within a xmgrace command disk-file:

```
s0 line type 1
s0 line linestyle 1
s0 line linewidth 1.0
s0 line color 3
s0 line pattern 1
```

In addition to the Grace documentation, the Chart::Graph::Xmgrace module from CPAN provides a good overview of those parameters most likely used when batch plotting a graph. Here's the colour table extracted from that documentation[12]. With this table, it is straightforward to see that "color 3" is actually "green":

```
+-------------------------------------------------------------------+
|                              COLORS                               |
+-------+-----+-------+-----+--------+-----+----------+---------+
| COLOR |VALUE| COLOR |VALUE| COLOR  |VALUE| COLOR    | VALUE   |
| white | "0" | blue  | "4" | violet | "8" | indigo   | "12"    |
| black | "1" | yellow| "5" | cyan   | "9" | maroon   | "13"    |
| red   | "2" | brown | "6" | magenta| "10"| turquoise| "14"    |
| green | "3" | grey  | "7" | orange | "11"| dark green| "15"   |
+-------+-----+-------+-----+--------+-----+----------+---------+
```

**Plotting using Chart::Graph::Xmgrace**

As mentioned earlier, Perl provides an interface to the Grace functionality by way of the Chart::Graph::Xmgrace module from CPAN. This provides the Perl programmer with a programmable plotting environment. This module works by populating various arrays with a set of commands to be included in a batch command disk-file. These commands are ultimately processed by Grace. The Chart::Graph::Xmgrace module greatly simplifies the formatting of command disk-files.

To use Chart::Graph::Xmgrace, the xvfb program needs to be installed. This is the X Windows Virtual Frame Buffer, which provides a dummy X Windows server used for the non-display of graphics on the command-line during automated processing. For more details (including instructions for downloading and installing xvfb), visit this web-site:

http://www.xfree86.org/4.0.1/Xvfb.1.html

Here is an extract from a Perl program used to produce a command disk-file to plot the "Absorbance" graph:

---

[12] http://search.cpan.org/~caidaperl/Chart-Graph-2.0/Graph/Xmgrace.pm.

### 438  Data Visualisation

```
use Chart::Graph::Xmgrace qw( xmgrace );
    .
    .
    .
xmgrace( {    "title"              => "Absorbance Data",
              "subtitle"           => "Thanks to Scott Prahl and
                                           Gary Deschaines",
              "type of graph"      => "XY graph",
              "output type"        => "png",
              "output file"        => "sample56_grace.png",
              "x-axis label"       => "Time",
              "y-axis label"       => "Relative Intensity",
              "grace output file"  => "xmgrace1.agr",
         },
         [ { "title"               => "data",
             "options"             =>
                 {  "line"         =>
                      { "color"    => "4",
                        "linewidth" => "3.5"
                      },
                 },
             "data format"         => "file"
           },
           "sample54.dat"
         ],
);
```

which, when executed, results in the visualisation shown in Figure 19.12 on page 439. On closer inspection, there are three problems with the resulting image:

1. The image size – It was not possible to set the image of the visualisation to anything other than 640 by 480 pixels, which is the default size.

2. The line colour – Despite setting the line colour to "3" (green), no colour is set.

3. The line thickness – Any specified line thickness is ignored by the module, which appears to set a maximum thickness of 1.

Inspecting the `xmgrace1.agr` disk-file revealed that all of the linestyle formatting information was missing. This may be a *bug* in the module[13].

> **Technical Commentary:** An alternative Perl interface to the functionality provided by Grace exists in the `Chart::GRACE` module, available from CPAN. This module is more orientated towards the display of data stored in arrays or Perl Data Libraries (PDLs).

---

[13] And it may well be fixed by the time you read this, as popular CPAN modules are always being updated.

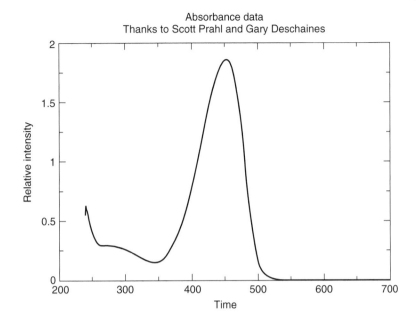

**Figure 19.12** The "Absorbance" image as produced by `Chart::Graph::Xmgrace`.

As suggested by the final maxim in Chapter 5[14], it is important to realise the level of functionality provided by a CPAN module. The `Chart::Graph::Xmgrace` module has some excellent features, but some functionality is (currently) missing.

## Where to from Here

Data visualisation is one of the most active areas of Bioinformatics because of the ever-increasing size, internal complexity and range of information that must be presented to and by researchers. The ultimate aim of any data visualisation is to make the data *mean* something, so the viewer is in a position to better understand the data.

Data visualisation is an area of great creativity, both scientifically and computationally. Take the time to explore what it has to offer.

## The Maxims Repeated

Here's a list of the maxims introduced in this chapter.

---

[14] *Always take the time to test downloaded CPAN modules for compliance to specific requirements.*

## 440   Data Visualisation

- *People are processing tools, too, especially when it comes to processing visual information.*
- *Producing plans avoids problems before problems surface.*

# Exercises

1. Amend the `Mer_Table.pl` program to automatically insert HTML hyperlinks to on-line database entries.
2. Develop a strategy to remove those trailing commas at the end of each of the lists produced by the `Mer_Table.pl` program. Implement the strategy.

# 20
# Introducing Bioperl

*Towards the Bioinformatics Perl programmer's nirvana.*

## 20.1 What is Bioperl?

*Bioperl* is a very useful set of Perl modules for Bioinformatics that handle common (and some not so common) Bioinformatics analysis tasks. These tasks include:

- Accessing databases.
- Aligning sequences.
- Accessing literature references.
- Accessing sequence annotation.
- Performing and parsing sequence similarity searches.

In many cases, Bioperl is just a convenient interface to existing software. For example, storage of the advanced database annotations uses a relational database (usually *MySQL*), the graphical output uses `gdlib` and multiple sequence alignment is performed by ClustalW or *T-Coffee*. Much of Bioperl's under-the-hood functionality is also borrowed from other Perl modules, such as the `libwww-perl` library when working with the web and `xpat` when working with XML. There are also interfaces to fully featured software packages such as *EMBOSS* and *Ensembl*. More on this later.

Bioperl is an open-source project. As such, it is possible to use Bioperl immediately after downloading the Bioperl modules from the project's web-site:

http://www.bioperl.org

The Bioperl modules are also on CPAN, but the version found there may lag behind the latest release from the project[1].

When using Bioperl, it is always a good idea to cite the project's web-site address and the current official publication that can be found in the web-site's FAQ. At the start of September 2003, the citation is:

Stajich, J. E., Block, D., Boulez, K., Brenner, S. E., Chervitz, S. A., Dagdigian, C., Fuellen, G., Gilbert, J. G. R., Korf, I., Lapp, H., Lehvaslaiho, H., Matsalla, C., Mungall, C. J., Osborne, B. I., Pocock, M. R., Schattner, P., Senger, M., Stein, L. D., Stupka, E. D., Wilkinson, M. and Birney, E., The Bioperl Toolkit: Perl modules for the Life Sciences, Genome Research, 2002, October 12(10): 1161-8.

## 20.2 Bioperl's Relationship to Project Ensembl

The Bioperl modules are the main interface engine behind the excellent *Ensembl Genome Browser*, available from this web-site:

http://www.ensembl.org

Originally, the Ensembl project was designed to be a very capable and reliable annotated data source for the Human Genome, but has since been used for other genomes, such as the Murine (Mouse) and *C. elegans* (worm). One of the strengths of Ensembl is the stability of the automatic annotation it provides, which makes it easy to link to either a local Ensembl system or a remote system via the Ensembl web-site.

Another strength is the ease of searching the information contained in Ensembl. Text, sequence and physical location are all supported on the web-site – the user literally clicks on the map of the chromosome to zoom in or out on that part which is of interest. A programmer interface (or *portal*) provides a mechanism to perform automated queries. Alternatively, it is possible to download the underlying database as either flat data-files or as a MySQL dump for import into a local Ensembl system or other analysis system.

## 20.3 Installing Bioperl

Installing the core Bioperl modules is, thankfully, straightforward. Download both the `current_core_stable.tar.gz` and `current_run_stable.tar.gz` disk-files,

---

[1] Release 1.2.2 of Bioperl is described in this chapter.

as these contain the latest stable release from the Bioperl web-site. Once downloaded, decompress them, as follows:

```
tar   -zxvf   current_core_stable.tar.gz
tar   -zxvf   current_run_stable.tar.gz
```

Change into each of the newly created directories[2] using the cd command, then issue the usual Perl module installation commands:

```
perl   Makefile.PL
make
make   test
su
make   install
<Ctrl-D>
```

Note the requirement, as usual, to issue the make install command as the superuser (*root*). Note, too, that the issuing of this set of installation commands in each of the directories takes a little while to complete. There may be a requirement to set the paths to installed programs that Bioperl relies on (such as ClustalW). Refer to the Bioperl documentation for details on how to do this, if necessary.

Bioperl, probably more than any other CPAN module, has a large number of extra dependencies. A number of other programs may or may not be required, depending on what it is Bioperl is going to be used for. Installing everything that Bioperl could potentially need is quite a task. For this introduction, your authors confined themselves to a small subset of all that's possible. The general advice is to install those modules needed and no more[3].

There is also a special bundle of Bioperl modules, called Bundle::BioPerl (note the non-standard capitalisation), which can help with installation and is also available on CPAN. When used with the CPAN module, the bundle automates the installation of Bioperl, as well as the most common dependencies. Here are the commands to use:

```
su
perl   -MCPAN   -e   "shell"
cpan>   install   Bundle::BioPerl
cpan>   quit
<Ctrl-D>
```

Again, note the requirement to issue these commands while logged in as *root*. The assumption in this chapter is that the *external modules* referenced during the Bundle::BioPerl installation have *not* been installed, except for the

---

[2] These are named bioperl-1.2.2 and bioperl-run-1.2.2 respectively.

[3] Bear in mind that it is possible to install modules later without too much difficulty.

libwww-perl library. To try out the *RemoteBlast* example, the `String::IO` module is needed, which (as usual) is available on CPAN.

## 20.4 Using Bioperl: Fetching Sequences

Bioperl comes in one of two flavours, object-oriented and non-object orientated (by less functional). The non-object based interface is easier to work with, so it is initially used within this chapter. Later examples use the object-oriented flavour.

Even though the vast majority of the Bioperl functionality is missing as a result of excluding the external modules/packages, a basic installation of Bioperl can still be applied to a number of problems. One of the simplest tasks is to use the `get_sequence` sub-routine (provided by Bioperl) to download sequences entries from remote databases such as SWISS-PROT, EMBL and GenBank.

When supplied with either a SWISS-PROT accession number (P00392) or a SWISS-PROT ID (MERA_PSEAE) on the command-line, the following program, called `simple_get_sequence.pl`, downloads the SWISS-PROT sequence and creates two disk-files. The first contains the verbose SWISS-PROT format together with a full annotation, and the second contains the sequence in FASTA format. The Bioperl-supplied `write_sequence` sub-routine creates the disk-files:

```perl
#! /usr/bin/perl -w

# simple_get_sequence.pl - a simple Bioperl example program, which
#                         downloads SWISS-PROT sequences.

use strict;

use Bio::Perl;

my $ID = shift;

my $Sequence = get_sequence( 'swiss', $ID );

write_sequence( ">./seqs/$ID.swp", 'swiss', $Sequence );
write_sequence( ">./seqs/$ID.fsa", 'fasta', $Sequence );
```

There's not much to the `simple_get_sequence.pl` program. After switching on strictness and using the `Bio::Perl` module, the $ID scalar is assigned a value from the command-line (thanks to the `shift` sub-routine). This scalar is immediately used in a call to Bioperl's `get_sequence` sub-routine, which goes off to the Internet and downloads the SWISS-PROT entry associated with the value of $ID. The entry is assigned to (or *associated with*) the $Sequence scalar. It is then used in two calls to Bioperl's `write_sequence` sub-routine, which each take the downloaded SWISS-PROT entry in $Sequence and create a disk-file from it. The first call creates a disk-file containing the verbose format, while the second creates a disk-file containing the FASTA format.

To execute this program, create a directory called `seqs` in the current directory, make the `simple_get_sequence.pl` program executable, and run it, as follows:

```
mkdir   seqs
chmod   +x   simple_get_sequence.pl
./simple_get_sequence.pl   P00392
```

If everything works as expected, the `seqs` directory now contains two disk-files, `P00392.swp` and `P00392.fsa`. The Bioperl modules are flexible enough to work out whether a human-friendly ID or an accession code is supplied on the command-line and download the correct sequence, as appropriate. Hence, these two invocations of the program download and create the same disk-files:

```
./simple_get_sequence.pl   P00392
./simple_get_sequence.pl   MERA_PSEAE
```

The `diff` utility is used to quickly check that any two disk-files are identical. These command-lines produce no output[4], which is `diff`'s way of indicating that the disk-files are the same:

```
diff   seqs/P00392.fsa   seqs/MERA_PSEAE.fsa
diff   seqs/P00392.swp   seqs/MERA_PSEAE.swp
```

The `get_sequence` sub-routine works equally well with DNA sequences from the GenBank and EMBL databases. To learn more about the other functionality included in the `Bio::Perl` module, consult the on-line documentation using this command-line:

```
perldoc   Bio/Perl.pm
```

or browse HTML-formatted version available on the Bioperl web-site.

## 0.4.1  Fetching multiple sequences

Let's download the sequences used in the ClustalW alignment example from Section 18.9.1 on page 407 in this book. Note that it's not an absolute requirement to use Bioperl to determine the list of required sequences, as it is possible to use the interactive Sequence Retrieval System (SRS) web service to download the list.

From SRS, extract a list of SWISS-PROT IDs by searching for `mera` and `merp`, copying (using cut 'n' paste) the results into two disk-files: one containing the `MerA` sequences and the other containing the `MerP` sequences. The list of determined proteins should be similar to these:

---

[4] They are the strong, silent type.

MERP_ACICA
MERP_ALCSP
MERP_PSEAE
MERP_PSEFL
MERP_SALTI
MERP_SERMA
MERP_SHEPU
MERP_SHIFL
```

and:

```
MERA_ACICA
MERA_ALCSP
MERA_BACSR
MERA_ENTAG
MERA_PSEAE
MERA_PSEFL
MERA_SERMA
MERA_SHEPU
MERA_SHIFL
MERA_STAEP
MERA_STRLI
MERA_THIFE
```

The lists are stored in the `Merp_IDs.1st` and `Mera_IDs.1st` disk-files respectively. Once the list is determined, the `Multi_Seq_Get.pl` program is used to download each list of sequences from the on-line databases with these command-lines:

```
./Multi_Seq_Get.pl   Merp_IDs.1st
./Multi_Seq_Get.pl   Mera_IDs.1st
```

The above command-lines create a collection of SWISS-PROT data files in the `seqs` directory. Here's the source code to the `Multi_Seq_Get.pl` program:

```perl
#! /usr/bin/perl -w

# Multi_Seq_Get.pl - when provided with a list of IDs, download them.

use strict;

use Bio::Perl;

while ( my $ID = <> )
{
    chomp( $ID );

    print "Fetching Sequence: $ID.\n";

    my $Sequence = get_sequence( 'swiss', $ID );

    write_sequence( ">./seqs/$ID.swp", 'swiss', $Sequence );
    write_sequence( ">./seqs/$ID.fsa", 'fasta', $Sequence );
}
```
```

The Multi_Seqs_Get.pl program simply wraps a while loop around the source code from the simple_get_sequence.pl program.

## 0.4.2 Extracting sub-sequences

For the phlyogenetics investigation, a requirement exists to align sequences of the HMA domain that exist as the major part of MerP and the first part of MerA.

In this section, Bioperl is used to download the sequences and "cut out" the HMA domains from the MerA proteins, creating new sequence disk-files containing the sections of interest. In an earlier chapter, a text editor is used to do this. However, there is a better way. One of the fundamental modules that Bioperl has is called Sequence[5], which provides a mechanism to extract sub-sequences. This is very similar to Perl's in-built substr sub-routine, which could be used to program a parser/sequence extractor. The word "could" is used in the last sentence for a reason: it's always possible to *roll your own* with Perl, but not always advisable.

> **Maxim 20.1** *Don't reinvent the wheel.*
> *Use Bioperl whenever possible.*

As most of the annotation is missing, all Bioperl is asked to do is create a new sequence with:

1. A particular sequence
2. A supplied ID
3. An accession code (which is actually based on the ID in this case)

Bioperl fills in the basic annotation fields automatically, producing a properly formatted SWISS-PROT file, which includes the sequence length, the molecular weight and the CRC checksum. This is all part of the Bioperl service. A program called Seq_Crop.pl extracts the first 92 amino acids from each MerA sequence ID passed from STDIN and writes the output to a disk-file called MerAs_cropped.swp:

```
#! /usr/bin/perl -w

# Seq_Crop.pl - extract a subsequence example program.

use strict;

use Bio::Perl;

use constant AMINO_COUNT => 92;

while ( my $ID = <> )
```

---

[5] This is often referred to as *Seq*.

448   *Introducing Bioperl*

```
{
    chomp( $ID );

    print "Processing sequence: $ID.\n";

    my $Org_Seq = read_sequence( "./seqs/$ID.swp", 'swiss' );

    my $Protein_Code = $Org_Seq->subseq( 1, AMINO_COUNT );

    my $Cropped_Seq = new_sequence( $Protein_Code, "$ID_CROPPED", $ID );

    write_sequence( ">>MerAs_cropped.swp", 'swiss', $Cropped_Seq );
}
```

When executed against the list of MerA sequences, the resultant disk-file, which is called MerAs_cropped.swp, contains a set of concatenated SWISS-PROT records. Here's an extract from the disk-file produced:

```
ID    MERA_ACICA_CROPPED      STANDARD;      PRT;    92 AA.
AC    MERA_ACICA;
DE
SQ    SEQUENCE     92 AA;  9363 MW;   F2AA11A3B589F55A CRC64;
      MTTLKITGMT CDSCAAHVKE ALEKVPGVQS ALVSYPKGTA QLAIEAGTSS DALTTAVAGL
      GYEATLADAP PTDNRAGLLD KMRGWIGAAD KP
//
ID    MERA_ALCSP_CROPPED      STANDARD;      PRT;    92 AA.
AC    MERA_ALCSP;
DE
SQ    SEQUENCE     92 AA;  9119 MW;   BD0F5CDA5FB699DC CRC64;
      MYLNITGMTC DSCATHVKDA LEKVPGVLSA LVSYPKGSAQ LATDPGTSPE ALTAAVAGLG
       .
       .
       .
```

Of course, this can also be accomplished using SRS to extract FASTA-formatted entries, which can then be manually edited. Doing so (manually) shows that the HMA domain is listed as the first 66 amino acids of the MerA proteins in the annotation, not 92 as assumed, on the basis of the length of the MerP proteins. Alternatively, change the AMINO_COUNT constant from 92 to 66 in the Crop_Seq.pl program and re-run it. It should be clear that even for the simple sequence processing tasks such as demonstrated above, the power of Bioperl is clear.

To produce an easy to upload disk-file, simply concatenate the MerP SWISS-PROT protein disk-files with the pre-concatenated MerA fragments as stored in the MerAs_cropped.swp disk-file.

## 20.5 Remote BLAST Searches

One of the most useful and interesting sub-routines provided by Bioperl is blast_sequence. As its name implies, this sub-routine performs a *sequence similarity search* using the Basic Local Alignment Search Tool (BLAST) from the NCBI web-site. There is a complementary sub-routine, called `write_blast`, which creates a disk-file from the results of the search. Typical usage within a program is:

    $blast_result = blast_sequence( $seq );

    write_blast( $filename, $blast_result );

Note that `write_blast` assumes that the `String::IO` module is installed within the local Perl environment. The Bioperl module provides a whole host of sub-routines for parsing any produced output. As these sub-routines operate on standard BLAST reports, they can process results generated locally using the `Bio::Tools::Run::StandAloneBlast` module and/or from a locally installed BLAST system.

While preparing this book, these sub-routines initially worked well and then – abruptly – stopped working. It is very likely that the NCBI BLAST service was (unintentionally) changed in a way that Bioperl could no longer use it. More than likely, this particular problem will be fixed with the next release of Bioperl[6]. Anyway, your authors wanted to show how to use other programs with Bioplerl!

An alternative strategy is to use the official NCBI remote BLAST solution, which is downloadable from the NCBI FTP site[7], which highlights an important maxim.

**Maxim 20.2** *Combine Bioperl with other tools to get your work done.*

## 0.5.1 A quick aside: the `blastcl3` NetBlast client

The `blastcl3` program is extremely flexible, and provides many command-line parameters. The most important are:

**-p** – the program name as a string (e.g., *blastp*)

**-d** – the database name as a string (e.g., *swissprot*)

**-i** – the query file (which defaults to `STDIN`)

**-e** – the expectation value (which defaults to 10.0)

**-m** – the alignment view options.

For example, to BLAST the sequence just downloaded back against the SWISS-PROT database, use the program `blastp`, which performs protein-to-protein comparison. The output can then be redirected into a disk-file, which is called `Swiss-Prot.NetBlast_P00392.res`. Use the following command-line:

---

[6] And by the time this book is published, this particular problem may no longer be an issue.

[7] Use this Internet address: `ftp://ftp.ncbi.org`.

```
blastcl3 -i ./seqs/P00392.fsa -d swissprot -p blastp         \
   > Swiss-Prot.NetBlast_P00392.res
```

The above invocation of `blastcl3` submits the request into the *job queue* on the NCBI web-site, uploading the disk-file containing the sequence via the Internet in the process. This can take some time (often many minutes) to complete.

When performing less than 20 sequences (perhaps as part of a small project or prototype analysis system), the advice is to use the on-line interactive web-based BLAST servers. Note that the BLAST client returns a flat data-file containing text only and it lacks features such as clickable hyper-links that reference the public databases, a feature that is provided by the web-site.

## 20.5.2 Parsing BLAST outputs

The structure of the BLAST output from either the on-line services, stand-alone systems, network clients or the `RemoteBlast` module follows the same basic format. There are three main sections:

1. The BLAST copyright message.

2. A summary of the similar sequence and measures of the amount of similarity[8].

3. The alignments between the query and database sequences.

Some of the most useful modules in Bioperl are those that parse BLAST output. One of the more recent and most general is the `Bio::SearchIO` module. This is a fully object-orientated module containing a wide variety of sub-routines (that are called *methods*).

Let's discuss a modified version of one of the example programs from the Bioperl documentation. This processes the output from the NCBI BLAST programs, and parses the disk-file passed to it on the command-line:

```
./Blast_parse.pl   Swiss-Prot.NetBlast_P00392.res
```

Here's the source code to the `Blast_parse.pl` program[9]:

```
#! /usr/bin/perl -w

# Blast_parse.pl - post-process/parse BLAST output.

use strict;
```

---
[8] Inspecting the output in detail, it is unsurprising that the scoring hit is the P00392 entry itself, which is what we expect: this was the query sequence after all! Also, the other MerA examples score above anything, so we can conclude that our BLAST search is behaving as expected.

[9] Look at the size of this program.

```perl
use Bio::SearchIO;

my $bls_report = shift;

my $in = new Bio::SearchIO( -format => 'blast',
                            -file   => $bls_report );
while ( my $result = $in->next_result )
{
    while( my $hit = $result->next_hit )
    {
        print "Hit = ", $hit->name, "\n";
    }
}
```

After the usual first line, the switching on of *strictness* and the use of the Bio::SearchIO module, the name of the disk-file as passed in on the command-line is assigned to the $bls_report scalar. This is then used to create a new Bio::SearchIO object that is assigned to a scalar called $in. It is this scalar that provides mechanisms to extract useful information from the BLAST output. In this program, two while loops cycle through the BLAST data.

Here are the first ten lines of output produced by the above invocation of the Blast_parse.pl program:

```
Hit = sp|P00392|MERA_PSEAE
Hit = sp|P94702|MERA_ENTAG
Hit = sp|Q52109|MERA_ACICA
Hit = sp|P94188|MERA_ALCSP
Hit = sp|P08332|MERA_SHIFL
Hit = sp|Q51772|MERA_PSEFL
Hit = sp|P17239|MERA_THIFE
Hit = sp|Q54465|MERA_SHEPU
Hit = sp||P08662_1
Hit = sp|P30341|MERA_STRLI
```

To learn more about the Bio::SearchIO module, use this command-line to view its documentation:

```
perldoc Bio/SearchIO.pm
```

The documentation is quite involved and provides details on the extensive features of the module. For example, it is a straightforward matter to extract individual alignments or convert BLAST output into HTML.

## Where to from Here

Bioperl is a flexible, extensive, powerful and standard set of Bioinformatics analysis and control modules for Perl programmers. This chapter touches on a very small part of Bioperl. There are lots more, including excellent modules for sequence analysis and data format translation, which are useful in Bioinformatics and bioscience in general. The real power of Bioperl comes from the external packages that it presents, in that they are neat, clean interfaces. This allows programmers to incorporate sophisticated functionality into their own programs without too much difficulty.

Readers are encouraged to take as much time as necessary to learn about the extensive features provided by the Bioperl modules. Feel free to get involved in working for the project, too. Volunteers are always welcome!

## The Maxims Repeated

Here's a list of the maxims introduced in this chapter.

- *Don't reinvent the wheel. Use Bioperl whenever possible.*
- *Combine Bioperl with other tools to get your work done.*

## Exercises

1. Download, install and configure Bioperl onto your computer. Take the time to explore its many features.

# *Appendix A*

# *Installing Perl*

The opening chapters of *Bioinformatics, Biocomputing and Perl* assume that Perl is installed on the computer being used. Perl is available in ready-to-run versions for many operating systems. For the complete list of supported platforms, visit this web-site:

    http://www.perl.com/CPAN/ports/

The good news is that if you are already running some version of Linux or Mac OS X, then Perl is – most likely – already installed, as it is included in these operating systems as an integrated component. The section below, entitled *Installing Perl The Hard Way*, contains instructions for installing Perl from its source code distribution/download. The remainder of this section provides instructions for getting and installing Perl on other platforms.

## Mac OS 7, 8 and 9

If the Apple Macintosh is not running Mac OS X (which includes Perl 5.6.0 as standard), then go to:

    http://www.macperl.com

to download MacPerl, which is based on the official 5.6.1 release of Perl. The MacPerl folk provide ongoing support for this version of Perl (which is tailored to work with "older" Macintoshes), as well as providing the full, on-line text to a book called *MacPerl: Power and Ease*. MacPerl is free to download and use.

## Windows 9x/ME/NT/2000/XP

A binary version of Perl, called ActivePerl, is available for the various Microsoft Windows platforms from *ActiveState* (a division of *SOPHOS*). Download ActivePerl from:

```
http://www.activeperl.com
```

The version of Perl available from ActiveState tends to lag behind the officially released version. For instance, at the time of writing, the Perl community is "shipping" Perl 5.8.2, whereas ActiveState's version is the 5.8.1 release. This is due to ActiveState taking the time to enhance the official Perl release with a large collection of Windows-specific functionality. ActivePerl, like MacPerl, is available for free download. ActiveState/SOPHOS also provide a collection of tools for use by professional developers. These tools are *not* provided free of charge.

## Installing Perl the Hard Way

Use the following command to check which version of `perl` is installed on a computer:

```
perl -v
```

This command results either in the display of a message detailing the version of Perl installed or in a "command not found" complaint from the operating system. Here's what is displayed on Paul's computer:

```
This is perl, v5.8.0 built for i386-linux-thread-multi
(with 1 registered patch, see perl -V for more detail)

Copyright 1987-2002, Larry Wall

Perl may be copied only under the terms of either the Artistic License or the
GNU General Public License, which may be found in the Perl 5 source kit.

Complete documentation for Perl, including FAQ lists, should be found on
this system using 'man perl' or 'perldoc perl'.  If you have access to the
Internet, point your browser at http://www.perl.com/, the Perl Home Page.
```

For purposes of illustration, the assumption is that users of UNIX or Linux (RedHat, Caldera, Debian, SuSe, Mandrake or whatever) do not have Perl installed, or that they intend to upgrade an existing implementation to the latest release. To begin, download this disk-file:

```
stable.tar.gz
```

from the following web address:

    http://www.perl.com/pub/a/language/info/software.html#sourcecode

Unpack the downloaded disk-file with this command-line:

    tar  zxvf  stable.tar.gz

which creates a new directory called `perl-5.8.2`, into which the entire collection of Perl source code is placed. Change into this new directory to begin the installation process:

    cd  perl-5.8.2

It may be worthwhile taking a few moments to read the supplied README and INSTALL disk-files. The installation begins with the forced deletion of two configuration commands:

    rm  -f  config.sh  Policy.sh

A script, specifically written to check the capabilities of the environment within which `perl` is to be installed, is then executed, as follows:

    sh  Configure  -de

Note that the `-de` command-line switches instruct the `Configure` script to execute with the most common and general options set as default. Note that failing to specify `-de` results in a long, interactive question-and-answer session with `Configure`, whereby the user is asked to confirm (or suggest alternatives for) the default selections that `Configure` is using. Unless you know what you are doing, the advice is to let `Configure` select defaults automatically. When `Configure` is complete, execute the next command to start the building of the Perl source code:

    make

This is a good point to take a break, put on the kettle and make a cup of tea/coffee. Even on the fastest of personal computers, building the Perl source code with `make` takes a number of minutes. Assuming all is well with `make`, the next step is to test the newly built `perl` before installing it. Use this command-line:

    make  test

This, too, takes a while. So, feel free to have another cup of tea/coffee. When the tests conclude, temporarily become the superuser (*root*) and install `perl` and its associated environment with these commands:

    su
    make  install

## Appendix A

Finally, while still logged in as superuser, issue the following command to convert your system's *header files* into something that `perl` can use, then logout using the familiar key combination[1]:

```
cd /usr/include && h2ph *.h sys/*.h
<Ctrl-D>
```

The Perl environment (including the `perl` program) is now installed and ready to use.

---

[1] It is not important that you know why this has to be done.

# Appendix B

# Perl Operators

Here is the list of Perl operators in precedence order, starting with those with the highest precedence.

- **->** - The infix dereference arrow operator, used when working with references (and objects).

- **++ and --** - The increment and decrement operators.

- **\*\*** - The exponential operator.

- **!, ~, \, + and -** - The logical negation (!), bit-wise negation (~), reference (\), numeric affirmation (+) and arithmetic negation (-) operators.

- **=~ and !~** - The binding operators (used when working with regular expressions and pattern matches).

- **\*, /, % and x** - The multiply, divide, modulus (%) and repetition (x) operators.

- **+, - and .** - The addition, subtraction and concatenation (.) operators.

- **<< and >>** - The left and right bit shifting operators.

- **<, >, <=, >=, lt, gt, le and ge** - The relational operators. There are two of each, one for working with numbers and the other for working with strings. Be careful to use <, >, <= and >= when comparing numbers, and use lt, gt, le and ge when comparing strings.

- **==, !=, <=>, eq, ne and cmp** - The equality operators. As with the relational operators, different versions exist for use with numbers and for strings. The <=> and cmp operators are used for comparison, and are typically used in conjunction with the inbuilt sort subroutine.

**&** – The bit-wise AND operator.

**| and ^** – The bit-wise OR and exclusive OR operators.

**&&** – The logical AND operator.

**||** – The logical OR operator.

**.. and ...** – the range operators.

**?:** – The ternary conditional operator.

**=, **=, +=, *=, &= and so on** – the assignment operators.

**, and =>** – The comma operator (typically used to separate list items).

**not** – A lower precedence alternative to !.

**and** – A lower precedence alternative to &&.

**or and xor** – Lower precedence alternatives to || and a logical eXclusive OR operator.

# Appendix C

# Perl's On-line Documentation

## A Short Guide to Perl's Documentation

Perl comes with a large and growing collection of on-line documentation. To access this documentation, use this command-line:

    man perl

The manual page displayed provides a long list of documents, themselves available as manual pages. Rather than present the entire list here, this appendix provides a short description of some of the more useful documents (and is based on the documents that come as standard with release 5.8.2 of Perl).

### perlfunc

This is a large document that provides a description of the collection of in built subroutines provided by `perl`. If the name of a subroutine is already known, the `perldoc` utility can be used to jump directly to the appropriate entry within `perlfunc` documentation. For instance, assuming a requirement to look up the `substr` subroutine, this command-line displays the appropriate `perlfunc` entry:

    perldoc -f substr

### perlreftut

Perl references are a special type of scalar variable that allow for and enable the creation of complex data structures (such as an array of arrays or a hash of hashes). This short guide provides a guide to the essentials of references.

## perldsc

Following on from the `perlreftut` document, this guide demonstrates how to use references with arrays and hashes to create complex data structures. Sample program code is provided for a number of useful data structures.

## perlretut

This is the Perl regular expression tutorial that covers most of what's required to be known about Perl's pattern-matching technology. It complements rather than replaces the material from Part I of this book.

## perldebtut

An excellent (and small) introduction to using the Perl debugger. A *debugger* is a software tool that can execute a program one statement at a time while the user (of the debugger) watches what's going on. When a program is not behaving the way it is expected to, running it through the debugger can help identify where things are going wrong. This manual page presents Perl's debugger to the new user.

# Appendix D

# Suggestions for Further Reading

In this appendix, those texts, journals, articles and other resources that supplement and support the material in this book are presented. These are ordered by chapter. Note that not all chapters have further reading suggestions.

There are many excellent Bioscience/Biochemistry/Molecular Biology/Genetics books available. Usually, one or more of these can be found on the Bioinformatician's bookshelf (or within their organisation's library). Here is a short list that Michael recommends and finds particularly useful:

- *Genetics: A Molecular Approach*, Brown, T., BIOS Scientific Publishers, 1992. This "older" book covers all of the fundamental concepts from a molecular perspective.
- *Genes VII*, Lewin, B., Oxford University Press, 2000. The "Genes" series of books are both comprehensive and comprehensible.
- *Biochemistry*, $5^{th}$ edition, Stryer, L., Berg, J. M. and Tymoczko, W. H. Freeman, 2002. This is a well established, standard Biochemistry textbook.
- *Molecular Cell Biology*, Darnell, J. E., W. H. Freeman, 2003. This is a very useful, modern and up-to-date molecular biology reference text.

## Setting the Biological Scene

- 1D66 is described in Marmorstein, R., Carey, M., Ptashne, M. and Harrison, S. C., *Nature*, **356**, 408, 1992.

- The *Mer Operon* is well characterized in Liebert, C. A., Hall, R. M. and Summers, A. O. , *Microbiology and Molecular Biology Reviews*, **63**, 507–522, 1999, as well as in Hobman, J. L. and Brown, N. L., *Metal Ions in Biological Systems*, Vol. 36, Marcel Dekker. (Sigel, A. and Sigel, H., editors), 1999.

## Setting the Technological Scene

- A good, modern introduction to the world of computers is *Computers in Your Future*, $4^{th}$ edition, by Bryan Pfaffenberger, Prentice Hall, 2002.

## The Basics

- The definitive, classic reference to Perl, now in its third edition, is *Programming Perl*, by Larry Wall, Tom Christiansen & Jon Orwant, O'Reilly & Associates, 2000. This book is often referred to as *The Camel* by practising Perl programmers. This text is the "Perl Bible",
- An excellent source of advice for new programmers is *Perl Debugged*, by Peter Scott & Ed Wright, Addison-Wesley, 2001. The emphasis in this book is on explaining how to avoid the most common problems encountered by the Perl programmer.

## Places to Put Things

The help files and on-line documentation included with Perl provide good descriptions of Perl variables and data structures. For the standard Perl variable types, use the `perldoc` utility to access `perldata`, as follows:

```
perldoc  perldata
```

For more complex data structures and representations, consider reading `perldsc` or `perllol`:

```
perldoc  perldsc
perldoc  perllol
```

## Getting Organised

- A thorough round-up of module design, creation and maintenance, as well as programmers' guide to CPAN, can be found in *Writing Perl Modules for CPAN*, by Sam Tregar, Apress (Springer-Verlag), 2002.

## About Files

A simple description of how to use and work with files and filehandles can be read by issuing this command-line:

    perldoc perlfaq5

An excellent introduction to the open subroutine and its various uses can be accessed as follows:

    perldoc perlopentut

## Patterns, Patterns and More Patterns

- To learn most, if not all, of what there is to know about regular expressions, study *Mastering Regular Expressions*, $2^{nd}$ edition, by Jeffrey E. F. Friedl, O'Reilly & Associates, 2002. This book is a great read.

## Perl Grabbag

- The perlfaq8 documentation is a good place to look for more details on interacting with the operating system (and its hardware) from within a Perl program.

- The ultimate Perl Grabbag is *Perl Cookbook*, $2^{nd}$ edition, by Tom Christiansen & Nathan Torkington, O'Reilly & Associates, 2003.

## The Protein Databank

- The Reference for 1LQT/U is Bossi, R. T., Aliverti, A., Raimondi, D., Fischer, F., Zanetti, G., Ferrari, D., Tahallah, N., Maier, C. S., Heck, A. J. R., Rissi, M., Mattevi, A. (2002), A covalent modification of NADP(+) revealed by the atomic resolution structure of FPRA, a mycobacterium tuberculosis oxidereductase, *Biochemistry*, **41**, 8807.

- The Reference for 1M7T is Dangi, B., Dobrodumov, A. V., Lousi, J. M., Gronenborn, A. M. (2002), Solution structure and dynamics of the human-Escherichia coli thioredoxin chimera: insights into thermodynamic stability, *Biochemistry*, **41**, 9376.

- Weissig, H., Shindyalov, I. N,, Bourne, P. E. (1998), Macromolecular structure databases: past progress and future challenges, *Acta Crystallographica Section D-Biological Crystallography*, **54**, 1085–1094.

- For an introduction to *Free R*, see Brunger, A. T. (1992), Free R-value – a novel statistical quantity for assessing the accuracy of crystal-structures, *Nature*, 355(6359), 472–475, as well as, Brunger, A. T. (1993), Assessment of phase accuracy by cross validation – the free R-value – methods and applications, *Acta Crystallographica Section D-Biological Crystallography*, **49**, 24–36.

- The PDB Data Uniformity Project is best described by Bhat, T. N., Bourne, P., Feng, Z. K., Gilliland, G., Jain, S., Ravichandran, V., Schneider, B., Schneider, K., Thanki, N., Weissig, H., Westbrook, J., Berman, H. M. (2001), The PDB data uniformity project, *Nucleic Acids Research*, 29(1), 214–218.

## Databases

- *MySQL*, $2^{nd}$ edition by Paul Dubios, SAMS, 2003, is the ultimate printed reference to the MySQL database management system.

- If you know another database system and want to apply your knowledge to MySQL, check out *The MySQL Cookbook*, also by Paul Dubios, but published by O'Reilly & Associates, 2002.

## Databases and Perl

- *Programming the Perl DBI* by Alligator Descartes & Tim Bunce, O'Reilly, 2000, is the classic DBI reference (although the module's included on-line documentation is also very good).

## Web Technologies

- A good, all-round text is *Programming the World Wide Web*, $2^{nd}$ edition by Robert W. Sebasta, Addison-Wesley, 2002.

- A book that is specific to Perl is *Writing CGI Applications with Perl* by Kevin Meltzer & Brent Michalski, Addison-Wesley, 2001.

## Tools and Datasets

- Sander, C. and Schneider, R., (1991), Database of homology-derived protein structures, *Proteins, Structure, Function & Genetics*, **9**, 56–68.

- Two original papers on dynamic programming sequence searching are: A general method applicable to the search for similarities in the amino acid sequence of two proteins, Needleman, S. B. and Wunch, C. D. (1970), *Journal of Molecular Biology*, **48**, 443–453; and Identification of common molecular

- subsequences, Smith, T. F. and Waterman, M. S. (1981), *Journal of Molecular Biology*, **147**, 195-197.
- Improved tools for biological sequence comparison, Pearson, W. R. and Lipman, D. J. (1988), *Proceedings of the National Academy of Sciences of the United States of America*, **85**, 2444-2448.
- The original reference for 1EHZ is Shi, H. and Moore, P. B., (2000), The crystal structure of yeast phenylalanine tRNA at 1.93 A resolution: a classic structure revisited, *RNA*, **6**, 1091.
- The original reference for the ClustalX program is Thompson, J. D., Gibson, T. J., Plewniak, F., Jeanmougin, F. and Higgins, D. G., (1997), The ClustalX windows interface: flexible strategies for multiple sequence alignment aided by quality analysis tools, *Nucleic Acids Research*, **24**, 4876-4882.
- Likewise for ClustalW, the reference is Thompson, J. D., Higgins, D. G. and Gibson, T. J., (1994), ClustalW: improving the sensitivity of progressive multiple sequence alignment through sequence weighting, positions-specific gap penalties and weight matrix choice, *Nucleic Acids Research*, **22**, 4673-4680.
- T-Coffee: A novel method for multiple sequence alignments, Notredame, C., Higgins, D. and Heringa, J., (2000), *Journal of Molecular Biology*, **302**, 205-217.
- COFFEE: A new objective function for multiple sequence alignment, Notredame, C., Holme, L. and Higgins, D. G., (1998), *Bioinformatics*, **14** (5), 407-422.
- Henikoff, S. and Henikoff, J. G., (1992), Amino acid substitution matrices from protein blocks, *Proceedings of the National Academy of Sciences of the United States of America*, **89**, 10915-10919.
- For a description of the original PAM matrices, see Dayhoff, M. O., Schwartz, R. M. and Orcutt, B. C., (1978,) in *Atlas of Protein Sequence and Structure*, Vol. 5, Supplement 3, Dayhoff, M.O. (ed.), NBRF, Washington, p. 345.
- Woodwark, K. C., Hubbard, S. J. and Oliver, S. G., (2001), Sequence search algorithms for single pass sequence identification: Does one size fit all? *Comparative & Functional Genomics*, 2(1), 4-9.

# Applications

- Durbin, R., Eddy, S., Krogh, A. and Mitchinson, G., (1998), *Biological Sequence Analysis*, Cambridge University Press, Cambridge.
- Borodovsky, M. and McIninch J., (1993), GeneMark: parallel gene recognition for both DNA strands, *Computers & Chemistry*, **17**, 123-133.

- Blattner, F. R., Burland, V., Plunkett, III, G., Sofia, H. J. and Daniels, D. L., (1993), Analysis of the *Escherichia-coli* genome .4. DNA-sequence of the region from 89.2 to 92.8 minutes. *Nucleic Acids Research*, **21**(23), 5408–5417.
- Borodovsky, M., McIninch, J., Koonin, E., Rudd, K., Medigue, C. and Danchin, A., (1995), Detection of new genes in the bacterial genome using Markov models for three gene classes, *Nucleic Acids Research*, **23**, 3554–3562.
- Burge, C. and Karlin, S., (1997), Prediction of complete gene structures in human genomic DNA, *Journal of Molecular Biology*, **268**, 78–94.
- Burge, C. B. and Karlin, S., (1998), Finding the genes in genomic DNA, *Current Opinion in Structural Biology*, **8**, 346–354.
- Schwede, T., Kopp, J., Guex, N., et al. (2003), SWISS-MODEL: an automated protein homology-modeling server, *Nucleic Acids Research*, **31**(13), 3381–3385.
- Rost, B. and Sander, C. (1993), Improved prediction of protein secondary structure by use of sequence profiles and neural networks, *Proceedings of the National Academy of Sciences USA*, **90**(16), 7558–7562.
- Rost, B. and Sander, C. (1994), Combining evolutionary information and neural networks to predict protein secondary structure, *Proteins: Structure Function and Genetics*, **19**(1), 55–72.
- Rost, B. and Sander, C. (1993), Prediction of protein secondary structure at better than 70-percent accuracy, *Journal of Molecular Biology*, **232**(2), 584–599.
- Mulder, N. J., Apweiler, R., Attwood, T. K., Bairoch, A., Barrell, D., Bateman, A., Binns, D., Biswas, M., Bradley, P., Bork, P., Bucher, P., Copley, R. R., Courcelle, E., Das, U., Durbin, R., Falquet, L., Fleischmann, W., Griffiths-Jones, S., Haft, D., Harte, N., Hulo, N., Kahn, D., Kanapin, A., Krestyaninova, M., Lopez, R., Letunic, I., Lonsdale, D., Silventoinen, V., Orchard, S. E., Pagni, M., Peyruc, D., Ponting, C. P., Selengut, J. D., Servant, F., Sigrist, C. J. A., Vaughan, R. and Zdobnov, E. M., (2003), The InterPro Database, 2003 brings increased coverage and new features, *Nucleic Acids Research*, **31**, 315–318.
- Zdobnov, E. M. and Apweiler, R., (2001), InterProScan – an integration platform for the signature-recognition methods in InterPro, *Bioinformatics*, **17**(9), 847–848.

## Data Visualisation

- A good book on visualisation is *Information Graphics: A Comprehensive Illustrated Reference* by Robert L. Harris, Oxford University Press Inc., USA, 2000.
- An excellent web resource is Ramana Rao's web-site: http://www.ramanarao.com.

# Appendix E

# Essential Linux Commands

To learn more about a particular command, view the *manual page* associated with it – simply type man followed by the name of the command you want to learn about. To exit from a manual page, press the q key (where "q" stands for "quit"). Most commands have options associated with them – don't try to guess the options... read the manual page!

## Working with Files and Directories

| | |
|---|---|
| cat | Type a disk-file to the screen |
| cd | Change directory (or return to %HOME directory) |
| chmod | Change the mode of a disk-file (e.g., to make it executable) |
| chown | Change the owner of a disk-file or directory |
| cp | Copy a disk-file/directory to a new location |
| find | Search for a disk-file on the system (see locate) |
| ftp | Transfer disk-files from one system to another |
| grep | Search for a text string in a group of disk-files |
| gzip/gunzip | Compress/uncompress a disk-file or group of disk-files |
| head | Display the first few lines of a disk-file on the screen |
| ispell | Spell-check a disk-file using the system dictionary |
| less | Type a disk-file to the screen one screen-full at a time |
| locate | Locate a specific disk-file on the system (see find) |
| ln | Create a symbolic link (alias/shortcut) to a disk-file |

---

*Bioinformatics, Biocomputing and Perl.* Michael Moorhouse and Paul Barry
© 2004 John Wiley & Sons, Ltd   ISBN 0-470-85331-X

| | |
|---|---|
| `ls` | List the contents of a directory to the screen |
| `mkdir` | Create a new directory |
| `mv` | Move or rename a currently existing disk-file/directory |
| `pwd` | Display the name of current working directory |
| `rm` | Delete one or more disk-files |
| `rmdir` | Delete a directory |
| `sort` | Sort a disk-file (using various techniques) |
| `tac` | Type a disk-file to the screen in reverse order (see `cat`) |
| `tail` | Display the last few lines of a disk-file on the screen |
| `wc` | Display the character, word, or line count of a disk-file |
| `zcat` | Type the contents of a compressed disk-file to the screen |
| `zmore` | Like `zcat`, only display the disk-file a screen-full at a `time` |

## Printing Commands

| | |
|---|---|
| `lpq` | Check the status of your entries on the print queue |
| `lpr` | Add an entry to the print queue |
| `lprm` | Remove an entry from the print queue |

## Networking Commands

| | |
|---|---|
| `netstat` | Show the network status for this system |
| `ping` | Is there anybody out there? Check a host for existence |
| `traceroute` | Show me how to get from here to there |

## Working with Processes

| | |
|---|---|
| `kill` | Stop a process (program) from running |
| `ps` | Report on the active processes |
| `top` | Who is doing what, and how much CPU are they using? |
| `w` | Display a summary of system usage on the screen |

## Working with Disks

| | |
|---|---|
| `df` | How much free disk space is there? |
| `du` | How is the disk space being used? |

## Miscellaneous Commands

| | |
|---|---|
| `cal` | Display a calendar on the screen |
| `clear` | Clear the screen |

| | |
|---|---|
| `date` | Display the current date and time on the screen |
| `echo` | Display a message on the screen |
| `man` | Read a manual page (type `man man` to learn more) |
| `passwd` | Change your password |
| `perl` | Run Perl (a great programming language ... ) |
| `su` | Create a shell under the ID of some other user |
| `telnet` | Log into a remote computer |
| `uname` | Display the machine and operating system name |
| `users` | List the current login sessions on the system |
| `vi` | Run `vi` (a great text editor ... ) |
| `whereis` | Locate a binary (executable), source, or manual page disk-file |
| `which` | List the path to a particular binary disk-file (executable) |
| `who` | Who is currently logged in |
| `whoami` | 'Cause I've forgotten ... |
| `Ctrl-D` | Signal end-of-file to running process (key combination) |

## Essential Systems Administrator Commands

Note that you will need to be logged in as *root* to use these commands effectively. Remember, as *root* you have complete power over Linux (so be careful).

| | |
|---|---|
| `cron` | Execute commands at scheduled times |
| `dmesg` | Display the system control messages |
| `e2fsck` | Check the health of a disk |
| `fdisk` | Fiddle with disk partitions (be very, very careful) |
| `fdisk` | You are being careful with `fdisk`, aren't you? |
| `ifconfig` | Configure your network interface card |
| `kill` | See `kill` above ... much more fun as `root` ... |
| `lilo` | Install the Linux Loader (read the `man` page) |
| `lpc` | Control a print queue |
| `mke2fs` | Create a disk-file system (i.e., format a disk) |
| `mount` | Add a disk into the active disk-file system (read the `man` page) |
| `reboot` | Reboot now! |
| `rpm` | The RedHat Package Manager |
| `shutdown` | Perform a nice safe, graceful, shutdown of the system |
| `tar` | Work with `tarred` disk-files (read the `man` page) |
| `umount` | Remove a disk from the active disk-file system |

# Appendix F

# vi Quick Reference

This quick reference will get you started. To learn more, from the Linux command-line, type `man vi`.

## Invoking the vi Text Editor

**vi** – Start the vi editor with an empty edit buffer

**vi** *file* – Edit a file called *file*

**vi +n** *file* – Edit a file called *file* and go to line *n*

**vi +/***pattern* *file* – Edit a file called *file* and go to the first line that matches the string *pattern*

## vi's Modes

vi can be in one of three modes:

**Edit mode** – Keys typed are added to the edit buffer

**Non-edit mode** – Keys typed adjust or move around the edit buffer

**ex mode** – Commands are executed within vi, which affect the edit buffer

To enter *edit mode*, press the Esc key, then type i
To enter *non-edit mode*, simply press Esc
To enter *ex mode*, press Esc, then type :

## Non-edit Mode Keystrokes

| | |
|---|---|
| ^ | Go to start of current line (first non-blank character) |
| 0 | Go to start of current line |
| $ | Go to end of current line |
| w | Go to next word |
| b | Go to previous word (back) |
| | |
| o | Insert blank line below current one, enter *edit mode* |
| O | Insert blank line above current one, enter *edit mode* |
| | |
| i | Enter *edit mode* by inserting text at current location |
| a | Enter *edit mode* by appending text after current location |
| A | Enter *edit mode* by appending to the end of the current line |
| J | Join the current line with that line immediately below it |
| | |
| Ctrl-G | Show current line number |
| $n$G | Go to line $n$ within the edit buffer |
| G | Go to bottom of edit buffer |

## Deleting Text (in Non-edit Mode)

| | |
|---|---|
| dd | Delete current line |
| dw | Delete next word |
| d^ | Delete to start of line |
| d$ | Delete to end of line |
| x | Delete a single character |

## Changing Text (in Non-edit Mode)

| | |
|---|---|
| cc | Change the current line, and enter *edit mode* |
| cw | Change the current word, and enter *edit mode* |
| r | Replace a single character |
| R | Replace characters until `Esc` is pressed |

## Cutting and Pasting (in Non-edit Mode)

| | |
|---|---|
| yy | Copy current line (the line is now *yanked*) |
| $n$yy | Copy $n$ current lines (multi-yank) |
| ye | Copy to the end of next word (little-yank) |

| | |
|---|---|
| p | Paste yanked text after or below cursor |
| P | Paste yanked text before or above cursor |

## Some ex Mode Commands

| | |
|---|---|
| :w | Write the edit buffer (i.e., save the file) |
| :w *file* | Write a copy of the edit buffer as *file* |
| :wq | Write the edit buffer, then quit |
| :q! | Quit without writing any changes (called "force quit") |
| :w! *file* | overwrite *file* with current edit buffer |
| :sh | Temporarily exit vi to access a Linux shell |
| :help | Access the vi on-line help |
| :help *cmd* | Access the on-line help for subject *cmd* |
| :set | Used to set and unset vi settings |
| :set all | Display the entire list of vi's current settings |

## Searching

| | |
|---|---|
| /*pattern* | Search forward in edit buffer for a match to *pattern* |
| / | Repeat last forward search |
| ?*pattern* | Search backward in edit buffer for a match to *pattern* |
| ? | Repeat last backward search |
| n | Repeat previous search (regardless of direction) |

# Index

++, increment operator, 34
-c, command-line switch, 22
-w, command-line switch, 24
<=> comparison operator, 157
<>, input operator, 41
=, assignment operator, 33
==, equality operator, 36
=~, binding operator, 44
%, modulus operator, 39
42, why forty-two?, 258

Aas, Gisle, 330
AceDB database, 307
ActivePerl, 275, 454
Adams, Douglas, 258
amino acids, 4, 173
    local structures, 173
Apache web server, 303
    preparing for Perl, 310
appending to disk-files, 116
applet, 409
ARGV array, 106
ArrayExpress sequence database, 342
arrays
    accessing elements, 51
    adding elements to, 51
    determining size, 51
    index values, 50
    introducing, 49
    naming, 49
    pictorial representation of, 50
    processing every element, 57

range of indexes, 54
referring to entire array, 51
referring to individual element, 51
removing elements from, 54
slicing, 54
splicing, 54
using foreach with, 58
associative array, see hashes 60
automatch program, 330
automating surfing, 329
    reasoning for, 329
    strategy, 335
    time saved, 334
Azimov, Dr. Issac, 414

backticks operator, 152
bacterial proteins, 380
Barioch, Amos, 341
base pairs, concept of, 3
bases program, 64
basic local alignment search tool, 340
BEGIN blocks, using, 241
Berners-Lee, Tim, 304
Bernstein, F. C., 206
Bernstein, H. J., 206
bestrict program, 148
binding operator, =~, 44
Bio::SearchIO module, 451
biodb2mysql subroutine, 139
bioinformatics tools, 339
bioinformatics, defined, xv
biomolecule structures, 174

---

*Bioinformatics, Biocomputing and Perl.* Michael Moorhouse and Paul Barry
© 2004 John Wiley & Sons, Ltd   ISBN 0-470-85331-X

Bioperl, 339, 365, 377, 408, 441
    `Bundle::BioPerl` module, 443
    citation, 442
    extracting subsequences, 447
    fetching multiple sequences, 445
    fetching sequences, 444
    installation, 442
    installation dependencies, 443
    not reinventing wheel, 447
    on-line documentation, 445
    relationship to BLAST, 449
    relationship to Ensembl, 442
    remote blast searches, 449
    two output formats, 444
BLAST, 299, 340, 357, 362, 378, 417
    `blastall` program, 366
    dealing with multiple hits, 379
    different algorithms, 392
    different types of searches, 369
    faster searching, 365
    `formatdb` program, 365
    general approach, 363
    identifying sequences, 382
    installation, 364
    interpreting alignment scores, 379
    interpreting results, 382
    interpreting scores, 379
    parsing output, 450
    PDB results summary, 389
    sequence similarity search, 378
    service provided, 449
    summary of outputs, 382
    variants on original, 363
`Blast_parse.pl` program, 450
`blastall` program, 369
`blastcl3` program, 449
`blastclust` tool, 370
blocks of code, 29
    using curly braces, 29
BLOCKS sequence motifs, 401
BLOSUM matrices, 361
BLOSUM62 matrices, 361
Bourne, P. E., 206
Boutell, Tom, xix, 422
Bragg's Law, 174
Bunce, Tim, 274

CD-HIT/CD-HI dataset, 213
Celera Genomics, 10

CGI, 311
    providing headers, 313
    relationship to HTML form, 316
    scripts, programs, CGIs, 314
    sending data to a program, 315
    the `cgi-bin` directory, 318
CGI module, 307, 418
    the `end_html` subroutine, 308
    the `param` subroutine, 316
    the `start_html` subroutine, 307
character classes, 127
    inverted, 127
    ranges of digits, 127
    ranges of letters, 127
`Chart::Graph::Xmgrace` module, 434
    plotting with, 437
`check_args` program, 108
`check_drivers` program, 275
chevrons, with disk-files, 117
`chkconfig` utility, 310
`chmod` command, 25
`chomp` subroutine, 118
`chop` subroutine, 118
CIFTr program, 206
clobbering, 116
`close` subroutine, 109
ClustalTree.class, 409
ClustalV, *see* ClustalW 358
ClustalW, 340, 357, 385, 398, 408
    command-line parameters, 361
    installation, 360
ClustalX, *see* ClustalW 358
`cmp` comparison operator, 156
codons, start/stop, 7
coffee
    when to drink a cup, 455
COFFEE program, 360
command-line arguments, 107
comments, in Perl, 28
Common Gateway Interface,
    *see* CGI 311
concatenation, 124
condition, explanation of, 30
constants, in Perl, 28
contact maps, *see* PDB 192
`Contact_map.pl` program, 193
    example output, 195
`convert_pdb` program, 208

CPAN, 96
    installing modules
        automatically, 99
        manually, 98
        testing, 100
    module building and testing, 98
    searching, 97
`cron` utility, 171
`Ctrl-C`, interrupting programs, 27
`Ctrl-D`, signalling end-of-file, 42

database
    case study, 227
    definition of, 219
    EMBL, *see* EMBL 228
    linking tables, 222
    MER, list of tables, 231
    metadata, 220
    programming
        avoiding enhancements, 274
        fetching rows of data, 282
        guarding against changes, 285
        potent combination, 292
        preparing Perl for, 274
        using placeholders, 287
        why do it?, 273
    RDBMS, 224
    relating tables, 220
    relating two tables, 254
    schema, 220
    single table problems, 222
    structure, 219
    SWISS-PROT, *see* SWISS-PROT 227
    type information, 220
database system
    Access, 225
    DB2, 225
    dBase, 225
    definition of, 224
    enterprise, 225
    FileMaker, 225
    Informix, 225
    Ingres, 225
    InterBase, 225
    MySQL, 226
    open source, 225
    Oracle, 225
    Paradox, 225
    personal, 225
    PostgreSQL, 226
    SQL Server, 225
    types of, 224
database table
    definition of, 219
Dayoff, Margrate, 341
`db_match_embl` program, 289, 320
`db_match_emblCGI` program, 321, 329
`DBD::mysql` module, 274
`DBI` module, 274
    checking installed drivers, 276
    data source, 277
    database handles, 277
    statement handles, 278
    using placeholders, 287
    variable naming conventions, 278
`DbUtilsMER` module, 279, 323
    using with CGI, 325
DeepView, 200, 388, 391, 392
    further investigations, 398
    structural alignment, 396
default array, 76
`delete` subroutine (with hashes), 62
Descartes, Alligator, 274
`determine_args` program, 106
`die` subroutine, 107
`diff` utility, 445
diffraction patterns, 175
displaying literal text, 117
DNA
    introduction, 1
    protein sequence example, 12
    relationship to RNA, 1
    transcription, 7
    translation, 7
double helix, 2
    direction, 3
    strands, 3
downloading datasets, 165
    local download advantages, 165
    local download disadvantages, 166
downloading raw DNA sequences, 377
downloading subsets, 169
drawing boxes
    problems with, 83
    properly, 84
`drawline` program
    final version, 89
    first version, 75

DSSP program, 197, 199
Dubios, Paul, 273
dynamic programming, 355

e-mail, of authors, xviii
each subroutine, 67
EBI's SRS, *see* SRS 297
elsif statement, 37
EMBL, 228, 299, 329, 340
    example entry (abridged), 229
    flat-file format, 377
    on-line manual, 228
    processing complications, 231
    understanding entries, 343
EMBL/GenBank, 341
Embl_plot.pl program, 427
EMBOSS, 8, 166, 299, 339, 377
    Transeq form, 8
Ensembl, 339, 442
eof subroutine, 112
European Bioinformatics Institute, 227
eval subroutine, 154
example heat map, 425
exception handling, 153
    using die with eval, 154
    using eval, 154
executing programs, how to, 24
exit subroutine, 106
ExPASy web-site, 388

false negative, 348, 386
false positive, 348, 387
false positive rate, 387
    problems with, 401
fast step algorithms, 355
FASTA data format, 355, 363, 368, 378
fastacmd program, 370
Felsentein, Joe, 358
Feng, Zukang, 206
file test operators, 108
filehandles, 109
    naming convention, 109
find utility, 275
fivetimes program, 35
foreach statement, 58
forever program, 26
formatdb program, 365
    command-line parameters, 365

Free R factors, 176
Free Software Foundation, 167
free_res program, 184
FreeType fonts, 422
Frishman, Dmitrij, 196

gawk utility, 203
GD module, 422
    installation, 423
    other uses, 429
    testing, 423
    using image canvases, 423
GD::Graph module, 431, 432
GD::Graph3d module
    plotting with, 432
GD::Text::Align module, 432
gdlib graphics library, 422
gene ontology, 342
GeneMark, 380
    accessing through EBI, 381
    summary of predictions, 388
generalised models, 5
    good enough, 5
    lie-to-children, 5
genes program, 66
genome sequencing, 10
    how it works, 10
GenScan program, 380
get_citations program, 265
get_dna_crossrefs program, 257
get_dnas program, 249
    invoking, 252
get_protein_crossrefs program, 256
get_proteins program, 239
    invoking, 245
getlines program, 41
global alignment algorithms, 363
global_scope program, 85
GNU software, 167
gnuplot program, 431
GO Consortium, 407
Grace, 431
    batch plotting, 435
    colour value chart, 437
    GUI version, 434
    interactive plotting, 434
    plotting with, 434
    scatter plots, 434

Grace program, *see* Grace 431
grep utility, 151
grouping program, 135
grouping of dihedral angles, 203
grouping2 program, 136
gzip utility, 167

handle, meaning of, 109
hard coding, best avoided, 106
hardware, 15
   relationship to software, 15
hashes
   accessing entries, 61
   accessing every entry, 65
   adding entries to, 60, 62
   alternative assignment notation, 62
   complete example, 64
   determining number of entries, 61
   introducing, 60
   major restriction, 60
   naming, 60
   pictorial representation, 60
   processing every entry, 66
   removing entries from, 62
   slicing, 63
   using foreach with, 67
   using the keys subroutine, 61
HERE documents, 159
   assigning to scalars, 159
   using with print, 160
high scoring segment pairs, *see* HSSP 363
homology modelling, 389, 401
   alternatives to, 390
HSSP, 342, 363
HTML, 304
   common graphic formats, 304
   creating manually, 305
   creating visually, 305
   creating with a program, 305
   forms and input, 316
      action attribute, 317
      avoiding the *Back* button, 319
   hypertext references, 314
   learning HTML, 305
   on-line tutorial, 306
   static vs. dynamic pages, 308
   table generation, 325
   textareas, 317

HTTP, 304
httpd.conf disk-file (Apache), 310
   editing, 310
   enabling CGI, 311
hybrid_scope program, 88

ImageMagick tools, 429
*in silico* testing, 348
*in vitro* testing, 348
*in vivo* testing, 348
increment operator, ++, 34
indentation, proper use of, 38
input operator, <>, 41, 104
input/output, 103
   modules, 119
Institute of Genome Research, 10
InterPro metadatabase, 406
introns, 387
iterateF program, 58
iterateW program, 57
iteration, 26

Java applet, 409
Java programming language, 300
Jigsaw web server, 303
Joy, Bill, xix

Kabsch, Wolfgang, 198
KDE Konqueror, 303
KEGG pathway databases, 407
keys subroutine, 61
kilobyte, explained, 182
Krieger, Elmar, 198

Lac Operon, 375
LAMP platform, 304
Lamport, Leslie, xix
last command, 36
layers of technology, 16
Lester, Andy, xix, 330
libpng library, 422
libwww-perl library, *see* LWP 330
Lincoln D. Stein, 311
Linux, xviii, 232, 469, 471
   command-line, 98
   essential commands, 467
Lion Bioscience, 297, 299
list context, 52

lists
    assigning to an array, 50
    convenient notation for, 59
    introducing, 50
`local` command, 89
local alignment algorithms, 363
`localhost`, 312
`locate` utility, 275, 310
loop, explanation of, 27
`ls` utility, 153, 168
LWP library, 171, 330
Lynx web browser, 303

Mac OS X, 453
MacPerl, 453
maintaining programs, 71
making programs executable, 25
markov models, 402
Martin, David, 167
`match_emblCGIbetter` program, 319
`match_emblCGI` program, 315
`match_embl` program, 144, 289, 315, 320
matrices, 362
megabyte, explained, 182
Mer Operon, 231, 365
    HMA domain examples, 407
    protein/function summary, 376
    scientific background, 374
        function, 374
        genetic structure/regulation, 374
        mobility, 375
    web-site, 13
`Mer_Table.pl` program, 418
    example output, 422
`merge2files` program, 106, 110
    problems with, 112
`merge2files_v2` program, 112
    problems with, 113
`merge2files_v3` program, 113
merging files, 105
    strategy, 105
messenger RNA, 353
metadatabases, 340
    InterPro, 347
    SRS, 347
micro array database, 342
Microsoft Internet Explorer, 303
Microsoft's IIS web server, 303

mirroring a dataset with `wget`, 168
missing proteins, 386
mmCIF, 340
    automating conversion, 208
    converting to PDB, 206
    data format, 179
    data-file format, 205
    general usage advice, 208
    how differs to PDB, 179
    introduction, 205
    performing the conversion, 206
    problems with conversions, 208
    relationship to PDB, 205
modelling with SWISS-PROT, 390
modules, 93
    blank template, 93
    creating, 94
    export variables, 94
modulus operator, 39
`mogrify` program, 429
    examples, 429
molecular biology
    central dogma, 6
Mozilla web browser, 303
multi-way selections, 37
`Multi_Seq_Get.pl` program, 446
multiple sequence analysis
    algorithms and methods, 359
`my` command, 86
MySQL, 232
    adding/creating tables, 235
    configuring, 233
    creating a database, 233
    creating users, granting rights, 234
    describing tables, 237
    importing data, 238
        specific example, 245
    installing on Linux, 232
    interpreting messages, 233
    LOAD DATA command, 246
    manual, 232
    monitor program, 233
    problems with sub-select, 253
        workarounds, 254
    showing databases, 233
    showing tables, 237
    web-site location, 232
    working at the prompt, 233

NCBI-BLAST, 364, 378
Net::MySQL module, 274
Netscape Navigator, 303
newline, explanation of, 22
NMR, 174
    constraints, 176
    introduction, 176
    major difference to X-ray, 189
    meaning of remarks, 184
    when to choose over X-ray, 177
non-DNA coding, 387
non-redundant algorithms, 213
non-redundant datasets, 211
non-redundant protein structures, 213
not equal to operator, 40
nuclear magnetic resonance,
    *see* NMR 174

objects, example usage, 331
ODBC, 274
oddeven program, 36
oddeven2 program, 38
oddeven3 program, 40
one-liners, 149
    simple calculator, 150
    with loops, 150
    with loops, printing, 151
ontologies, 407
open subroutine, 109
Open Rasmol program, 392--394
open reading frames, 7
    getting it wrong, 8
opening named disk-files, 108
Opera web browser, 303
operators
    list of, 457
or, && operator, 109
our command, 88
output redirection, 245
over predicted genes, 387
Oyama, Hiroyuli, 274

package statement, 94
PAM250 matrices, 362
Pascal programming language, 29
pattern-matching, introduction to, 44
patterns program, 44
patterns, *see* regular expressions 121
PAUP, 358

PDB, 174
    accessing annotation data, 183
    accessing data items, 182
    contact maps, 192
    data format explained, 183
    data-file formats, 179
    database cross-referencing, 186
        different types, 187
    downloading data-files, 181
        using EBI, 181
    example structures, 180
        1LQT, 180
        1M7T, 180
    extracting 3D coordinates, 191
    flat text data-files, 177
    growth of, 178
    how structures are determined, 178
    introduction, 173
    introduction and history, 177
    legacy structures, 180
    non-redundant dataset, 182
    plotting results, 186
    processing DBREF lines, 188
    shortcomings, 179
    understanding ATOM lines, 190
    understanding remarks, 184
    working with coordinates, 188
PDB-Select dataset, 212, 213
pdb2cif program, 206
pdbselect program, 169
PDL, Perl Data Libraries, 438
perl versus Perl, 17
Perl programming language
    case sensitivity, 29
    documentation, 459
        perldebtut, 460
        perldsc, 460
        perlfunc, 459
        perlreftut, 459
        perlretut, 460
    installation instructions, 453
        from source code, 454
        on Linux, 454
        on Macintosh, 453
        on Windows, 454
    on-line documentation, 90
        using perldoc utility, 90
    statements, 22

perlretut documentation, 146
pfscan program, 403
PHYLIP program, 358
phylogenetics, 407
pinvoke program, 152
PIR database, see SWISS-PROT 341
polypeptides, see amino acids 4
pop subroutine, 56
predictions, 348
    balancing errors, 351
    improving performance, 352
    problems of, 348
    using multiple algorithms, 352
prepare_embl program, 143
preparing data files, 365
print subroutine, 21
printf subroutine, 150
PRINTS sequence motifs, 401
prior art, implications of, 343
private_scope program, 87
processing data files, 41
produce_simple program, 306
produce_simpleCGI program, 307
PROSITE, 342
    concluding comments, 407
    downloading instructions, 403
    example invocations, 403
    patterns and matrices, 402
    sequence motifs, 401
protein crystals
    requirement for, 175
protein databank, see PDB 173
protein structure methods, 177
protein synthesis, 353
proteins, 4
    determining function, 173
    ligands, 4
    residues, 4
    structure of, 4
    tags, 4
PSI BLAST, 364
PubMed, 342
push subroutine, 56
pushpop program, 56

qq operator (generalised quote), 287
qw, quote words operator, 59
qx operator, 152

R Factors, 175
ramachandran plots, 203
range operator, .., 55
ray tracing, 389
reading from disk-files, 105
redirecting standard error, 117
redirecting standard output, 117
redundancy
    non-representative data, 212
    reducing, 212
    why is it there?, 211
reflections, 175
regular expressions, 43, 457
    after-match variables, 135
    alternation, 126
    anchors, 132
        to end-of-line, 133
        to start-of-line, 133
        word boundaries, 132
    applying globally, 333
    character classes, 127
    concatenation, 124
    dealing with greed, 137
    EMBL data example, 142
    escaping metacharacters, 125
    grouping parentheses, 135
    match any character, 131
    modifiers, 141
    nested grouping parentheses, 137
    optional patterns, 130
    pattern metacharacters, 124
    remembering matches, 135
    repetition, 124
    repetition counts, 130
    search and replace, 140
    shorthand, 127
    substituting whitespace, 141
    substitution, 140
    the slash-*something* shorthands, 129
    using alternative delimiters, 138
    what are they?, 122
    why use them?, 122
    with HTML tables, 333
remote blast searches, see Bioperl 448
return command, 84
reverse subroutine, 156
RGB notation, 424

RNA
    introduction, 1
    messengers, 353
    relationship to DNA, 1

sample56.pl program, reformatted, 434
Sander, Christian, 198
scalar subroutine, 52
scalar context, 52
scalars, 31
    correctly formed, 31
scope, 85
    global, 86
    more on globals, 88
    private, 86
scoring matrices, 361
searching for installed modules, 275
sed utility, 368
select_filter program, 215
    used with grep, 217
selection
    explanation of, 34
    multi-way, 37
    statement, 34
semantics, explanation of, 23
Seq_Crop.pl program, 447
sequence alignment
    with ClustalW, 358
sequence assembly, 11
sequence databases
    how they relate, 341
    primary, secondary, tertiary, 340
    summary, 347
    three main examples, 341
    three types defined, 341
sequence motif databases, 401
sequence motifs
    relationship to protein, 401
sequence re-assembly, 12
sequence retrieval system, see SRS 297
sequence scoring profiles, 402
sequence, in programs, 22
shell script, 393
shift subroutine, 56, 78
show_tables program, 276
show_tables2 program, 280
signalling end-of-file, 42
simple_coord_extract program, 191
simple_get_sequence.pl program, 444

simplepat program, 134
sleep command, 30
slice or splice, which?, 55
slices program, 55
slicing arrays, 54
slow step algorithms, 355
slurper program, 115
slurping disk-files, 114
    warning, 116
software, 15
    category: applications, 16
    category: operating systems, 15
    category: tools, 16
sort subroutine, 67, 156
sort utility, 158
sortexamples program, 155
sortfile program, 157
sorting, 155
    alphabetically, 156
    numerically, 157
specific_crossref program, 285
splice subroutine, 54
SQL
    defining data, 226
    manipulating data, 227
    SELECT query, 246
        ORDER BY clause, 247
        WHERE clause, 247
    sub-select, 253
SRS, 297, 343, 367
    accessing and using, 298
    data formats, 299
    download formats, 377
    downloading Mer Operon, 377
    what is cannot do, 299
    why is it important?, 298
Stallman, Richard, xix
standard error, 42
standard input, explanation of, 41
standard modules, 96
standard output, 42
start codons, 7
statement qualifier, 40
STDERR, 42, 104
STDIN, 42, 104
STDOUT, 42, 104
Stein, Dr. Lincoln D., xix, 307, 422
stop codons, 7

streams, 103
    standard error, 104
    standard input, 103
    standard output, 104
    why two output streams?, 104
strictness, 147
STRIDE, 196
    extracting amino acids, 204
    how it works, 198
    identifying hydrogen bonds, 198
    installing, 197
    parsing output, 200
    using, 197
strings, explanation of, 42
structural alignment, 396
structural genes, 374
structural refinement, 184
Structured Query Language, *see* SQL 226
subroutines, 73
    calling/invoking, 73
    creating, 74
    default parameters, 78
    indentation styles, 74
    list of in-built, 90
    named parameters, 80
    other names for, 73
    processing parameters, 76
    returning results, 84
substitution, 140
substitution matrices, 361
    amino acid, 361
Sun Microsystems, 300
Swiss Institute of Bioinformatics, 227
SWISS-MODEL, 388
    structural prediction, 388
    what is cannot do, 400
SWISS-PROT, 227, 340--342
    alignmaster output, 390
    example entry, 228
    modelling techniques, 390
    on-line manual, 227
    output from ProMod, 391
    processing complications, 229
    understanding entries, 346
syntax, 23
    checking with -c, 22
    explanation of, 23
`system` subroutine, 152

tab-delimited disk-files, 238
table row
    definition of, 219
taint mode, importance of, 309
TCOFFEE, *see* COFFEE 360
tea, *see* coffee 455
`tentimes` program, 32
`terrible` program, 37
testing
    *in silico*, 348
    *in vitro*, 348
    *in vivo*, 348
testing for self-consistency, 348
the three C's, 46
threshold scores, 351
timing, implication of, 399
TMTOWTDI, Perl's motto, 25
Torvalds, Linus, xix
transposons, 375
TreeView program, 358
TrEMBL, 340, 341
tRNA gene prediction, 349
tRNA package, 340
tRNAScan-SE, 340, 351
    case study, 353
    use with predictions, 347
true negative, 349
true positive, 349
true positive rate, 402
trueness, checking for, 41
TrueType fonts, 422
`ttf.pl` program, 423

`undef` subroutine, 63
undefined, 63
UniProt Consortium, 341
`unique_crossrefs` program, 258
`unless` statement, 109
`unshift` subroutine, 56
`use lib` directive, 96
`use strict` directive, 148
    with subroutines, 149
    with variables, 148
`use subs` directive, 149
`UsefulUtils` module
    first version, 95
    using, 95
    using with `slurping`, 115

validating a model, 348
validation, 348
variable containers, 31
variable interpolation, 117
variable, explanation of, 31
variables
    scalar, 31
vi text editor, 471
visibility, 85
visualisation
    comparing senses, 413
    creating high-quality images, 422
    displaying genes, 426
    importance of, 414
    introduction, 413
    plotting graphs, 431
    reducing complexity, 425
    resizing vs. resampling, 430
    SWISS-PROT identifiers, 417
    using HTML, 415
    why use HTML?, 415

Wall, Larry, xix
warn subroutine, 104
warnings, turning on, 24
wc utility, example usage, 144
web databases, 320
web development infrastructure, 303
    backend database, 304
    client-side programming, 304
    content, 304
    server-side programming, 304
    transport protocol, 304
    web client, 303
    web server, 303
web server errors, 313

web-enabling a program, recipe for, 315
web-site, for book, xviii
welcome, the first program, 21
welcome2 program, 22
welcome3 program, 25
Westbrook, John, 206
wget utility, 167
    deep copying, 168
    documentation, 167
    downloading PDBs, 167
    mirroring a dataset, 168
    useful parameters, 169
whattimeisit program, 308
    testing with Apache, 312
whereis command, 17
which command, 17
which_crossrefs program, 282
which_crossrefs2 program, 284
which_crossrefs3 program, 284
while loop, 26
whoops program, 23
Wiedmann, Jochen, 274
world wide web, 303
writing to disk-files, 116
    appending, 116
WU BLAST, 363
WWW
    downloading datasets, 165
WWW::Mechanize module, 329
    example of use, 332

x, the repetition operator, 67
X-ray Crystallography, 174
    introduction, 174
    problems with, 175
xmgrace program, *see* Grace 431